T0141355

ABOUT ISLAND PRESS

Island Press, a nonprofit organization, publishes, markets, and distributes the most advanced thinking on the conservation of our natural resources—books about soil, land, water, forests, wildlife, and hazardous and toxic wastes. These books are practical tools used by public officials, business and industry leaders, natural resource managers, and concerned citizens working to solve both local and global resource problems.

Founded in 1978, Island Press reorganized in 1984 to meet the increasing demand for substantive books on all resource-related issues. Island Press publishes and distributes under its own imprint and offers these services to other nonprofit organizations.

Support for Island Press is provided by The Geraldine R. Dodge Foundation, The Energy Foundation, The Charles Engelhard Foundation, The Ford Foundation, Glen Eagles Foundation, The George Gund Foundation, William and Flora Hewlett Foundation, The James Irvine Foundation, The John D. and Catherine T. MacArthur Foundation, The Andrew W. Mellon Foundation, The Joyce Mertz-Gilmore Foundation, The New-Land Foundation, The Pew Charitable Trusts, The Rockefeller Brothers Fund, The Tides Foundation, and individual donors.

FIRE ECOLOGY OF PACIFIC NORTHWEST FORESTS

FIRE ECOLOGY OF PACIFIC NORTHWEST FORESTS

James K. Agee

ISLAND PRESS

WASHINGTON, D.C. ❑ COVELO, CALIFORNIA

The author is grateful for permission to include the following previously copyrighted materials in redrafted form, except where noted: Figures 4.5 and 11.11 are from *Forest Ecology and Management*, Elsevier Science Publishers BV. Figure 5.5 is courtesy of Yale University. Figures 4.19, 7.7, and 10.12 are from *Ecology*, and Figure 11.4 is from *Ecological Monographs*, published by the Ecological Society of America. Figures 10.5 and 10.11 are from *Northwest Science*, published by Washington State University Press. Figure 11.2 is from the *Journal of Forestry*, published by the Society of American Foresters. Figure 11.5 is original artwork from *Scientific American*, published by Scientific American, Inc. Figure 12.11 is from the *Journal of Range Management*, published by the Society for Range Management.

Library of Congress Cataloging-in-Publication Data

Agee, James K.
Fire ecology of Pacific Northwest forests / James K. Agee.
p. cm.
Includes bibliographical references (p.) and index.
ISBN 1-55963-230-5
1. Forest management—Northwest, Pacific. 2. Fire ecology—Northwest, Pacific. 3. Forest fires—Environmental aspects—Northwest, Pacific. I. Title.
SD144.A13A55 1993
574.5'264—dc20
93-2071
CIP

Printed on recycled, acid-free paper.

Manufactured in the United States of America

10 9 8

This book is dedicated to my parents,
Carroll and Lee Agee,
who supported a young boy's dream to learn about forests,
and to
Professor Harold Biswell,
pioneer fire ecologist, who helped the dream come true.

CONTENTS

PREFACE

THIS BOOK BEGAN as a source book for natural area managers interested in restoring or maintaining fire in the natural areas of the Pacific Northwest. It grew to encompass a broader charge: to provide a natural baseline that wildland managers, or those interested in wildland management, could use in understanding the effects of natural or altered fire regimes in the western United States. This ecological perspective about fire is not a prescriptive guide, since prescriptions must include management objectives. The management emphasis is on the role of fire in natural areas, but such information is also useful in fire applications for other management purposes.

The structure of most virgin forests in the American West today reflects a past disturbance history that includes fire. Although media reports of the 1988 Yellowstone fires treated the scene as an ecological catastrophe, these forests were born of fire in the 1700s and are now being reborn in the 1990s. Knowledge of the natural and often inevitable disturbances likely to affect forests, including fire, is essential to any forest management plan, whether the objective is timber production, wildlife conservation, or wilderness management. Creating desirable forest stand structures in the future for these objectives may not require simulation of past fire activity. Such efforts, however, will be successful only if we understand the processes responsible for desirable structures we see today before undertaking future stand manipulation.

The geographical coverage of this volume is applicable to much of the western United States, although the focus is on forest types found in Oregon, northern California, and Washington. Where those types occur in adjacent regions, information on them has been included. I chose to exclude those forest types endemic only to other areas of the West, such as giant sequoia forests. Nonforest vegetation is included where it is transitional to forest, as in oak and juniper woodlands or subalpine environments. I did not attempt to provide detailed discussion of fire weather and fire behavior. These are the subjects of other monographs and could only be touched upon in this one. I purposely downplayed the coverage of shrub and herbaceous vegetation so as to keep the book from becoming too long. Although the book features natural history, political events at the regional and national level have influenced the use of fire in our forests, and a cultural history of fire is included to place the information in context.

The organization of the forest chapters primarily follows a "potential vegetation" concept. This means that some tree species, notably Douglas-fir, are discussed in several chapters. For example, Douglas-fir is discussed in chapters 7 and 8 as a major seral dominant, and in chapter 10 as a seral and climax dominant. This organization helps clarify the role of a particular species within the successional stages, or sere, of the various forest zones. The index will help readers locate all discussions of widely distributed and important species, such as Douglas-fir.

A decade ago, I reviewed Wright and Bailey's book *Fire Ecology* for *Science* magazine and noted their claim that their book was a "progress report" in a rapidly expanding field. The treatment I have chosen focuses on the West and includes more material on stand development than was covered in their book. I hope that this book stimulates more research and progress reports on the fascinating and complex subject of fire ecology.

ACKNOWLEDGMENTS

THIS BOOK COULD NOT HAVE BEEN WRITTEN without the support and encouragement of institutions, friends, and colleagues. The National Park Service financially and logistically supported several fire ecology projects described in this book, under Cooperative Agreement CA-9000-8-0007 Subagreement 7. Those projects could not have been completed without the help of my graduate students over the past 15 years, and I thank all those students.

Reviews of chapters were graciously provided by the following individuals: Stephen Arno, Mark Finney, Richard Fonda, Jerry Franklin, Charles Grier, Charles Halpern, Mark Huff, J. Boone Kauffman, Bruce Kilgore, Robert Martin, Philip Omi, David Parsons, David Peterson, Lois Reed, John Stuart, Frederick Swanson, Peter Teensma, Donald Theoe, Stephen Veirs, Jan van Wagtendonk, and Ronald Wakimoto. Their suggestions and information led to substantial improvements. Of course, the responsibility for final changes, and for what was judiciously added and subtracted, remains in my hands.

James G. Darkow Design, Seattle, redrafted many of the illustrations. Lys Ann Shore copyedited the book and significantly improved the presentation. I would also like to thank Barbara Dean and Barbara Youngblood of Island Press, who supported the concept of the book from my first correspondence with the press and kept my spirits high near the finish line.

Finally, I owe a debt of gratitude to my wife, Wendy, and children, Jules and Suzie, for their patience and understanding for the two years while I worked on this manuscript over seemingly endless evenings and weekends.

FIRE ECOLOGY OF PACIFIC NORTHWEST FORESTS

CHAPTER 1

THE NATURAL FIRE REGIME

DISTURBANCE IS AN INTEGRAL PROCESS in natural ecosystems, and management of forest ecosystems must take into account the chance of natural disturbance by a variety of agents. In some situations, such as park or wilderness management, natural disturbance may be required by law or policy to maintain natural ecosystems. In others, natural disturbance may wreak havoc with specific management goals, such as wood production or maintenance of a specific wildlife habitat. Fire is a ubiquitous disturbance factor in both space and time, and it cannot be ignored in long-term planning. Its effects can be integrated into land management planning through an understanding of how fire affects the site and the landscape.

Today's plant communities reflect species assemblages in transition, each reacting with different lag times to past changes in climate, and each migrating north or south, up or downslope. Many species have not closely coevolved with the other species they are found growing with today, because of differential rates of migration over past millennia. Each species, however, may have coevolved for much longer periods with particular processes associated with it.

Fires have been associated with most species of angiosperms and gymnosperms through much or all of their evolutionary development.

THE PALEOPYRIC IMPERATIVE

Fire is by no means a recent phenomenon. As long as plant biomass has been present on the earth, lightning has ignited fires, and the myriad ecological effects have been repeated time and again. The history of fire extends well back into the Paleozoic Era, hundreds of million years before the present and long before the angiosperms existed on earth. The Carboniferous Period, so named because of the extensive coal deposits formed during that time, have extensive amounts of fusain (Komarek 1973, Beck et al. 1982). Fusain is a fossil charcoal produced by fires that is almost completely inert, allowing it to survive through the geologic eras (Harris 1958). Fusain has little volatile content and glows on combustion, in contrast to coalified plant tissue, which burns with a smoky flame (Harris 1981). Wildfire was probably a regular occurrence on the earth during and since the Mesozoic (Cope and Chaloner 1985), when gymnosperms dominated the earth and angiosperms developed.

Fire may have been associated with the extinction of dinosaurs. A catastrophe following a large meteorite striking the earth is now a widely accepted theory for the significant deposition of iridium at the Cretaceous-Tertiary (K-T) boundary, also associated with significant peaks in carbon content (10^2–10^4 above background levels; Wolbach et al. 1988). The carbon is mostly soot, and the ejecta from the hypothesized collision lie on top of the soot, implying that the soot was created rapidly and was deposited before the weeks- to months-long deposition of the remainder of the mineral ejecta. A single massive global fire or a series of forest fires occurring around the globe would have been necessary to explain the amount of carbon found in these deposits (Wolbach et al. 1988). Whether such fires were simply another effect of the meteorite impact or whether they were in fact a co-primary cause of biological extinction is a question that may be debated for decades. The magnitude of such a potential event makes the Yellowstone fires of 1988 seem no more than a minor spark on the landscape.

Ecosystems with substantial presence of fire almost always contain species that are able to take advantage of it to survive as individuals or species. Plant adaptations, which will be discussed further in chapter 5, such as thick bark, enable a species to withstand or resist recurrent low intensity fires while less well-adapted associates perish. Some pine species have serotinous (late-opening) cones, which have changed little since the mid-Miocene (Axelrod 1967). While closed, these cones hold a viable seedbank in the canopy that remains protected until the trees burn. After a fire, the cone scales open and release seed into a freshly prepared ashbed. Other species maintain a similar seedbank in the soil, which lies dormant until heated. Many species have the ability to sprout after being burned, either from the rootstock or from the stem. The adaptations of plant species to fire are more widespread and common than animal adaptations, but they are less spectacular than the adaptation of the *Melanophila* beetle.

The *Melanophila* beetles are flat-headed borers, found worldwide, which usually breed in fire-damaged pines. Eggs are deposited below the bark, where in larval form the beetles feed on the cambium of newly killed trees and later emerge as adult beetles. Adults are known to be stimulated by heat and/or attracted to smoke. Linsley (1943, p. 341) noted that at University of California football games, with 20,000 or so cigarettes ablaze at any time (remember, this was the 1940s), a haze of tobacco smoke would hang over Memorial Stadium. Melanophila beetles would "annoy patrons by alighting on the clothing or even biting" during a big game, which was more disturbing to fans than a Stanford touchdown. Linsley found that the beetles had sensory pits on their bodies and could somehow sense heat or smoke. Later these pits were determined to be infrared detectors that allowed beetles to find burned areas where newly damaged trees were likely to be found and where the highest probability existed of successfully rearing a brood. This adaptation can only be interpreted as a direct attraction to the presence of fire to increase species fitness, an adaptation that must have taken millennia to evolve.

The earth has long been a fire environment. The fires of Indonesia in 1982 (Davis 1984) and northeastern China in 1987 (Salisbury

1989), each of which burned millions of hectares, are a testament that earth is still a fire environment. Smaller episodes like the Yellowstone fires of 1988 (570,000 ha) are not the first nor will they be the last to strike the northern Rocky Mountains (Christensen et al. 1989). Our approach to fire management in North America must accommodate fire (Pyne 1989a); we cannot be so bold as to think we can eliminate fire from the landscape. It has been with us so long precisely because it is an inevitable part of our environment.

THE RECENT QUATERNARY

Our knowledge of fire on the Pacific Northwest landscape improves as we approach the present, although much remains unknown. In particular, evidence since the last glaciation suggests a substantial interaction among vegetation, climate, and fire that continues to the present. Climate directly affects vegetation and influences the probability that the vegetation will burn. During periods of climatic change, when conditions at many sites will favor establishment of new species combinations, burning will increase the rate of expansion of shade-intolerant vegetation and decrease the spread of shade-tolerant, late successional species (Brubaker 1986).

Changes in species composition on a site may be inferred from pollen analysis of cores, usually drawn from peatland areas. Ages of the sequences within a core are determined from radiocarbon dating, and an index to fire activity can be determined from charcoal in the same layers. Pollen analyses can be used to reconstruct regional vegetation patterns during the Holocene (Fig. 1.1). In the western Cascades, the relationship between vegetation and fire activity was very dynamic (Tsukada et al. 1981, Cwynar 1987). Retreat of the Fraser glaciation resulted in a forest dominated by spruce and lodgepole pine (*Picea* and *Pinus contorta*) in the Puget Trough between 15,000 and 12,000 ybp. Western hemlock (*Tsuga heterophylla*) entered at that time, suggesting a warming that was associated with increasing summer drought. Douglas-fir (*Pseudotsuga menziesii*) and bracken fern (*Pteridium aquilinum*) became dominants, with red alder (*Alnus rubra*) dominant in riparian settings (Barnosky et al. 1989). (Appendix B lists the common and scientific names of plants mentioned in the text.)

The period between 10,500 and 7,000 ybp was warmer and drier than today. Samples from that time period contain the greatest charcoal peaks, implying that fire was more prevalent during that dry, warm period. Over the past 5,000 years, charcoal peaks have declined from their maxima, and a more stable lowland vegetation, increasingly dominated by western hemlock and western redcedar (*Thuja plicata*), has persisted to the present. Although fire may have

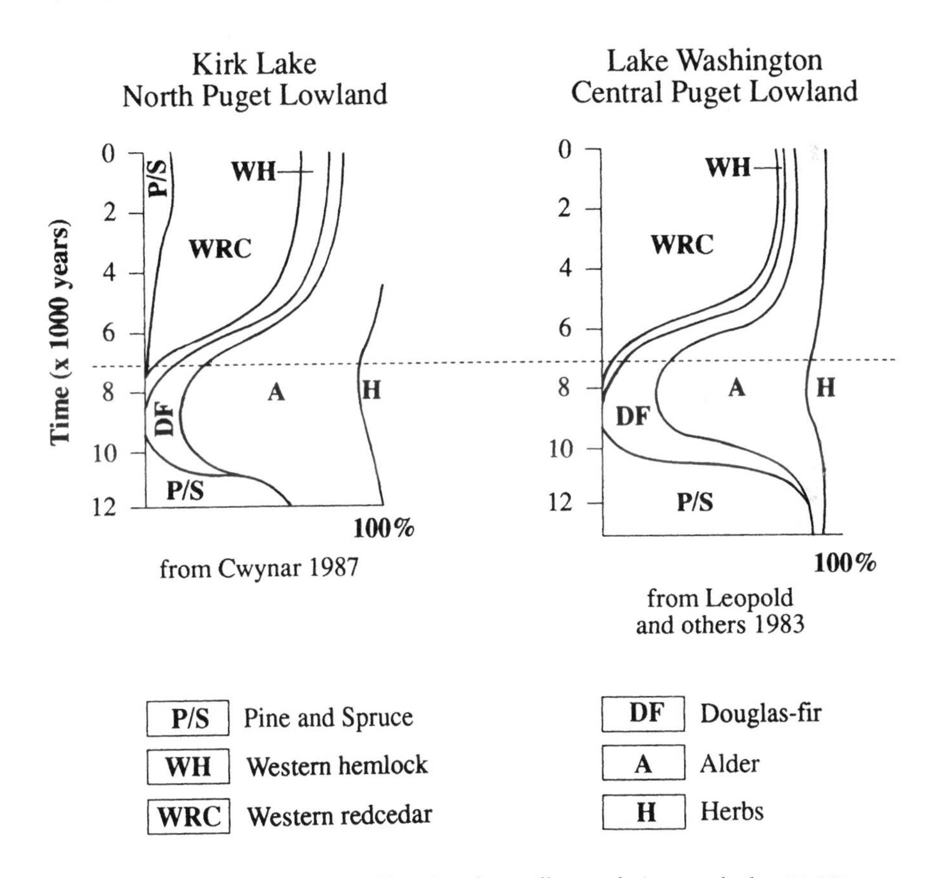

FIG. 1.1. *Trends in species composition, based on pollen analysis, over the last 12,000 years in the Puget Trough, Washington. Dashed line is date of Mazama ash layer, about 6,600 ybp.* (From Brubaker 1991)

interacted with species such as Douglas-fir for hundreds of generations, it appears to have interacted with the species mix common to today's mesic old-growth, Douglas-fir forests for perhaps 10–20 generations (5,000 years) of Douglas-fir (Brubaker 1991).

THE CURRENT MILLENNIUM

Fire evidence in the Pacific Northwest for the current millennium becomes more obvious, since many of the tree species can live for 500–1,000 years (Franklin and Waring 1979). These trees may provide, through forest age structure or fire scars, a direct record of fire activity (see chapter 4). Almost every forest type has experienced a fire in the current millennium, and some may have burned more than a hundred times. Although the evidence of fire is visible on today's landscape, presence alone is an insufficient criterion by which to understand the effects of fire in forested ecosystems. Not only is there variability in fire frequency between forest types, but this frequency has varied over time (Fig. 1.1). It is also necessary to understand other characteristics of fire before fire effects can be holistically interpreted.

THE PROCESS OF DISTURBANCE

Traditional theories of natural disturbance have embraced two concepts that are now discarded: it must be a major catastrophic event, and it originates in the physical environment and therefore is an exogenous agent of vegetation change (White 1979). We now embrace a much broader concept of disturbance, recognizing a gradient from minor to major and the endogenous nature of many disturbances (due either to biotic agents or ecosystem states that encourage an agent). As we accept this broader concept, we thereby create a fuzzier image of disturbance.

Disturbance is difficult to define in ecological terms. The simple dictionary definition of the word is to "interrupt" or to "break up a quiet or settled order." We are well aware that forest ecosystems are not quiet or settled orders, whether recently burned by crown fires or the oldest of old-growth stands. How can we define disturbance in the context of a dynamic ecosystem? When do the "normal" rhythms of the system oscillate to the point where they become "abnormal" or a disturbance? There is no clear answer, as shown by the definition of White and Pickett (1985, p. 7): "A disturbance is any relatively discrete event in time that disrupts ecosystem, community, or population structure and changes resources, substrate availability, or the physical environment." They note that this definition is purposely generalized and place Harper's (1977) "disasters" (a frequent disturbance likely to be repeated in the life cycles of successive generations) and "catastrophes" (a rare disturbance unlikely to be repeated as a selective force) as subsets of "disturbance." Disturbance in forest ecosystems, however, need not be either a disaster or a catastrophe in the normal sense of these words.

Disturbance effects can be ordered to some extent by several characteristics (Table 1.1). These characteristics may be used to describe a single event or a series of events. Disturbance type includes but is not limited to fire, wind, ice and freeze damage, water, landslides, lava flows, insect and disease outbreaks, and effects caused by humans (White and Pickett 1985).

The characteristics in Table 1.1 are not wholly sufficient to describe disturbance effects. Effects of wind, for example, will depend on local topography and forest structure. Blowdown is more important on poorly drained soils (Gratkowski 1956), in wide valleys, and where the area is oriented along the direction of the prevailing winds (Moore and Macdonald 1974). Species' tolerances to wind may be site-specific. Western hemlock is generally prone to windthrow, western redcedar and Sitka spruce (*Picea sitchensis*) may at times be windfirm, and Douglas-fir has been described as both wind-tolerant and wind-sensitive (Boe 1965, Moore and Macdonald 1974). Dominants in a stand are often more windfirm than intermediate crown-class trees (Boe 1965, Gordon 1973). The characteristics of the disturbance are a starting point to understand ecological effects.

Sometimes simple descriptors may disguise the processes creating the major ecological effects. At Mount St. Helens, a layer of tephra now called the Wn erupted in 1480, depositing a meter or more of tephra near the mountain (Yamaguchi 1986). Yet the falling clasts

TABLE 1.1 CHARACTERISTICS OF DISTURBANCE REGIMES

Descriptor	*Definition*
Type	The agent of disturbance: fire, wind, etc.
Frequency	Mean number of events per time period, expressed in several ways: Probability—Decimal fraction of events per year Return interval—The inverse of probability; years between events Rotation/cycle—Time needed to disturb an area equal in size to the study area
Predictability	A scaled inverse function of variation in the frequency
Extent	Area disturbed per time period or event
Magnitude	Described as intensity (physical force, such as energy released per unit area and time for a fire, or windspeed for a hurricane), or severity (a measure of the effect on the organism or ecosystem)
Synergism	Effect on the occurrence of the same or other disturbances in the future
Timing	The seasonality of the disturbance, linked to differential susceptibility of organisms to damage based on phenology

SOURCE: Adapted from White and Pickett 1985.

of the Wn tephra were apparently warm rather than hot, since twigs and leaves at the base of this layer are not carbonized, and many Douglas-fir survived the event with a meter or more of tephra around their bases. In contrast, the 1980 eruption of the volcano left much less tephra in many locations but blew down many trees with an earthquake-triggered explosion of hot and rapidly moving rock debris (Lipman and Mullineaux 1981). The thickness of the tephra, as a single measure of the disturbance, may not be well correlated with its ecological effects.

In the alluvial flats of the redwood region of California, coast redwood (*Sequoia sempervirens*) is adapted to periodic disturbance by silt deposition from periodic flooding events around the tree bases. While the other conifers are usually killed after such deposition, coast redwood can produce a new root system to replace its buried root system and take advantage of the resources in the new substrate (Stone and Vasey 1968). However, such trees are not adapted to coarse-textured deposits associated with rapid logging of unstable watersheds (Agee 1980). The "drought fickle" deposits are not capable of supporting a new root system, when water tables rise and flood the buried roots. Thus, a change in the quality of the flooding disturbance, rather than a simple measurement of its depth, adversely affected the ability of individuals of a species to survive an event they had survived for previous centuries.

Disturbance is not an easy process to characterize, as these examples attest. Fire effects have been investigated less thoroughly than aspects of fire behavior related to fire control. We now have good capability for prediction of fire behavior, linking fuels, weather, and topography information into behavioral characteristics of a single fire. The next step, predicting the effect of a fire, or of a series of fires varying in frequency, intensity, seasonality, and extent, is just beginning.

FIRE AS A DISTURBANCE FACTOR

Fire is a classic disturbance agent within the criteria of White and Pickett (1985). It is a relatively discrete event in time, although it may vary from seconds in a grass fire to weeks or months in a peat fire. Fire changes ecosystem, community, and population structure, either by selectively favoring certain species or creating conditions for new species to invade. It usually favors early successional species but sometimes can "accelerate" succession to favor late successional species. Fire also changes resource availability. It usually increases mineral elements, such as calcium or magnesium, and temporarily reduces total site nitrogen while at the same time increasing available nitrogen. The physical environment is also altered. The removal of the organic layer covering the soil can deepen permafrost thaw levels, and a blackened soil surface and loss of tree canopy after intense fires can increase maximum and decrease minimum surface temperatures. Such effects may be specific to a fire or an ecosystem, however, so that gross generalizations about fire effects are not possible. In a particular forest type, the ecological effects of fire can be better understood by knowledge of the species present and their relative competitive abilities, as well as of their reaction to the complex process called fire.

The disturbance descriptors in Table 1.1 are a good start to describe the effects of fire on Pacific Northwest forests. Each forest type tends to have unique effects when burned, due to variation in fire characteristics, such as frequency and intensity. These effects are also linked to the species present and to the adaptations of each to the specific combination of descriptors, which can be called a fire regime.

Fire Frequency

The presence of fire on most landscapes is observed through fire-scarred trees, or age classes of trees or shrubs that regenerated after the last fire. Using several techniques, we can calculate from such records an average return interval for fire (Table 1.2). Techniques to measure fire frequency, and the limitations of each, will be discussed in chapter 4. For now, the data in Table 1.2 should be interpreted with the knowledge that fire frequency estimates were made using a variety of methods and periods and probably do not reflect the actual fire record of any single decade or century of the current millennium.

The fire-return intervals were developed for Oregon and Washington using forest-cover type areas and definitions drawn from the first regional forest survey of the region (Andrews and Cowlin 1940, Cowlin et al. 1942). "Fire cycle" data were drawn from available literature. Both regional and forest type differences are evident. On average, fire return intervals in Oregon forests (42 years) are about half as long as in Washington forests (71 years), reflecting the somewhat cooler, moister weather of Washington, particularly on the west side of the Cascades (Franklin and Dyrness 1973). There also exists a difference of more than an order of magnitude in fire frequency by forest type, with the ponderosa pine type burning almost every decade and subalpine and spruce/hemlock forests burning perhaps once in the current millennium. From these data alone, a remarkable presence and variation of fire in the region is obvious.

TABLE 1.2. FIRE-RETURN INTERVALS FOR OREGON AND WASHINGTON OVER THE PAST FEW CENTURIES

FOREST TYPE	OREGON			WASHINGTON		
	Area in Type[a] *(1,000 ha)*	*Fire Cycle (yr)*	*Area Burned per Year (1,000 ha)*	*Area in Type (1,000 ha)*	*Fire Cycle (yr)*	*Area Burned per Year (1,000 ha)*
Cedar/spruce/hemlock	292	400[b]	0.7	1,291	937[g]	1.4
Douglas-fir	4,444	150[c]	29.6	3,068	217[j]	14.1
Mixed conifer	399	30[d]	13.3	504	50[k]	10.1
Lodgepole pine	757	80[e]	9.4	211	110[l]	1.9
Woodland	1,001	25[f]	40.0	12	25	0.5
Subalpine	1,075	800[g]	1.3	935	500[m]	1.9
Ponderosa pine	3,142	15[h]	209.4	1,438	15	95.8
Other	2,397	133[i]	18.0	1,949	298	6.5
Total/average	13,507	42	321.7	9,408	71	132.2

[a] Andrews and Cowlin 1940, Cowlin et al. 1942.
[b] Agee (unpublished analysis from Andrews and Cowlin 1940).
[c] Means 1982, Morrison and Swanson 1990.
[d] McNeil and Zobel 1980, Atzet and Wheeler 1982, Agee 1991b.
[e] Gara et al. 1985.
[f] Martin and Johnson 1979, Bork 1985, Atzet and Wheeler 1982.
[g] Extrapolated from Fahnestock and Agee 1983.
[h] Weaver 1959, Bork 1985.
[i] A weighted average of other types.
[j] Agee et al. 1990, Fahnestock and Agee 1983.
[k] Finch 1984, Wischnofske and Anderson 1983.
[l] Agee et al. 1990.
[m] Fahnestock 1976, Agee et al. 1990.

PREDICTABILITY

Predictability, or variation in fire frequency, helps explain the presence or absence of some species in ecosystems. Let's assume that an ecosystem burns intensely about every 30 years, and that shrubs are adapted to fire by one of two mechanisms: they sprout, or they have seeds that are able to survive the fire and help replace the mature individuals killed by the fire. Let's say that the seeding-type shrubs need five years before they reach maturity and set seed. In a fire regime where fire frequency is exactly 30 years (very predictable) or even 20–40 years, both types of species will be found, since the disturbance allows each type of species to complete its life cycle. Now, let's assume in a portion of the landscape (maybe south aspects), one fire cycle is missed but is "made up" by two fires in close succession, perhaps two or three years. At the time of the second fire, the population of seeder-type shrubs, all of which were killed by the first fire, has no mature individuals. The seedbank created by the previous generation has germinated, resulting in a population of seedlings two to three years old, none of which is capable of setting seed. After the second fire, which kills the immature individuals, the shrub population will consist primarily of sprouting species, a result not of the average fire frequency, which may have stayed the same, but rather of the predictability of the event.

Another example illustrates the other end of the predictability spectrum. Let's assume an ecosystem where the average fire occurs on a 10 year "cycle" and where there are two tree species: one that develops thick, woody bark early in life and one that has much thinner bark until it is 20–30 years old. The thinner barked species is better able to survive in thick, suppressed stands than the thick-barked species. In this scenario, a 10 year fire cycle (or shorter) selects for the species with thicker bark, although many individuals of the thicker-barked species are also killed by these recurrent fires. If an unusual 30 year period goes by without a fire in one section of the forest, the thinner barked species may grow large enough to survive the next fire and may eventually dominate this patch of forest. Hundreds of years later it may be difficult, without a detailed fire history, to recognize that this patch is the result of the variation in fire frequency rather than its mean.

Magnitude

The magnitude of a disturbance event can be described as intensity or severity. Here only intensity will be used; severity will be defined later in the context of fire regimes. Fireline intensity is derived from the energy content of fuel, the mass of fuel consumed, and the rate of spread of the fire. The units of fireline intensity reflect energy release (kW) per unit length (m) of fireline: energy release along a linear fire front. The length of the flames of a fire can be related to its fireline intensity:

$$I = 258\ FL^{2.17}$$

where I = fireline intensity (kW m^{-1}), and FL = flame length (m). A shortcut to a rough estimate of fireline intensity is provided by Chandler et al. (1983):

$$I = 3\ (10\ FL)^2$$

From observation of the flame length as it passes a point (Fig. 1.2), its rate of energy release can be measured.

The units of measurement at first seem rather odd, but wildland fires generally move along a well-defined front as a line phenomenon, so that the rate of energy release along the line is a logical descriptor of the intensity of the event. Fireline intensity can be directly linked to some ecological effects. For example, crown scorch is highly correlated to fireline intensity (with adjustments for ambient air temperature and windspeed).

Three general categories of fireline intensity can be related to

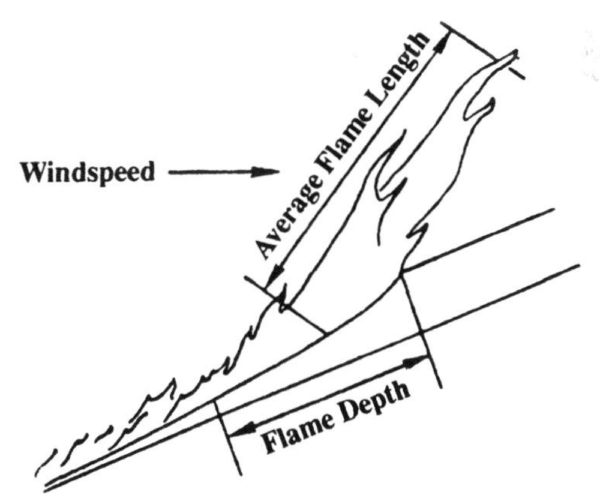

FIG. 1.2. *Flame dimensions shown for a wind-driven fire on a slope.*
(From Rothermel and Deeming 1980)

flame length and fire behavior (Table 1.3, Fig. 1.3). The surface fire is the lowest intensity category, with flame lengths <1 m, and ecology crews can generally work around the head of the fire. The understory fire (flame lengths 1–3 m) is of intermediate intensity and must be viewed from a distance, as heat and smoke can be excessive and erratic fire behavior is possible. The crown fire is the highest intensity category (flame lengths >3 m), and can range over 30,000–40,000 kW m^{-1} (Stocks 1987). Most prescribed fires, ignited by managers, are in the first category, while natural fires in the Pacific Northwest can span the entire range.

Extent

The areal extent of a fire has been difficult to link to fire effects. If a fire consumes or kills mature individuals of a species over a wide area, and its seedbanks are destroyed, recolonization may be delayed, particularly if the species has large, heavy seeds. The new community may be dominated by survivors and light-seeded spe-

TABLE 1.3. FIRELINE INTENSITY INTERPRETATIONS

FIRELINE INTENSITY		FLAME LENGTH	INTERPRETATION OF THE FIRE
SI units (*kW m^{-1}*)	*English units* (*BTU sec^{-1} ft^{-1}*)	(*m*)	
0–258	0–75	0–1	Surface fire: Most prescribed fires in this range. Control by hand line probable. Fire-monitoring information can generally be collected near head of fire.
258–2,800	75–812	1–3	Understory fire: Fires too intense for direct attack. Equipment usually successful. Fire monitoring usually remote, by camera, thermocouples, etc.
>2,800	>812	>3	Crown fire: Erratic fire behavior (crowning, spotting, major fire runs) probable. Fire monitoring must be remote for safety reasons.

SOURCE: Partly adapted from Rothermel 1983.

FIG. 1.3. *Fireline intensity, ranging from intense crown fire behavior* (A) *exceeding 50,000 kW m^{-1} to understory fires* (B) *around 5,000 kW m^{-1} to very low intensity fires <50 kW m^{-1}* (C).
(Photos [A *and* B] *courtesy of Canadian Forestry Service; author photo* [C])

cies. In marginal environments for a species, such as in subalpine areas, availability of seed source year after year may be critical in postfire reestablishment (Agee and Smith 1984). Large areal extent may also explain long regeneration delays in lowland environments (Franklin and Hemstrom 1981). In chaparral ecosystems, extent has the opposite effect. Very small fires act as attractants for heavy browsing by deer and rabbits, and may result in the conversion of brush patches to stable grasslands (Biswell et al. 1952).

Seasonality

Season of burning can be very important in determining fire effects. Spring burning occurs at a time when buds are flushing and are very susceptible to damage. Late in the season, buds have hardened and are much more capable of withstanding heat. In a recent fire-effects model, season and bud size were incorporated into the algorithm that predicted crown damage from fireline intensity (Peterson and Ryan 1986).

In the spring shrubs often are at yearly lows in terms of carbohydrate reserves in the roots because of the demand to produce new shoots, leaves, and flowers. Burning in spring will weaken and can occasionally kill sprouting shrubs, and burning when the soil is moist can kill seeds of native herbaceous perennials (Parker 1987). Spring burning can also kill fine conifer roots (Swezy and Agee 1991), which may predispose trees to moisture stress during the coming dry season. These examples of seasonality effects are primarily oriented to out-of-season burning. Prescribed burning in the spring may be preferred for air quality purposes or for ease in controlling fire, but it is not generally a preferred season from a natural ecological point of view in the Pacific Northwest.

Synergism

Fire may interact with other disturbances in a synergistic manner. By creating open landscapes when it burns intensely, fire may increase the magnitude of rain-on-snow events, and it can increase shallow soil-mass movements on the landscape by reducing fine root biomass (Swanson 1981). It can open a stand to increased windthrow by reducing crown resistance through tree mortality. It can encourage creation of infection courts for fungi (Gara et al. 1985) on roots that are scarred.

The effect of fire on trees can include excessive heat transmitted to roots, cambium, or crown. Each effect, separately or in combination, can reduce the tree's resistance to insect attack. The probability of successful insect attack increases with increasing fire damage (Fischer 1980).

The overall effect of a forest fire, then, is the sum of its direct, indirect, and synergistic effects. These synergistic effects may be difficult to quantify (for example, root infection courts) or may be tied to even more complex disturbances, such as regional droughts in the case of insect attacks. If a stand is already drought-stressed, the added stress effects of fire may greatly increase the chance of an insect outbreak.

FIRE REGIMES OF THE PACIFIC NORTHWEST

A fire regime is a generalized description of the role fire plays in an ecosystem. Systems for describing fire regimes may be based on the characteristics of the disturbance, the dominant or potential vegetation of the ecosystem in which ecological effects are being summarized, or fire severity based on the effects of fire on dominant vegetation. Each system has been applied to Pacific Northwest forests, and each is generalized enough to be useful for some applications. None has substantial advantages over the others, so all these systems may continue to find use in the region.

The first system of fire-regime definition is based on the nature of the disturbance (Table 1.4). This system was developed by Heinselman (1981) for forest ecosystems in the lake states and was applied to Pacific Northwest forests by Agee (1981a) (Fig. 1.4). This system uses combinations of frequency and intensity to describe six fire regimes, ranging from 2 (frequent, low intensity fire regime) to 6 (very infrequent, intense fire regime). Of course, some fire regimes

TABLE 1.4. FIRE REGIMES DEFINED BY THE NATURE OF THE DISTURBANCE

Fire Regime Number	*Description of the Regime*
0	No natural fire (or very little)
1	Infrequent light surface fire (more than 25 year intervals)
2	Frequent light surface fires (1–25 year return intervals)
3	Infrequent, severe surface fires (more than 25 year return intervals)
4	Short return interval crown fires (25–100 year return intervals)
5	Long return interval crown fires and severe surface fires in combination (100–300 year return intervals)
6	Very long return interval crown fires and severe surface fires in combination (over 300 year return intervals)

SOURCE: Heinselman 1973.

seem to fit between those described or to be a combination of two of them. The strength of this system is its basis in the disturbance itself; it can be used widely to collectively describe ecosystem effects of fire. A variant of this system is the fire regime as defined by Johnson and Van Wagner (1985), which includes depth of burn (duff removal) in addition to measures of fire frequency and intensity.

The second system of fire-regime definition uses the characteristics of the vegetation (Table 1.5). In this example, the potential vegetation (in the sense used by Daubenmire 1968a) of a forest is used to define "fire groups," for which fire frequency, intensity, and effects can then be summarized. When a management system is based on similar vegetation units, such as habitat types, this fire-regime system is a useful way to catalog fire and ecological information. Such a grouping is not easily applied to areas with different vegetation. Some of the fire groups on the Lolo National Forest in western Montana (Davis et al. 1980), for example, are not present on the national forests east of the Continental Divide in Montana (Fischer and Clayton 1983). Disturbance may affect understory species composition, too, so identification of habitat type is not

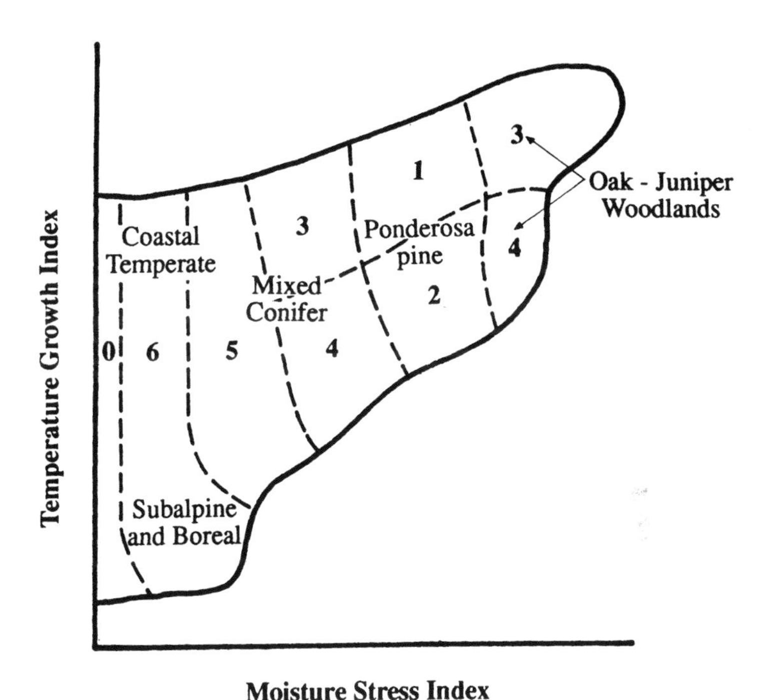

0 Little Fire Influence

1 Infrequent Light Surface Fires (Above 25 Yr.)

2 Frequent Light Surface Fires (1-25 Yr.)

3 Infrequent Severe Surface Fire (Above 25 Yr.)

4 Short Return Interval Crown/ Severe Surface Fire (25-100 Yr.)

5 Long Return Interval Crown/ Severe Surface Fire (100-300 Yr.)

6 Very Long Return Interval Crown/ Severe Surface Fire (300+ Yr.)

FIG. 1.4. *Fire regimes based on physical characteristics of the disturbance for the Pacific Northwest.*
(From Agee 1981a)

independent of fire history (Stewart 1988). The simplicity of the system begins to bog down when one considers the literally hundreds of fire groups (or habitat types) across the western United States. Although this system is best applied on a local basis, the site-

TABLE 1.5. FIRE GROUPS BASED ON POTENTIAL VEGETATION OF THE SITE

Fire Group	*Description of Fire Groups*
0	A heterogeneous collection of special habitats: scree, forested rock, meadow, grassy bald, and alder glade.
1	Dry limber pine (*Pinus flexilis*) habitat types.
2	Warm, dry ponderosa pine (*Pinus ponderosa*) habitat types. These sites may be fire-maintained grasslands.
3	Warm, moist ponderosa pine habitat types. These sites will support dense thickets of ponderosa pine in the absence of frequent fires.
4	Warm, dry Douglas-fir (*Pseudotsuga menziesii*) habitat types. These sites are dominated by ponderosa pine and experience understory thickets of Douglas-fir if protected from disturbance.
5	Cool, dry Douglas-fir habitat types. Douglas-fir is often the only conifer on these sites.
6	Moist Douglas-fir habitat types.
7	Cool habitat types usually dominated by lodgepole pine (*Pinus contorta*).
8	Dry, lower subalpine habitat types.
9	Moist, lower subalpine habitat types. Fires are infrequent but severe.
10	Cold, moist, upper subalpine and timberline habitat types. Fires are usually infrequent due to fuel limitations, but have long-standing effects.
11	Warm, moist grand fir (*Abies grandis*), western redcedar (*Thuja plicata*), and western hemlock (*Tsuga heterophylla*) habitat types. Fires are infrequent but severe.

SOURCE: Davis et al. 1980.

NOTE: These groups are defined for Lolo National Forest in Montana and would need to be redescribed for other forest areas.

specific nature of fire effects makes it appropriate in many cases, and it is the primary system of chapter organization in this book.

The third system of fire-regime classification is based on the effects of fires on dominant vegetation, from low to high severity (Fig. 1.5). It is the most generalized and simplest system of the three,

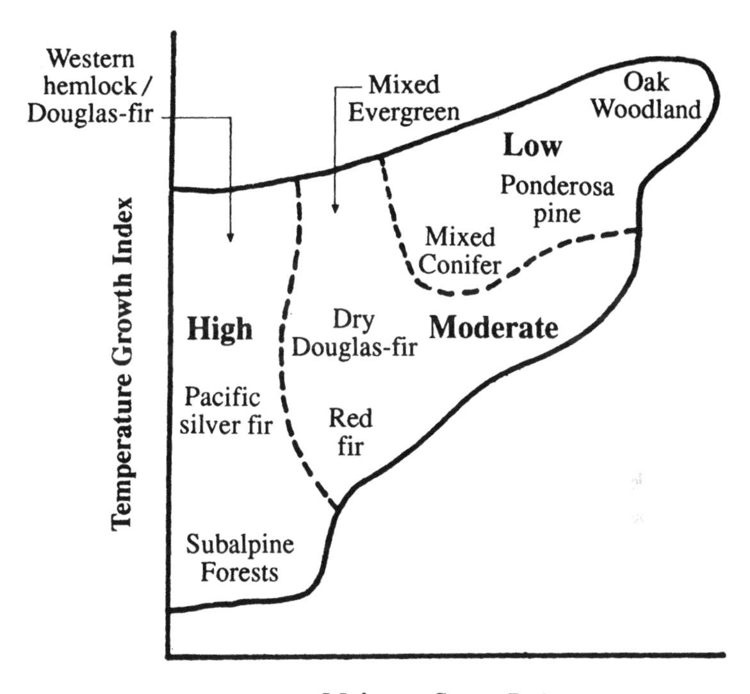

FIG. 1.5. *Fire regimes based on fire severity. Stands in low severity fire regimes have 20% or less of the basal area removed by a fire, while stands in high severity fire regimes have 70% or more of their basal area removed.* (From Agee 1990)

so it can be universally applied, but it loses discriminating power at a local level. Because almost every fire has very severe effects on some species, even while leaving other species unaltered, this system must be defined in terms of dominant tree species to be meaningful. Because some tree species may have unique fire adaptations to resist damage, fires of similar intensity may be classed as fires of different severity in different ecosystems. This system can be used more than others to recognize the variability in fire that occurs within or between fires on a site (Fig. 1.6). On any site, all levels of fire severity will be present over large scales of space and time, but characteristically in different proportions. The most complex fire regimes are in the moderate-severity fire regimes in the central part of Figure 1.6.

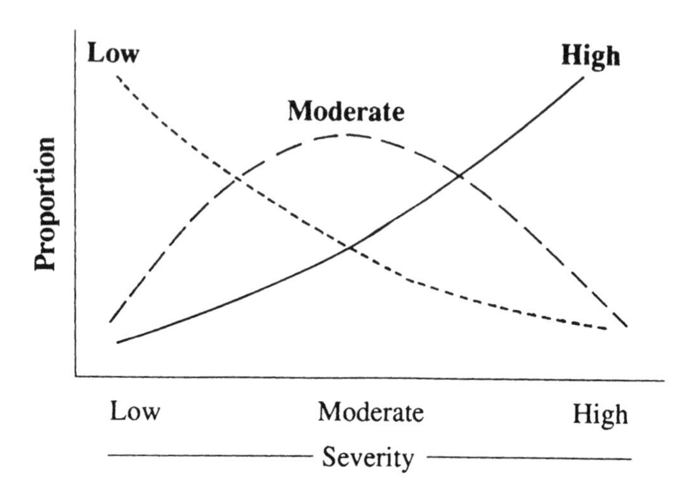

FIG. 1.6. *Variation in fire severity within a general fire-severity type. The general fire-severity types differ in the proportion of each fire-severity level present, with the moderate severity fire regime having the most complex mix of low, moderate, and high severity fires.*

Whether physical, floristic, or ecological criteria are used to define fire regimes, the concept of the fire regime is a useful way to catalog fire effects. In chapters 7–12 floristic groupings have been used to summarize ecological effects, but physical and ecological criteria are also used as appropriate.

CHAPTER 2

THE NATURAL FIRE ENVIRONMENT

THE MANY COMBINATIONS OF FIRE REGIMES and forest types in the Pacific Northwest are a product of repeated fires with variable spread rates and intensities. Many of the fires are ignited by natural sources, while others can be traced to human sources. The human component of the fire environment is discussed in chapter 3. This chapter focuses on the natural fire environment, covering those physical aspects of fire useful in understanding fire effects. This chapter is not intended to serve as a guide for predicting fire weather or behavior. For a more thorough analysis, the reader is strongly encouraged to consult the publications cited in the text.

THE PROCESS OF IGNITION

Fires can be ignited by various natural processes. Falling rocks may produce sparks that can ignite fine fuels. Spontaneous combustion can occur from thermal decomposition of well-insulated organic materials; this can be an oxidation process, as in a pile of oily rags, or a thermal decomposition process not requiring oxygen (Davis 1959). Volcanoes may create forest fires as superheated lava approaches or engulfs nearby forests (Smathers and Mueller-Dombois 1974). Superheated air can scorch forests surrounding volcanoes, as occurred at Mount St. Helens in 1980 (Means et al. 1982), and glowing avalanches can entrain and burn forests, as suggested by the buried, charred stems in pumice deposits around Crater Lake. All these sources, however, are responsible for less than 1 percent of naturally occurring forest fires.

Lightning is the primary natural source of forest fires in the world. As many as 44,000 thunderstorms occur daily over the earth (Trewartha 1968), and up to 1,800 such storms may be in progress at any one time (Taylor 1974). Thunderstorms are most common in the warmer latitudes of the earth, the warmer seasons of the year, and the warmer hours of the day, suggesting that surface warming is important in thunderstorm development. Thunderstorm activity in the Pacific Northwest is relatively mild compared to the southeastern United States, which averages 10 times as many thunderstorm days per year (Trewartha 1968). Portland and Tacoma average 5 thunderstorm days per year, while Baker, Oregon, and Walla Walla, Washington, east of the Cascades average between 10 and 15; Mobile, Alabama, averages 74 thunderstorm days per year (Alexander 1924). However, the average annual number of lightning fires is greater in the West because less precipitation accompanies the thunderstorms. The area around Mobile averages <25 lightning fires per million ha per year, compared to 25–100 in the Pacific Northwest. Less than 2 percent of annual fire occurrences in the Southeast, Northeast, and Midwest are lightning-ignited, while in the Rocky Mountains 57 percent and in the Pacific states 37 percent of all ignitions are from lightning (Taylor 1974).

A thunderstorm is a violent type of convectional shower associated with conditionally unstable air, a triggering mechanism to release the instability, and sufficient moisture in the air (Schroeder and Buck 1970). Surface heating of air or the introduction of cold air at high altitude will steepen the temperature lapse rate, or temperature change with altitude, of the lower atmosphere. A moist parcel of air at the surface that is lifted may then rise by free convection. As it rises, it cools and releases latent energy through the condensation process. The moisture condensed by this updraft is carried upward as raindrops and ice crystals, while air temperatures within the cell remain higher than those outside the cell. The puffy clouds formed at this time cause this to be known as the cumulus stage (Schroeder and Buck 1970). Next, in the mature stage, rain begins to fall from the thunderstorm cell. Downdrafts of air accompany the falling rain particles and can create strong surface winds. In the Pacific Northwest, thunderstorm tracks may be as much as 40 miles wide, essentially a moving system of thunderstorm cells covering the entire sky (Morris 1934).

Lightning potential is highest during the mature stage of the thunderstorm. Under normal conditions, the atmosphere is positively charged relative to the ground, but in a thunderstorm cell a negatively charged lower cloud layer induces a positive charge on the ground. Most lightning discharges are within-cloud, representing discharges between the negative charges of the lower cloud and positive charges of the upper cloud. Cloud-to-ground discharges, which are responsible for forest fires, occur between the positively charged ground and negatively charged lower cloud. In the Pacific Northwest, within-cloud discharges are a large percentage of total discharges (Morris 1934). Lightning is accompanied by thunder as air is heated and expanded along the path of the discharge. As the downdrafts continue, the thunderstorm cell weakens and eventually dissipates.

Not every lightning strike will cause a fire. As the lightning bolt strikes, probing its way rapidly from the cloud to the ground, a series of return strokes will flash upward to the cloud. Three or four return strokes are common, causing the bolt to flicker (Schroeder and Buck 1970). Sufficient energy is present in the discharge to melt metals, fuse soil particles, and ignite dead woody plant material (Taylor 1970). Where a long continuing current phase (>40 msec) is associated with the return strokes, the probability of an ignition increases (Fuquay et al. 1972). Ignition will also depend on the type of fuel encountered on the ground and its moisture content. In the Olympic Mountains, Pickford et al. (1980) correlated an increased probability of ignition with the number of days without significant rain previous to the thunderstorm.

Two common types of thunderstorms occurring in the Pacific Northwest are the cyclonic, associated with a regularly formed low pressure system, and the anticyclonic, consisting of a low pressure trough between two highs (Alexander 1927). Western Washington usually experiences the cyclonic type, which carry significant rainfall and are not often associated with forest fires. The anticyclonic type is common in July and August, and more than half of them cause fires (Alexander 1927). Within the Pacific Northwest, significant variation exists in thunderstorm track patterns across the region and from year to year (Morris 1934; Fig. 2.1). Although these tracks were defined only for national forest areas during the 1920s, they indicate that areas such as the Okanogan Highlands, Blue Mountains, and central Oregon Cascades are "hot spots." Similar patterns are described for the same area for the 1945–66 period by Komarek (1967).

Considerable variation was found from year to year and from storm to storm, some being widespread and others consisting of localized events (Morris 1934). The pattern of lightning in southern Oregon in 1987, which caused 600 fires, is an example of Morris's general pattern, which tends to have more cloud-to-ground discharges. These simultaneous fires burned 75,000 ha in Oregon alone (Helgerson 1988). Although the Olympic Mountains were omitted from the analysis shown in Figure 2.1, lightning does strike these mountains and cause forest fires (Pickford et al. 1980).

Lightning starts vary by elevation, aspect, and fuel type (van Wagtendonk 1986). In forests of California (Komarek 1967) and Idaho (Fowler and Asleson 1984), mid-elevation areas had the most starts. This is hypothesized to be the result of interaction between fuel moisture (increasing with elevation and reducing probability of ignition) and lightning strikes (increasing with elevation) (Fowler and Asleson 1984). With elevation zones, the upper one-third of slopes had the most ignitions, and south aspects had more starts than north aspects.

Lightning starts by vegetation type are difficult to assess. A widespread fuel type may have the highest number of ignitions, but when normalized on a unit area basis it may be no more significant than other vegetation types. In Idaho, Fowler and Asleson (1984) suggested there were significant differences in ignitions between vegetation types from the number expected if ignitions were proportionally distributed by area. In the Northern Region national forests, Bevins and Barney (1980) noted that the Douglas-fir type

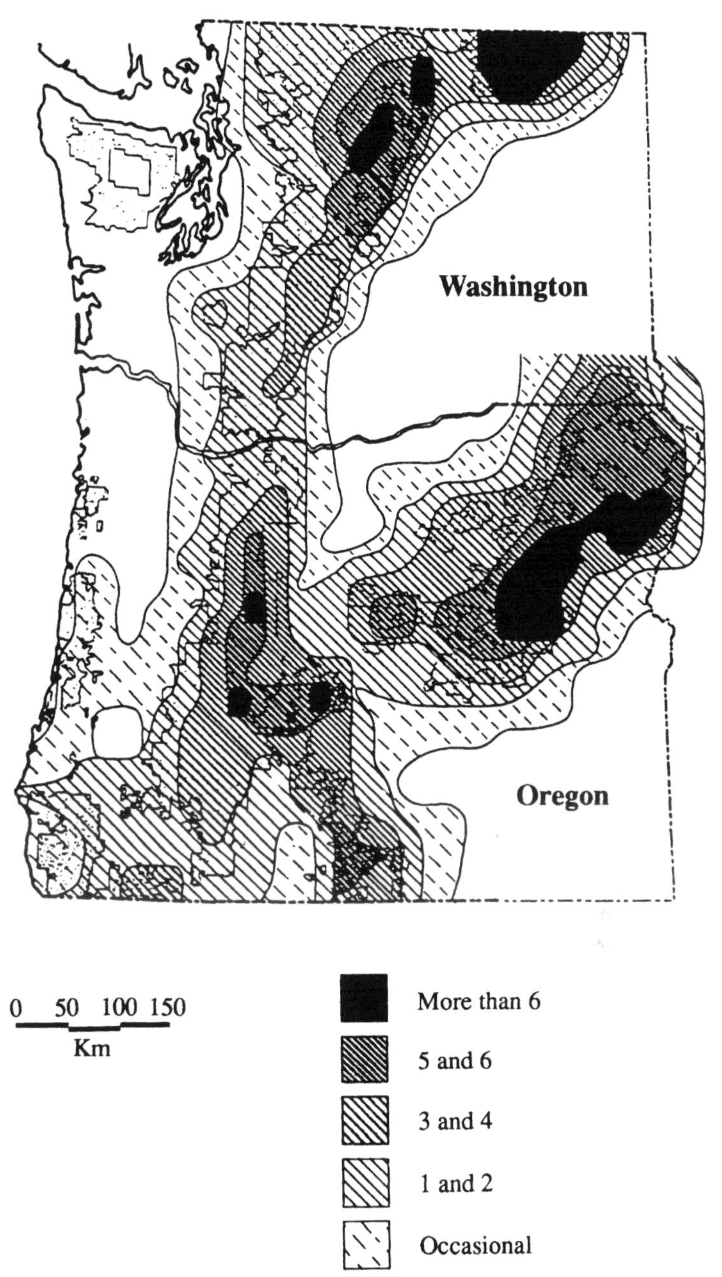

FIG. 2.1. *Regional annual lightning patterns on national forests of Oregon and Washington. These patterns are an index to lightning fire activity and show number of storms per 40,000 ha per year.*
(From Morris 1934)

had the highest number of fire starts, but when normalized to a basis of starts per unit area, ponderosa pine and western white pine had the highest number of starts, while Douglas-fir was seventh of the ten types identified. In California, van Wagtendonk (1986) found the red fir type to have the highest number of lightning fire starts per unit area, although upper and lower mixed-conifer types had a higher total number of fires.

These data suggest that a meteorological basis for natural fire has existed worldwide for thousands of millennia (Komarek 1967, 1968), and the Pacific Northwest is no exception. Although the frequency of lightning storms in the region is relatively low, it nevertheless is sufficient, given the summer/dry weather patterns, to explain much if not all of the historic fire activity recorded in the forests of the region.

THE PROCESS OF COMBUSTION

Sustained combustion from an ignition source requires three ingredients, often referred to as the fire triangle: fuel, heat, and oxygen. Every fire-control technician is taught that fire suppression involves removing one leg of that triangle; for example, application of water removes heat and oxygen, and fireline construction removes fuel. In forest fires, chemical energy stored in forest fuels is converted to radiant, kinetic, and thermal energy through the combustion process (Byram 1959). The chemical energy in the fuel has itself been transformed from the sun's radiant energy through photosynthesis. It can either be oxidized slowly, as microbes decompose organic matter in the forest floor, or rapidly through the process of combustion.

Combustion is thus a rapid oxidation process dependent on fuel, heat, and oxygen. The process can be described in four stages (Fig. 2.2): (1) preheating and distillation; (2) distillation and the burning of volatile fractions; (3) burning of the residual charcoal; and (4) cooling (Byram 1959). During the preheating stage, the process is endothermic, requiring heat to raise the temperature of the fuel particles above 250–300°C. Moisture is vaporized, and chemicals with low boiling points are volatilized. The transition to the second stage is the ignition process above 300°C, as the process becomes exothermic. No specific temperature is associated with the transition, as the flaming stage depends on the richness of the gas mix, the amount of water vapor within it, and the availability of oxygen (Chandler et al. 1983). Flame temperatures can exceed 500–600°C.

As the volatiles are consumed, the process of glowing combustion of the residual surfaces of carbon begins the third stage. The fourth stage is the cooling period, at the end of which ambient temperatures have again been reached. If the process is complete, ash will be the only material left, composed of minerals, such as calcium or magnesium, with boiling points above combustion temperatures. Normally, the combustion process is incomplete, leaving blackened carbonaceous materials behind that are relatively inert and resistant to subsequent microbial decay. Such combustion residues may persist for thousands of years even in oxygenated environments.

Four modes of heat transfer occur during the combustion process: conduction, convection, radiation, and mass transfer. Conduction is the movement of heat from one molecule to another. The particles themselves are not displaced in the process. Conduction is generally of little importance in fire spread but of major importance in fire

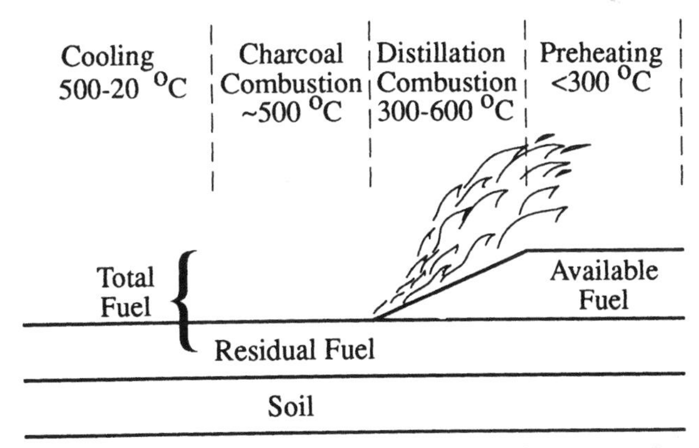

FIG. 2.2. *Four stages of the combustion process. The amount of fuel consumed is known as available fuel.*

effects. Heating of cambial tissues through bark, heat transfer along logs, and soil heating from burning surface organic matter occur primarily through conduction. Low conductivity can result in high conservation of energy, allowing a smoldering log fire to dry and slowly consume a log with relatively high moisture content (>40 percent).

Convection is the transfer of heat by currents in liquids or gases. Heated molecules move and are displaced in convection. Convection is an important heat transfer process in fire spread and effects. Crown scorch, for example, is modeled as a convective process (Van Wagner 1973). Radiation is the transfer of heat by electromagnetic wave motion. From a point source, it decreases with the square of the distance away from that source. Radiation can be important in preheating surface fuels, and in combination with convection it can be important in creation of fire scars (e.g., Gara et al. 1986). The fourth type of heat transfer, mass transfer, is also called spotting (Albini 1979, Chandler et al. 1983). Active firebrands may be carried by wind currents ahead of the main fire front, usually under conditions of very low fuel moisture. In the 1988 White Mountain wildfire complex in northeastern Washington, unusual spot fire distances of 30 km were noted under high winds and exceptionally low fine fuel moisture levels of about 2 percent (R. Dunnagan, Incident Commander, personal communication, 1988). Similarly long spotting distances have been recorded in Australia (Chandler et al. 1983).

The direct effects of fire on living organisms depend on critical temperatures being reached and maintained for a period of time. For example, a usual standard of plant tissue death is 60°C for one minute. Whether a tissue such as stem or root cambium reaches this critical temperature is not only a function of heat input at the bark or soil surface but also of the characteristics of the medium through which the heat is conducted. The density of the medium, its specific heat, and its thermal conductivity will determine for a given heat input the temperature rise at some distance into the medium. Water is a critically important variable in altering these characteristics. Its high specific heat and high conductivity relative to wood, litter, and soil can have significant effects on vegetation and soil heating (see chapters 5 and 6).

ELEMENTS OF FIRE BEHAVIOR

Fire behavior can be understood through the use of another fire triangle, the fire behavior triangle: weather, topography, and fuels. When specific characteristics of each of these factors are known, the behavior of a single fire at its leading edge can be predicted. Surface and understory fires are predictable within reasonable limits. The chances of firebrand spotting and crown fires can be estimated, but the behavior of crown fire is still relatively unpredictable.

WEATHER

Radiant energy from the sun is the engine that drives the earth's oceanic and atmospheric circulation. Such circulation patterns result in the local weather affecting forest fires. Spatial and temporal global patterns may be linked to forest fires much more than we have realized. In the Olympic Mountains of Washington, Henderson and Peter (1981) identified large-fire years of the past as occurring in 1309, 1442, 1497, and 1668. These years of large fire events are also within temporal envelopes of low sunspot periods of the past: 1280–1350, the Wolf minimum; 1420–1530, the Sporer minimum; and 1645–1715, the Maunder minimum (Stuiver and Quay 1980). The sunpot minima are thought to be periods of global cooling, and the Maunder minimum is associated with the beginning of the Little Ice Age. Beyond the seeming paradox of large fire events occurring under a global cooling scenario, the linkage of local fire history to global climate adds a fascinating layer of complexity to current scenarios of global change.

A better-documented example of regional fire patterns and global weather cycles is provided by the fire histories of the southwestern United States (Swetnam and Betancourt 1990). A synchrony of fire-free and widespread-fire years was found across the ponderosa pine forests of Arizona and New Mexico. These synchronies were linked to high (*La Niña*) and low (*El Niño*) phases of the Southern Oscillation, a measure of sea surface pressure around the equator that affects the position of the Intertropical and South Pacific convergence zones. For example, during 1982–83, an extreme *El Niño* event, little fire activity occurred across the southwestern United States due to above-normal rainfall carried northward from the tropics. At the same time, millions of hectares burned in Borneo (Davis 1984) and Australia (Rawson et al. 1983) due to below-normal precipitation there caused by the shift in the Intertropical convergence zones associated with a strong *El Niño* event.

Much more is known about general circulation patterns and the associated winds and moisture that are important in fire behavior. Unequal heating of the earth by the sun results in temperature differences across the earth's surface. Air in these warm zones, such as the belt around the equator, rises and moves both north and south, transporting energy with it and pulling in cooler, denser air at the surface from those areas. At about 30° latitude, this air descends and begins to move both north and south at the surface. Because the earth is turning, such air movement to north or south appears to occur at an angle, apparently "deflected" to the east as it moves north and to the west as it moves south. In the Pacific Northwest (30–60° latitude), such winds are known as *prevailing westerlies*, because they are generally transporting energy to the north but appear to be moving from the west. In Hawaii (between 30 and 0° latitude), the steady *trade winds* moving south appear to be from the northeast.

If atmospheric circulation were so simple, bands of high pressure would be found at about 30° and at the poles, while low pressure would exist at the equator and at 60° latitude. In fact, some semi-permanent cells of high and low pressure exist, which affect both the moisture content of the atmosphere and wind. One of these, the Pacific High, which tends to remain off the Pacific Coast during summer, deflects low pressure cells to the north, resulting in relatively dry summers in the Pacific Northwest. In winter, the Pacific High moves to the south, allowing the Aleutian Low to bring winter storms into the area. Several other, less dominant patterns exist, and some are associated with severe fire weather during the dry season.

The ocean is the major source of moisture in air masses. As the major precipitation-bearing storms in the Pacific Northwest move onshore from the ocean, they encounter mountain masses: the

Coast Range, the Olympics, and the Cascades. Each of these mountain systems forces the air masses to rise, cool, and precipitate moisture.

As an air parcel rises, the decrease in pressure with altitude causes it to cool at a rate known as the dry adiabatic lapse rate (−10°C per 1,000 m rise). As it cools, its absolute humidity stays relatively constant while its saturation vapor pressure decreases, because cold air can hold less moisture than warm air. Therefore, its relative humidity (the proportion of absolute to saturation vapor pressure) increases. As the parcel's relative humidity reaches 100 percent saturation, clouds form and precipitation may occur. Further rising of the air parcel will result in the temperature of the parcel decreasing at the wet adiabatic lapse rate (−5°C per 1,000 m rise). Once over the mountain, the parcel may decrease in altitude, and in so doing it adiabatically warms, quickly becoming unsaturated and stopping the precipitation process. It will then warm or cool at the dry lapse rate until it is again saturated. An example of the effect of this process is shown in Figure 2.3 for the central area of Oregon from the coast to the eastern Cascades. The west side of the Coast Range receives precipitation to the crest in this example, while the east side remains dry, and the Cascades receive moisture only above 1,000 m elevation. The eastern Cascades receive a cool, moist, but not saturated air parcel. If it warms moving across the plateau, it will not yield precipitation until it is forced well above 2,000 m elevation in mountains farther to the east. Each successive set of mountains becomes drier, thus making the regional pattern of annual precipitation (Fig. 2.4) more easily understood.

Summer drought is characteristic of the Pacific Northwest (Munger 1925). Even on wet sites, less than 10 percent of the total precipitation falls during the summer (Waring and Franklin 1979) due to the presence of the Pacific High offshore. Coastal and western portions of the region have more clouds and cool marine air, which lower the evaporative demand. At some forested sites fog drip can be an important additional precipitation mechanism (Harr 1980).

Gradient winds are created by air flowing between high and low pressure on a synoptic scale (Schroeder and Buck 1970). Near and at the surface, those winds will be affected more by landscape characteristics, such as valleys and mountain systems. Winds that affect fire behavior are the result of interactions between gradient

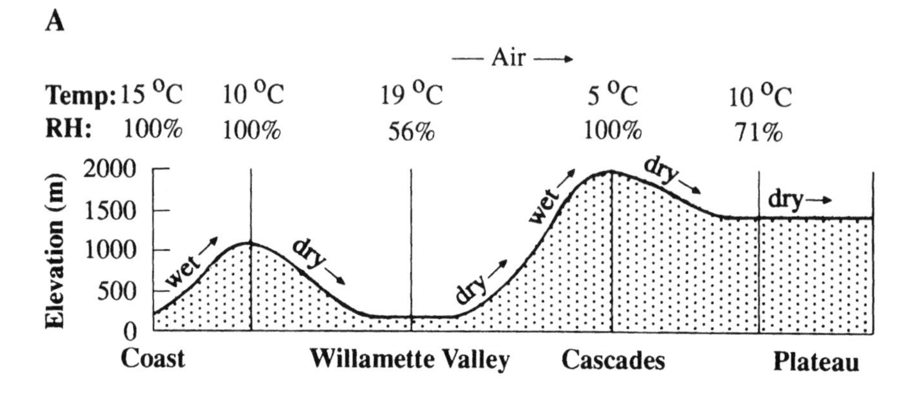

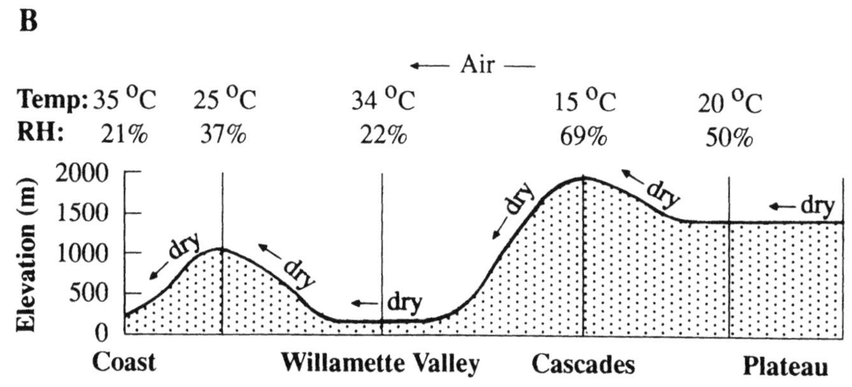

FIG. 2.3. *Changes in air parcel temperature and relative humidity during onshore flow* (A) *and offshore flow* (B). *Dew point lapse rates are ignored in this example. In offshore flow, called east wind, the air parcel arrives at the coast warmer and drier than it began inland.*

winds and local winds, created by local differential heating of the air and channeling due to local topography. The resultant wind may be stronger or weaker than either the gradient or convective wind, depending on whether those winds are blowing in the same or opposite directions.

A gradient wind important in the Pacific Northwest for conditions that favor wildfire behavior is the foehn wind, which can occur in two ways. First, air from higher elevations east of the Cascades can be forced by a strong high pressure system over and across the range

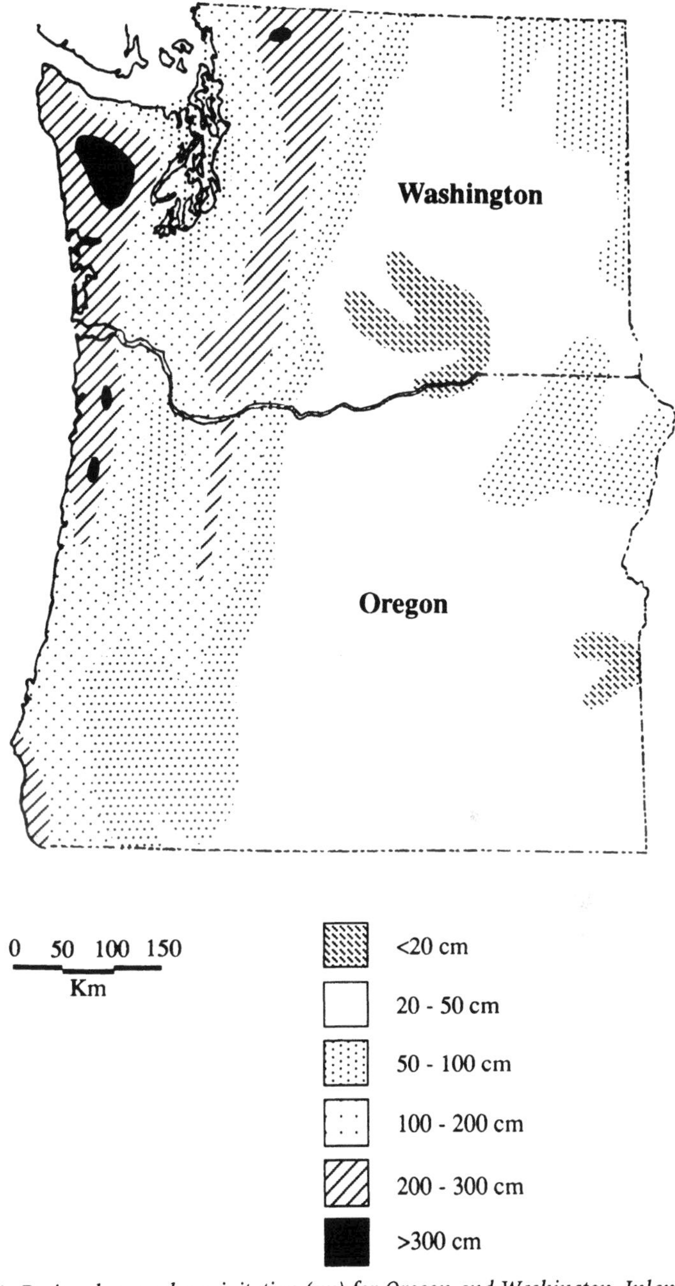

FIG. 2.4. *Regional annual precipitation (cm) for Oregon and Washington. Inland areas are much drier than the coast.*
(From Franklin and Dyrness 1973)

toward a low pressure system offshore to the west. The air parcel may arrive warmer and drier than it began (Fig. 2.3B) because of cooling and warming at different lapse rates. An offshore flow results in a much warmer and drier air parcel due to adiabatic warming as it moves toward the coast. If the pressure gradient between the inland high and coastal or offshore low is large, very strong winds can create severe fire weather. The second mechanism for foehn winds is warm air subsidence from a strong, stable high pressure system, a more common foehn phenomenon in California. The famed Santa Ana winds of southern California, Chinooks of the Rocky Mountains, and Mono winds of the Sierra Nevada are foehn winds. In the Pacific Northwest, east wind is the bland name we have chosen for this phenonemon. Many of the large fires of the Pacific Northwest (Yacolt of 1902, Tillamook of 1933) are associated with east winds or strong north winds.

Slope and valley winds and sea breezes are diurnal convective winds that occur because of differential heating of land, or sea and land, surfaces. They are generally shallow winds (in comparison to gradient winds) produced by local pressure gradients. Slope and valley winds (Fig. 2.5) can be thought of as a counterclockwise diurnal cycle. In the morning, heating of upper slopes by the sun causes upper-slope air to rise and draw cooler air upslope. As the general upper valley heats, the upslope flow moves to a stronger upvalley flow during the afternoon. Cooling of the upper slopes in late afternoon begins a shallow downslope flow, which becomes stronger as the night proceeds. If one were to stand at one point at mid-slope for a diurnal cycle, the wind would appear to shift in a counterclockwise direction. In some locations, larger scale phenomena will overwhelm the slope and valley winds. For example, in the Lake Chelan basin in the eastern Cascades, local thermal lows caused by surface heating in the sagebrush steppe to the east create downcanyon afternoon airflows during summer months. A similar situation occurs in east-facing canyons in many Pacific coastal regions (Schroeder 1961).

The sea breeze is associated with some downcanyon winds. During the day, inland air heats more rapidly than air over the ocean, and the pressure differential causes cooler marine air to move inland, reaching maximum penetration at about the same time of day as maximum temperature (Schroeder and Buck 1970). This air will

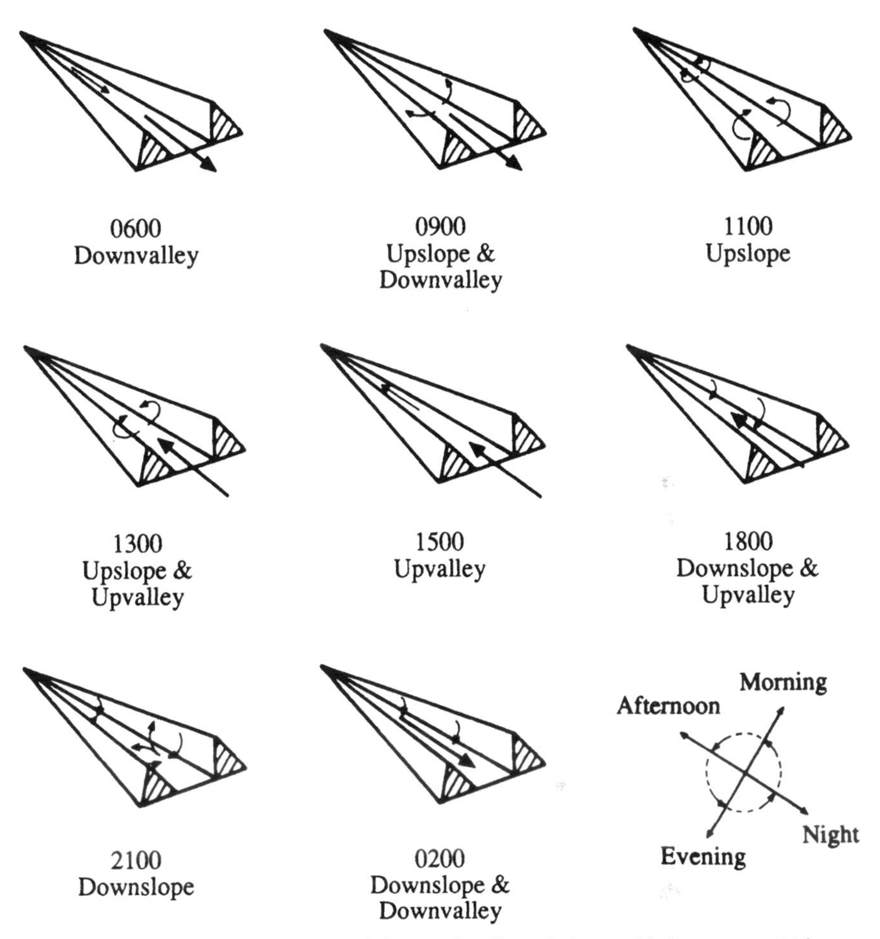

FIG. 2.5. *Counterclockwise pattern of slope and valley winds at mid-slope over a 24 hr period. Strength of wind is represented by length of arrow. Counterclockwise mid-slope pattern of wind direction during a diurnal cycle is shown at lower right.*

funnel along river valleys and sometimes will crest ridges and move downslope in east-facing canyons during midday.

Such wind patterns repeat year after year and help explain fire patterns on the landscape. In the eastern Olympic Mountains, many historical fires have left a chevron pattern on slopes of the major river valleys, reflecting the probable pattern of east winds augmenting the afternoon valley winds. In the western Olympics, the gradient east wind opposes the valley wind, so fires are blockier in shape and appear to move more easily upslope than upvalley. An afternoon downvalley wind pattern was noted on the 1970 Entiat fires (George Fahnestock, personal communication, 1985), and is consistent with historical fire scars on the downvalley (lee) sides of trees in the Stehekin Valley, the next drainage north of the Entiat.

Topography

Topography is the most constant of the three legs of the fire behavior triangle. Elevation, slope, aspect, and physiography all contribute to potential fire behavior. Elevation is important both regionally and locally. Regionally, temperature declines with elevation and tends to be an important environmental gradient affecting the distribution of major vegetation zones (Zobel et al. 1976). Temperature will affect the length of the fire season, particularly at high elevation where snowpack will limit both growing season and fire season.

Slope is an important direct input to fire behavior models. Steeper slopes cause fire to spread faster. Radiant heat is emitted closer to upslope fuel particles and can preheat those particles more effectively. Convective heat moving upslope will also increase heating of the fuel particles. Slope position affects fire behavior since fires starting at the top of a slope are more likely to be dominated by backing and flanking fire behavior, while those starting at the bottom of the slope are more likely to be dominated by heading fire.

Slope with aspect influences fire behavior indirectly by affecting available moisture. South aspects more perpendicular to the sun's rays receive higher solar radiation, and evaporation is higher. Stage (1976) provides a quantitative index to predict relative moisture availability using slope and aspect, and this approach, slightly modified, has been used locally to relate environmental gradients to vegetation patterns (Agee and Kertis 1987). Generally, steep south-facing aspects are driest, and northeast aspects are most mesic. In some situations, such as the San Juan Islands of Washington, steep north-facing aspects are more sheltered from dry southerly summer winds and exhibit more mesic vegetation than flat to gentle north-facing aspects (Agee 1987).

Physiographical effects on fire behavior are related to the shape of the country, either regionally or locally. At a regional level, low mountain passes may be areas where winds funnel across the Cascades. Fires may be more wind-driven in such areas. In areas along the Columbia Gorge, the typically strong upriver winds in summer favored by windsurfers can increase fire spread, and at other times dry, hot east winds funnel through the gorge in the other direction. The patterns of the 1902 Yacolt burn in the Wind River area suggest that east winds were prevalent and strong (Gray 1990). Patterns of fire behavior near Lake Crescent, Washington, where east winds are funneled through a narrow valley, are cigar-shaped as a result of wind acceleration; the most recent example was the 1951 Forks fire (Fig. 2.6). Winds in small, narrow drainages may increase fire intensity near the heads of canyons, and riparian areas may burn with higher intensities than the surrounding landscape because of this channeling effect (Agee 1988).

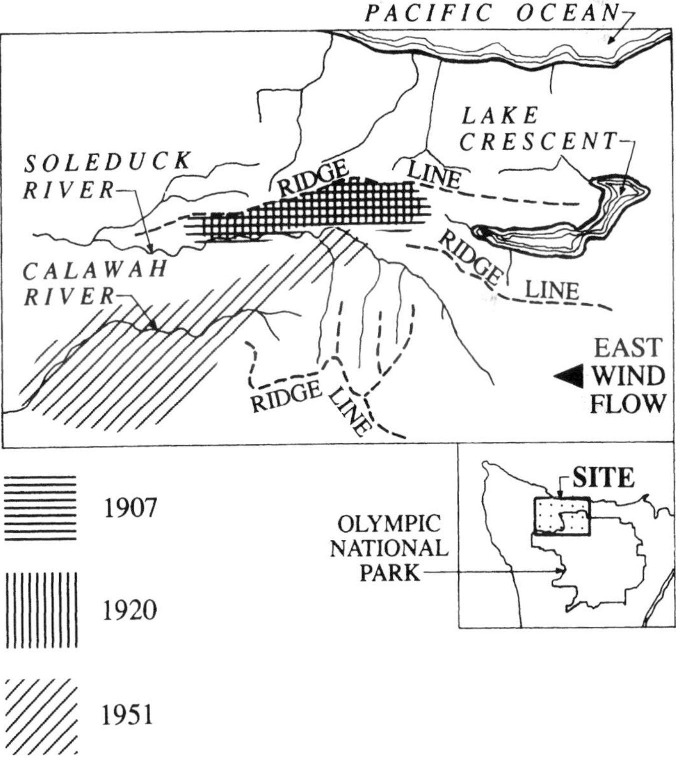

FIG. 2.6. *Funneling of east winds through a narrow corridor west of Lake Crescent, along the northern Olympic Peninsula. Elongated patterns suggest a high level of wind influence on fire patterns, created by topography.*

FUELS

Forest fires can potentially consume all the aboveground biomass on a site, as well as some roots. Normally this is not the case, because some fuels are more susceptible to burning than others. For example, a fire in old-growth, Douglas-fir forest may be intense enough to kill all the vegetation to ground level while consuming only 15 percent of the aboveground biomass (Table 2.1). The 15 percent consumed is known as available fuel, and this proportion of total fuel on the site will vary between vegetation types, and over time in the same types. In chaparral ecosystems available fuel may equal total fuel in some fires, but in most forest types a high proportion of biomass in live stems and large-diameter down logs does not burn, so that available fuel is less than half of the total fuel. Most fuel consumption is aboveground, but old stumps and associated roots will commonly burn out. In wet areas with substantial peat deposits, the ground itself can burn under unusually dry conditions.

Each category of fuel usually has a unique chemical makeup. The heat content of each fuel constituent will vary depending on its chemical composition, and this chemical makeup may vary significantly over a season (Table 2.2). Cellulose has a heat content about 65 percent as high as lignin, and extractive heat content (terpenes, resins, etc.) is about double that of cellulose. The noncombustible mineral content of fuels normally ranges around 5–10 percent of the dry weight; the heat content is usually expressed on an ash-free

TABLE 2.1. AVAILABLE FUEL AS A PERCENTAGE OF TOTAL FUEL

Biomass Category	*Amount in a Douglas-fir forest (tons/ha)* [a]	*Percent consumption by high intensity forest fire* [b]
Stems	645	5
Branches	53	10
Foliage	12	100
Understory vegetation	7	75
Snags, down logs	215	30
Forest floor	51	80

[a] Grier and Logan 1977.

[b] Flames >3 m length, percentages from Fahnestock and Agee (1983).

basis because the ash does not contribute to it. Some minerals, such as phosphorus, have an active dampening effect on pyrolysis and are used in fire-retardant compounds. In most models of fire behavior, heat content is treated as a constant of about 18.5 kJ/g.

The transfer rate of heat and moisture to and from dead fuels depends on the surface area/volume ratio of the fuel particle. Forest fuels are often modeled as cylinders of indeterminate length, so that the surface area/volume ratio can be determined from particle diameter. Smaller particle diameters are associated with higher surface area/volume ratios and with increased ability to gain and lose heat and moisture (this is why small sticks are used to start campfires, with larger pieces added later). Fuel size categories are usually defined by diameter classes representing different surface area/volume ratios. The loading (fuel quantities) within each size (surface area/volume) class is used as an input to models for prediction of fire behavior.

Each of the size classes of fuel is hygroscopic, having a unique pattern of moisture gain and loss with the atmosphere over time. Larger surface area/volume ratios of fuels of smaller size are associated with faster absorption and evaporation. In an environment of constant temperature and relative humidity, all fuel sizes will eventually come to the same moisture content, called the equilibrium moisture content (Table 2.3). Equilibrium moisture content is more sensitive to changes in relative humidity than to temperature. As the environment changes, the smaller fuels will be able to gain or lose moisture more rapidly and thus have a shorter lagtime to reach

TABLE 2.2. ENERGY CONTENT OF COMMON FUELS

Fuel Constituent	*Energy Content (kilojoules/g)*
Pitch	35.3
Oak wood	19.4
Mixed conifer litter	21.8
Chamise leaves (early season)[a]	21.1
Chamise leaves (late season)[a]	23.4
Snowbrush leaves (early season)[b]	19.4
Snowbrush leaves (late season)[b]	21.8

[a] Philpot (1969).
[b] Richards (1940).

TABLE 2.3. EQUILIBRIUM MOISTURE CONTENT OF WOOD AS PERCENT OF DRY WEIGHT, IN VARIOUS CONSTANT ENVIRONMENTS

Temperature		RELATIVE HUMIDITY (PERCENT)									
°C	°F	*90*	*80*	*70*	*60*	*50*	*40*	*30*	*20*	*10*	*5*
10.0	50	20.9	16.4	13.4	11.2	9.5	7.9	6.3	4.6	2.6	1.4
15.6	60	20.7	16.2	13.3	11.1	9.4	7.8	6.2	4.6	2.5	1.3
21.1	70	20.5	16.0	13.1	11.0	9.2	7.7	6.2	4.5	2.5	1.3
26.7	80	20.2	15.7	12.9	10.8	9.1	7.6	6.1	4.4	2.4	1.3
32.2	90	19.8	15.4	12.6	10.5	8.9	7.4	5.9	4.3	2.3	1.2
37.8	100	19.5	15.1	12.3	10.3	8.7	7.2	5.8	4.2	2.3	1.2
43.3	110	19.1	14.7	12.0	10.0	8.4	7.0	5.6	4.0	2.2	1.1

SOURCE: USDA 1987.

the new moisture equilibrium. Small diameter fuels have a short timelag (1 hr), while larger fuels may have timelags exceeding 1,000 hr (Table 2.4). Fuel moisture is assumed to change exponentially, so that within a unit time period moisture content will move 63 percent toward the new equilibrium moisture content (Fosberg and Deeming 1971). For example, a twig of 0.25 cm at 20 percent moisture content placed in a controlled environment with 10.8 percent equilibrium moisture (26.7°C, 60 percent relative humidity) will be at about 14.2 percent moisture (63 percent of the way to the new equilibrium) in one hour, moving another 63 percent of the way in the next hour, and so forth. A larger twig (1.5 cm) will, on average, take 10 hours to complete each of these steps.

In the field, with changing temperatures and relative humidities on a diurnal, synoptic, and seasonal basis, no equilibrium moisture is likely to persist, so that fuels of the various timelag classes are

TABLE 2.4. TIMELAG SIZE CLASSES OF FUEL

FUEL PARTICLE DIAMETER		TIMELAG CLASS
(cm)	*(in.)*	*(hr)*
0–0.62	0–0.25	1
0.62–2.54	0.25–1.0	10
2.54–7.62	1.0–3.0	100
7.62–20.32	3.0–8.0	1,000
>20.32	>8.0	>1,000

continually adjusting to a variable equilibrium moisture content. The smaller size classes, such as the 1 hr timelag fuels, vary significantly on a daily basis and reflect immediate fire weather, while the 1,000 hr timelag fuels react more slowly and reflect more seasonal trends (Fig. 2.7).

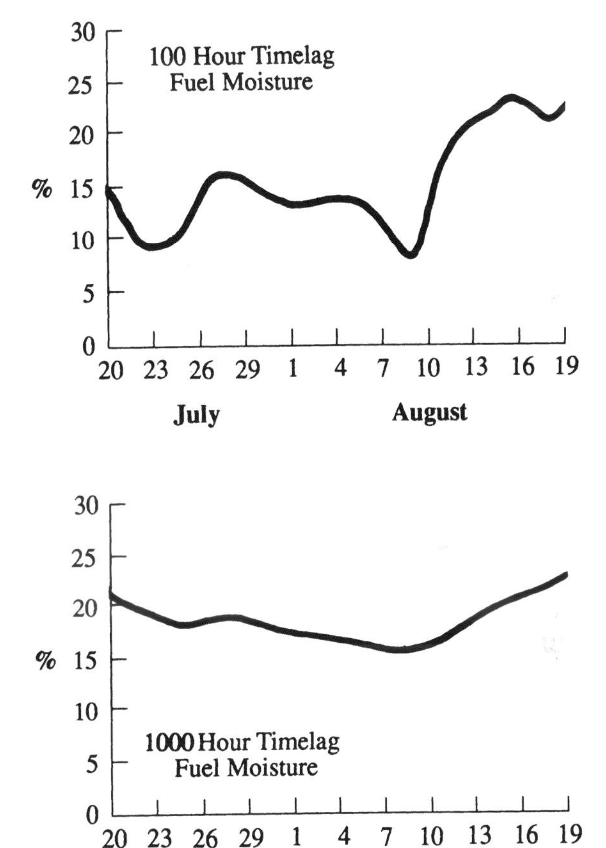

FIG. 2.7. *Moisture content of large dead fuel over a summer month in the Olympic Mountains from four weather stations. Precipitation occurred on 26 July and beginning 8 August. The 100 hr timelag fuels are more responsive than the 1,000 hr timelag fuels; moisture content of the 1 hr and 10 hr fuels, not shown, would fluctuate even more than the 100 hr timelag fuels.*
(From Huff and Agee 1980)

Moisture patterns for live fuel are more complex. Because only the smallest live fuels are normally involved in combustion, leaf moisture is synonymous with live fuel moisture. The moisture content of new leaves may exceed 300 percent on a dry weight basis, decreasing over the season to the level of the older foliage and small twigs, which in the Pacific Northwest may fluctuate around 100 percent (Fig. 2.8). Live fuel moisture is a function of the balance between potential transpiration loss and replacement of that loss from stem water storage or from root uptake of soil moisture. When live fuel moisture dips below 100 percent, crown fire behavior is likely (Woodard et al. 1983) if the crown density is optimum or surface fire intensity is high enough (Van Wagner 1977). In chapar-

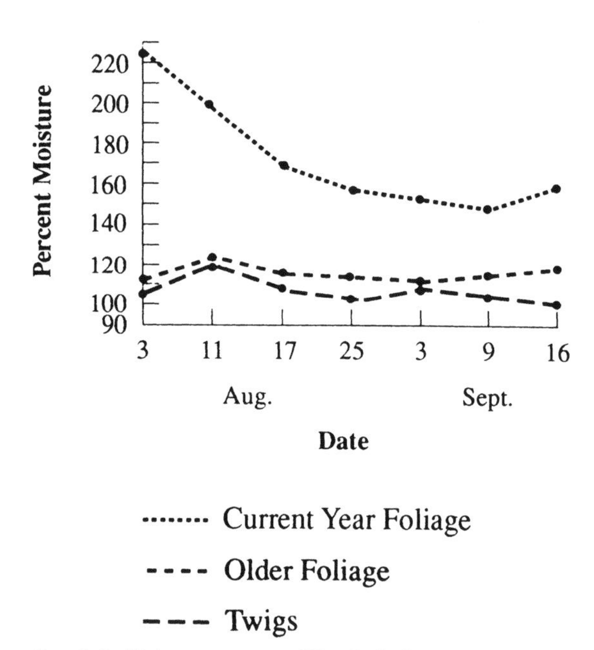

FIG. 2.8. *Moisture content of live fuel of subalpine fir needles and twigs from mid-season on, showing little trend for older needles and twigs, and declining moisture for current-year (new) foliage. Crown fire behavior would be more likely late in the season with declining live fuel moisture, in combination with draped lichens mixed with the foliage.*
(From Huff et al. 1989)

ral ecosystems, live fuel moisture may dip to below 60 percent, and fuels become very flammable.

As moisture content increases in live or dead fuels, flammability decreases because more energy is needed to heat and vaporize water. Above a certain moisture content, called the moisture of extinction, flaming combustion no longer occurs. For most dead fuels, this is in the range of 12–40 percent (Albini 1976, Chandler et al. 1983). For live fuels, with generally higher oil and resin content, the moisture of extinction may exceed 120 percent (Chandler et al. 1983); the moisture of extinction for smoldering duff may be in the same range (Sandberg 1980).

The arrangement of fuel particles in the fuelbed is another important input to models for prediction of fire behavior. Fuelbed density (mass per unit volume) affects the porosity of the fuelbed and its ability to supply oxygen to the fuel. When particle densities are known, a dimensionless ratio, called the packing ratio, can be calculated as the proportion of the fuelbed volume occupied by fuel particles. Along a fire spread gradient from high to low, typical packing ratios might be: grass (.001), loose litter (.01), tightly packed litter (.1). Packing ratios can affect fire behavior by being too low (too much aeration) or too high (fuel too tightly packed).

Fuels are a dynamic component of ecosystems. The amount present at one time will be a function of site productivity, decomposition rates, and past disturbance. Forest fires can reduce total fuel loads but may increase fuels in the dead category by killing live vegetation. Fire regimes are not solely fuel-driven; they depend on the interaction of topography and weather with fuels.

MODELS FOR PREDICTION OF FIRE BEHAVIOR

Fire behavior integrates the effects of the components of the fire behavior triangle on the rate of spread and heat release rate of wildland fires. Development of these models took two decades of intensive work, primarily by U.S. Forest Service scientists (Rothermel 1983). The first decade was spent producing a robust model of surface fire behavior. The second decade was spent obtaining and interpreting outputs for fire managers. This work still continues. The best summary of predictions of fire spread and fire intensity is that of Rothermel (1983), who cites several other publications as being especially helpful in different areas: weather (Schroeder and Buck 1970), fuels (Anderson 1982), calculations (Albini 1976, Burgan 1979), spot fire distance (Chase 1981), interpretation (Andrews and Rothermel 1982), and verification (Rothermel and Rinehart 1983).

The information flow of the model (Fig. 2.9) indicates that the fire behavior model requires up to five inputs, each of which has been briefly discussed in this chapter. The fuel model may be one of thirteen "fire behavior fuel models" defined by Albini (1976; Table 2.5) or a "roll yer own" model for local application based on site-specific fuel information (Burgan and Rothermel 1984). Once the fuel model is specified through experience or a key, the other inputs must be specified. Dead fuel moisture is entered by timelag class, up to 100 hr timelag fuels. Larger fuels are not considered; although they are important in later energy release and fire effects, they do not significantly interact with the leading edge of the fire, which is the focus of the model. Live fuel moisture is estimated for those models with significant live fuel, and slope and windspeed are

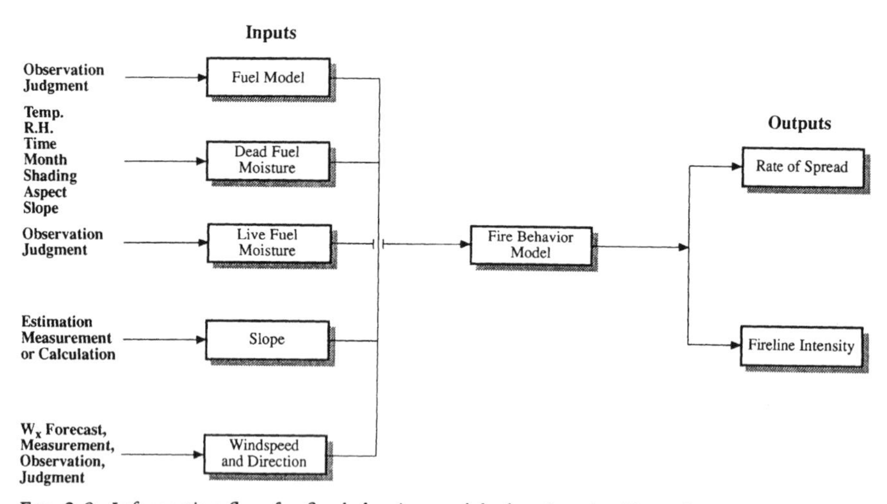

FIG. 2.9. *Information flow for fire behavior model, showing significant inputs and outputs applicable to fire effects.*
(Adapted from Rothermel 1983)

entered. Wind is a critical input to the model, and care must be taken to accurately estimate windspeed at estimated midflame height. The standard 6 m weatherstation windspeed must be adjusted downward by 40–90 percent to account for friction near the

TABLE 2.5. FUEL MODELS FOR FIRE BEHAVIOR

Model Number	*Description*
Grass and grass-dominated complexes (cured fuels only)	
1	Short grass (0.3 m)—western grassland, not grazed
2	Timber (grass and understory)—open pine grassy understory, wiregrass/scrub oak
3	Tall grass (0.6 m)—bluebunch wheatgrass, bluestems, broomsedge, panicgrass
Chaparral and shrub fields	
4	Chaparral (2 m)—mature (10–15 yr) chaparral, manzanita, chamise
5	Brush (0.6 m)—salal, laurel, vine maple, alder, mountain mahogany, young chaparral
6	Dormant brush, hardwood slash—Alaska spruce taiga, shrub tundra, low pocosins
7	Southern rough—southern rough (2 yr), palmetto-gallberry
Timber litter	
8	Closed timber litter—tightly compacted short-needled conifer litter, not much branch/log fuel
9	Hardwood litter—loosely compacted long-needled pine or hardwood litter
10	Timber (litter and understory)—larger dead fuel as well as litter, some green understory, old-growth hemlock-fir.
Logging slash (western mixed conifers)	
11	Light logging slash (<90 t/ha)—most needles fallen, compact; partial cuts and clearcuts
12	Medium logging slash (90–270 t/ha)—most needles fallen, slash somewhat compact
13	Heavy logging slash (>270 t/ha)—most needles fallen, slash somewhat compact

SOURCE: Albini 1976.

ground and sheltering by overstory vegetation. Rothermel (1983) summarizes input refinement; he notes that procedures must be learned and practiced repeatedly if results are to be reliable. Model outputs are made in English units, which can be easily converted to SI units (see appendix A) for purposes of fire effects. The usual interpretive uses are designed for management groups, which typically work in English units.

Fire behavior outputs can be obtained from hand-held computers or microcomputers. The first hand-held system was the Texas Instruments TI-59 calculator augmented with a custom-designed chip (Burgan 1979). Later the Hewlett-Packard HP-71 was similarly adapted with a custom chip. In 1984 a series of interactive computer programs for fire behavior were developed. The overall system is known as BEHAVE and has two major subsystems designed to build and test fuel models and to predict fire behavior (Burgan and Rothermel 1984, Andrews 1986). BEHAVE can be used on microcomputers and is designed to be—and actually is—user-friendly.

An example of some fire behavior output for a standard model for fire behavior prediction (Table 2.5) is shown in Table 2.6. Other custom fuel models could have been built and used in this example. The first step is to define the character of the fuelbed. Loads of fuels by timelag class are needed. Surface area/volume ratios are set in the standard models and need to be estimated for custom models; a combined (weighted) surface area/volume ratio is then calculated. Other fuel input variables also need to be specified. The second step is to define the environmental conditions in which the fire will burn the fuelbed. Fuel moisture by timelag class can be imported from standard "low," "medium," or "high" classes, or the values can be individually defined. One environmental parameter can be specified to take three values. The third step is to run the model; the fire behavior results are calculated and displayed in graphic or tabular form.

The fire behavior results in Table 2.6 show a fire with a predicted rate of spread up to 0.41 km hr^{-1} (0.25 mi hr^{-1}), and flame lengths of 1.5–2.4 m. It would be classified as an understory fire (see chapter 1), and control strategies would be based on indirect or flank attack rather than a direct attack at the head of the fire. Monitoring would likely be remote, using cameras or thermocouples, rather than direct observation. The prediction is valid only for the conditions

TABLE 2.6. FIRE BEHAVIOR OUTPUT FROM BEHAVE

LOAD (T/AC)		S/V RATIOS		OTHER	
1 HR	3.01	1 HR	2000.	DEPTH (FEET)	1.0
10 HR	2.01	LIVE HERB	190.	HEAT CONTENT (BTU/LB)	8000.
100 HR	5.01	LIVE WOODY	1500.	EXT MOISTURE (%)	25.
LIVE HERBS	0.00	SIGMA[a]	1764.	PACKING RATIO	0.01725
LIVE WOODY	2.00	S/V = (SQ FT/CU FT)		PR/OPR[b]	2.35

ENVIRONMENTAL DATA[c]		FIRE BEHAVIOR RESULTS[d]			
		FIRE VARIABLE[e]	MIDFLAME WIND 4.	7.	10.
1 HR FM	6.				
10 HR FM	7.				
100 HR FM	8.	ROS (FT/M)	7.	14.	22.
LIVE HERB FM	120.	FL (FT)	5.	6.	8.
LIVE WOODY FM	120.	IR (BTU/SQFT/M)	5735.	5735.	5735.
		H/A (BTU/SQFT)	1248.	1248.	1248.
SLOPE (%)	30.	FLI (BTU/FT/SEC)	154.	293.	460.

NOTE: In this run one of the standard NFFL fuel models, Model 10 (Timber, Litter, and Understory), is used. Because the program produces English units only, they are presented here.

[a] SIGMA is a weighted surface area/volume ratio for the fuel complex.
[b] PR/OPR is the packing ratio divided by the optimum packing ratio.
[c] "Medium" standard environmental data were used in this example, but a wide range of environmental data can be used.
[d] Wind was chosen at the environmental parameter to vary in this example, but other environmental factors (slope or fuel moisture) could have been chosen.
[e] ROS = rate of spread; FL = flame length; IR = reaction intensity; H/A = heat per unit area; FLI = fireline intensity.

specified in the inputs, so as fuel type, slope, or time of day change, a different prediction will result.

At present, BEHAVE has limited application for predicting fire effects (Andrews 1986) because it focuses on the flaming front and does not consider residual combustion, which is critical for duff consumption or soil and root heating. Such flaming front predictions are useful, however, in effects such as crown scorch (Van Wagner 1973).

Fire behavior models can be adapted to increase their utility for predicting fire effects. Some of the fire effects models summarized later in the book use these fire behavior models as a start and then adapt them for predictions of heating and tree mortality. Over time, the strength of current models for fire behavior should spill over into more applications for predicting fire effects.

CHAPTER 3

THE CULTURAL FIRE ENVIRONMENT

FIRE HAS BEEN PART OF THE NATURAL ENVIRONMENT for 350 million years, about as long as plants have inhabited the earth (Cope and Chaloner 1985). Yet it is also a cultural phenomenon, since human beings use fire (Komarek 1971). No natural history of fire can be complete without full recognition of the symbolism and cultural meaning of fire. This cultural heritage influences our moral vision of fire, and the ways we control and use fire, from slash burning to natural-fire policies. The forests on the landscape today, while reflecting past geological and climatological events, also reflect the influence of Native Americans, European American settlers, and the inhabitants of today. We have much to learn from the smokes of the past, including the prehistoric effects of fire in the region. This chapter emphasizes the role of human beings in the development and implementation of fire policy. It is organized chronologically rather than topically, since the evolution of culture has shaped and will continue to shape the development of fire policy.

THE PREHISTORIC PERIOD

Although fire has been a potent natural force for millions of years, the "celebrated moisture" of the Pacific Northwest has masked the importance of fire in Pacific Northwest ecosystems (Pyne 1982). The relevance of fire to Oregon and Washington today is limited to perhaps the last 10,000 years. Before then, the vegetation was markedly different in many parts of the region; modern species assemblages have appeared only in the last 5,000 years (Barnosky et al. 1989, Brubaker 1991). Pollen profiles in several areas of Washington (Tsukada et al. 1981, Leopold et al. 1982, Heusser 1978) suggest a warmer, drier period about 6,000–9,000 ybp, with significant charcoal fragments mixed with the pollen of that period. In many lowland areas Douglas-fir was a dominant, but there was much less western hemlock and western redcedar than we see today.

Beginning about 5,000 years ago and continuing to this century, a cooler, moister climate has increased the relative importance of western redcedar and western hemlock at lower elevations. There is a strong correlation between the expansion of the range of western redcedar and the development of woodworking technology by native cultures about 5,000–2,500 ybp (Hebda and Mathewes 1984).

Native American Burning

Anthropogenic ignitions are known to have been important over past millennia in certain forest and grassland types of the Pacific Northwest. The dry Douglas-fir and ponderosa pine forests of the eastern Cascades were burned quite frequently by Native Americans (Barrett and Arno 1982). However, the role of aboriginal burning is less clear west of the Cascades for wetter coastal Douglas-fir forests. Native American burning appears to have varied across the region, as a result of the wide range of natural and cultural environments.

Native cultures clearly managed fire to meet their resource needs. For many years, analysts dismissed the possibility of sophisticated resource use of fire activity by Native Americans: "It would be difficult to find a reason why the Indians should care one way or another if the forest burned. It is quite something else again to contend that the Indians used fire systematically to 'improve' the forest. . . . yet this fantastic idea has been and still is put forth time and again" (Clar 1959, p. 7).

Perhaps the most detailed documentation of the relationship between aboriginal cultures and fire is found in southwest Australia (Hallam 1975). Hallam shows that the natives there used fire in a sophisticated manner; the somewhat patchy evidence for the Pacific Northwest suggests similar resource-oriented relationships between native cultures and fire. Pyne (1982) carries this argument to its illogical extreme, claiming that the fire regime of the Olympic Mountains and the entire Northwest was shaped by anthropogenic fire, and that all stands are certainly human-caused. The role of human beings in the ignition and spread of forest fires is important, but this importance varies from place to place and from culture to culture.

At least three common patterns of Native American burning were found in the Northwest: frequent burning in westside prairies and adjacent dry Douglas-fir forest, maintenance of small patches of open prairie for agriculture or hunting by coastal or mountainous tribes, and widespread burning by inland or "plateau" tribes east of the Cascades (cf. Knudson 1980).

Native American burning was a common practice in the Willamette Valley grasslands and oak woodlands (Boyd 1986). Boyd's detailed study of the burning strategies of the Kalapuya Indians of the Willamette Valley contains many historical accounts. David Douglas, the English botanist who explored much of the Pacific Northwest in the 1820s and for whom Douglas-fir is named, traveled up the valley in 1826 and recorded in his diary: "Country undulating; soil rich, light with beautiful solitary oaks and pines interspersed through it . . . but being all burned. . . . Most parts of the country burned; [it is] only on little patches in the valleys and on the flats near the low hills that verdure is to be seen" (Davies 1980, p. 94).

Slightly to the south, in September 1841 Charles Brackenridge, a botanist with the Wilkes expedition, recorded these observations in the Umpqua Valley: "the flats between the rising ground is rich deep soil with Clumps of Ash and Dogwood, the grass had all been burnt up by a fire which we saw rageing ahead of us and were compelled to urge our horses through it. . . . Atmosphere so dense we could not see more than 1/4 mile ahead" (Wilkes 1845, p. 72).

At the base of the Siskiyou Mountains, in what appears to have been more forested terrain, the Wilkes travelers observed: "In passing through the woods we suddenly came on to an Indian woman who was blowing a brand to set fire to the woods probably, we stoped to speak to her, but she was sullen, & dogged, & made no reply, & we passed. . . . We found the woods in fire in several places & at times had some difficulty in passing some of the gullies where it was burning" (Henry Eld, in Boyd 1986, p. 73).

Reasons for such apparently frequent burning included deer hunting through the use of herding fires; harvesting of tarweed (*Madia* spp.); and collection of grasshoppers, hazelnuts, acorns, berries, and roots (such as camas [*Camassia quamash*] and bracken fern rhizomes [*Pteridium aquilinum*]) (Boyd 1986). Similar uses for prairie fires in Washington are documented for grassy prairies of the Chehalis-Centralia area (Boyd 1986) and Whidbey Island (Norton 1979, White 1980). Native Americans seem to have burned the valley bottoms of at least some of the major tributaries of the Willamette River (Teensma 1987), but Burke (1979) found little evidence to support such burning in the more mountainous regions. Botanical evidence (Habeck 1961, Thilenius 1968, Kertis 1986) and pedological data (Ugolini and Schlicte 1973) suggest that prairie-forest ecotones frequently burned and that forest area has expanded over the last century at the expense of prairie.

In the moist forests of the more mountainous areas of the Coast Range, northern Cascades, and Olympic Mountains, there is little convincing evidence that aboriginal ignitions were a widespread ignition source. The Tillamook, Siuslaw, and Quileute tribes are known to have maintained forest openings as yarding areas for deer and for berry production (Boyd 1986). When the forest burned, fires were often of high intensity and uncontrollable. The tribes of coastal Oregon were the victims of some of these fires, having been driven to the waters of the Pacific Ocean to survive (Morris 1934b). The source of these fires is never clearly identified or placed in a cultural context. Native American legends tell of two other fires, both in Washington montane forests, that occurred in prehistoric times. The Quinault tribe has a legend of a great fire that swept down from the Olympic Mountains perhaps 500 years ago, pushing the people into the sea, but the ignition source is not identified (Anonymous 1983). In an area roughly bounded by Mount St. Helens, Mount Rainier, and Centralia, a 500,000 ha fire is rumored to have occurred about 1800. This so-called "Big Fire" was supposedly set by the Cowlitz tribe against the Nisqually tribe, or was set by the Nisquallies as a means of generating rain during a drought (Clevinger 1951). The existence of two stories about the same fire makes both suspect: either one, or both, could be true. At present, the case for widespread aboriginal fires throughout the wetter part of the Douglas-fir region is not convincing.

The plateau tribes lived in classic "fire environments" capable of widespread underburns almost every year. The inland tribes burned forest for several reasons: (1) to maintain open stands for travel; (2) to improve hunting by stimulating growth of desirable plants, to facilitate stalking, or to surround game animals; (3) to produce certain food or medicinal plants; (4) to clear campsites to reduce the risk of wildfire and avoid ambushes; (5) to improve grazing for horses (obtained in the early 1700s); and (6) to communicate with other groups (Shinn 1980, Barrett and Arno 1982). Similar anecdotes to those of the Willamette Valley are found for the Klamath Indians of southeastern Oregon (Boyd 1986) and for tribes inhabiting the dry coniferous forests of California (Lewis 1973).

Settler Fires

During the 1830–60 period, visitors and settlers recorded considerable evidence of fire and smoke in the Pacific Northwest (Norton 1979), extending the smoke-filled diary accounts of the Wilkes expedition of 1841. Many large forest fires occurred during the nineteenth century in mountainous terrain, the history of which is summarized by Morris (1934). Some of these were human-caused, but not all were settler-related.

Two notable nineteenth-century Americans had firsthand observations of fire in Puget Sound country. In 1895, Mark Twain greeted a smoke-choked Olympia in August with the comment, "I regret to see—I mean to learn—I can't see, of course, for the smoke—that your magnificent forests are being destroyed by fire. As for the smoke, I do not mind. I am accustomed to that. I am a perpetual smoker myself." His patience was less evident weeks later as the ship he was to take to Australia to continue his world tour ran aground because of the smoky haze (Glickstein 1987, p. 46). When John Muir first visited the Puget Sound area in the 1890s, he remarked on the solitary snags remaining from old fires, showing that "a century or two ago the forests that stood there had been swept away in some tremendous fire at a time when rare conditions of drought made their burning possible" (Weaver 1974, p. 305).

The advent of modern forest fire management in the Douglas-fir region began after the disastrous regional fires of 1902. These fires together form one of the ten largest and most destructive fires in U.S. history (Davis 1959). Severe forest fires occurred in nearly every county west of the Cascades in Washington and Oregon (McDaniels 1939). Most fires seemed to be a combination of settler clearing fires and "slashing" fires, set purposely after logging to prevent subsequent fires. The largest of these was the Yacolt fire in southwestern Washington, which covered 434,000 acres and resulted in 16 deaths.

Organized forest fire protection began soon after. Forest landowners in Linn County, Oregon, formed the Linn County Fire Patrol Association in 1904 (Davis 1959). The Washington Forest Fire Association (WFFA) was formed in 1908, and the Pacific Northwest Protection and Conservation League, renamed the Western Forestry and Conservation Association (WFCA), was formed in 1909 (Allen 1911a, Cowan 1961, Rakestraw 1979), primarily for forest fire prevention.

THE BEGINNING OF MANAGEMENT, 1910–30

The story of modern fire management in the Pacific Northwest is a seeming paradox. In forest types where fire had been historically infrequent, mandatory fire use became institutionalized (the "westside story"), while in forest types where fire had frequently underburned the forest, fire use was outlawed (the "eastside story").

These stories begin with a large fire episode that occurred in 1910 in northern Idaho/northwestern Montana, covering 1.2 million ha and resulting in the deaths of 85 people. These fires stimulated a national concern over protection from forest fires. The political connections of Gifford Pinchot of the Forest Service helped to pass the Weeks Act in 1911, not long after Pinchot's firing in 1910 (Rakestraw 1979). This allowed purchase of lands necessary to adequately protect navigable streams from fire.

Foresters believed that before forest management could become effective, control over forest fire, particularly underburning, was imperative (Steen 1976). The continued use of fire in the infrequent but high intensity fire regimes of the Pacific Northwest contrasted with the approach adopted in California and the eastern Cascades in low severity fire regimes, where frequent surface fires had burned through mixed conifer and ponderosa pine forests for centuries.

The Westside Story

The "westside story" was driven by the forest industry's recognition that protection of virgin timber was imperative to maintain a timber supply for its mills. By 1910 industrial leaders also clearly recognized the relationship between slash fires (known as slashing fires at the time) and wildfires. George S. Long (1910) of Weyerhaeuser noted that the public had a right to regulate the operations of slashings, and the goal was to burn off the quickly inflammable slash at times when there was no danger of the fire spreading to the adjacent forest. The forest industry pushed for and obtained state legislation that regulated the disposal of slashings. In Oregon a permit was required to burn slash during summer months, and all snags higher than 25 ft had to be cut. In Washington similar rules were adopted but with less discretionary language for suspension of rules (Allen 1911b). Fire control organizations began to appear in California, too (Clar 1969a).

Regulation of slash burning to protect forests expanded the public perception of forests. Up to that time forests had been seen as a resource to be "bottled up" by government or exploited for the use of "timber barons" (Allen 1911b). In the first year of operation, the Washington regulations resulted in 30 convictions, which were lauded by the WFCA and WFFA (Anonymous 1912). Slashing fires were still a major cause of wildfires (Elliott 1911), and the national attention on forest protection focused all slash-burning regulation toward the abatement of fire hazard. Long (1911, p. 12) questioned the requirement for compulsory burning but concluded, "In the bottom of my heart I think every slashing should be burned."

Recommendations for slash-disposal techniques began to dot the literature in 1912. Most articles defended the need to burn slash, so the arguments mainly concerned season and ignition technique. In 1912 over 12,000 ha of slash was burned in western Washington (Allen 1912). The position of the WFFA and the Forest Service was that spring and fall burning were equally satisfactory (Allen 1912, Ames 1912); the WFCA advocated fall burning, after a series of escaped slashing fires in May of that year (Bridge 1912).

The typical technique was to fall snags within 30 m of green timber and use a backing fire, which cost about 45 cents per hectare (Elliott 1911). Hot fires were preferred to cool ones, in the belief that high initial consumption reduced the probability of reburns. In flat terrain, center firing with subsequent perimeter ignition was recommended, while on slopes blacklining the top and then lighting around the bottom was recommended (Joy 1913).

By 1917 additional laws were passed in both Oregon and Washington that helped to institutionalize compulsory slash burning (Larson 1966, Cramer and Graham 1971). Landowners were held financially accountable for fires from any cause on their land unless the slash-fire hazard was "abated," which was interpreted to mean "burned."

The 1920s began with two events that helped shape future slash-fire policy for decades. In late 1921 a tremendous windstorm blew down large tracts of timber on the Olympic peninsula. Landowners responded by burning 28,000 ha of slash on the peninsula alone to ward off potential wildfire effects (Cowan 1961); much of this was older slash with advance regeneration. As no holocaustic fire resulted from the blowdown, this event appeared to validate the need of burning all slash for hazard reduction. The second event marked the beginning of the end of spring burning. An east wind in May 1922 caused slashing fires to burn 40,000 ha of cutover and virgin timber (Cowan 1961); 15 of the 28 large wildfires were escaped slashing fires (Joy 1922). WFFA began to change its earlier recommendation of spring burning by recognizing potential dangers of holdover fires and recommending no burning during east winds, "without going into the merits and demerits of spring burning" (Joy 1922, p. 30).

The 1920s also marked the first research work applicable to slash burning. The association of critical fire weather and low relative humidity by Julius Hofmann (1922), a silviculturist at the Wind River Experiment Station in southern Washington, initiated a more scientific approach to the burning of slash. He suggested burning at night, when higher relative humidity would allow better control of the fire. However, there is little indication that night burning was ever widely adopted. Hofmann's bulletin on Douglas-fir regeneration (1924) recommended spring burning with mopup. One of his most significant, though incorrect, findings with relevance to slash burning was the statement that "seed remains viable in the forest floor much longer than two years and still produces good stands of young growth" (1924, p. 17). This statement was first published in 1917 (Hofmann 1917a) and was used by opponents of slash burning in the next few years, although it was refuted by Isaac years later (Isaac 1935). The research of the 1920s, as sporadic as it was, largely focused on methods, and most statements were based on speculation.

Divided opinion over slash burning began to surface by 1925. In 1924 passage of the Clarke-McNary Act had increased federal help to states for fire control; slashings were still a major source of wildfire acreage in Oregon and Washington. A classic dialogue spanning the opinions of the day was heard at the Pacific Logging Congress in 1925 (Lamb 1925, Joy 1925, Allen 1925). Representing the proburning position was George Joy, once fire warden of WFFA but now state forester of Washington. Taking the middle ground was E. T. Allen, venerable WFCA spokesman. Representing the opposing view was Frank Lamb, a respected industry spokeman for two decades.

Lamb led off the discussion by noting that most cutover land was being reburned up to a dozen times, and that the line of least resistance was to promiscuously burn slashings in spring and fall. At first such areas were small and isolated, but with the rapidly extended logging operations of 1915–25, most operations were now contiguous, and many timbered areas were bordered by cutover land 10–30 mi in width. The object of most fire protection was to protect only virgin timber, so cutover areas were repeatedly overrun by fires. Lamb referred to seed in the forest floor being destroyed, perhaps a reference to Hofmann's study of 1924. Nature's way, in his opinion, was to cover the landscape with new growth, which quickly fireproofed the area. Lamb claimed that fire hazard on logged and unburned areas remained high only for five years, rapidly decreasing after that. Furthermore, the annual summer pall of smoke was causing Washington to lose much tourist business. He concluded his presentation with what must be considered a classic takeoff of *Hamlet*: "To Burn, or Not to Burn" (Fig. 3.1).

Lamb's presentation must have been a hard act to follow, but George Joy gamely presented the state view. He noted that slash remained a menace for 15–20 years. The state could interpret the regulations to waive mandatory slash burning (although waivers seem to have been rarely granted, and liability for subsequent fires was not waived). He concluded by suggesting that April and May were the most hazardous months to burn, because the relative humidity was lower then than in September (the data to back this up were, to be generous, rather sketchy).

The moderating opinion was presented by E. T. Allen, who sagely noted that slash burning was right in its place and wrong outside of its place. Blanket rules, such as mandatory slash burning, were good in most cases and bad in a few cases. Because a lot of hard work went into getting blanket rules, agencies didn't want them compromised by providing any exceptions. Allen concluded that there were no good blanket rules, a phrase that was often repeated in later decades.

By 1928 there was some opinion that Lamb's recommendations of no burning did seem to fit coastal areas (Alexander 1929). Broadcast burning was recognized as no panacea; wildfires could still pass through areas where slash had been burned, and sometimes control problems were harder in burned areas. In Grays Harbor County, Washington, these ideas were overwhelmed by statistics showing cutover areas to be the site of two-thirds of wildfire area burned, and spring burning to be the primary cause of slash-burn escapes (WFCA 1929). By the end of the 1920s, spring burning had been largely abandoned.

TO BURN, OR NOT TO BURN

(With apologies to Hamlet)

To burn or not to burn; that is the question:
Whether 'tis better in the mind to suffer
The rules and laws of an outrageous custom,
Or to take arms against the forest's troubles,
And by opposing, end them. To fire, to burn;
To burn, perchance to destroy; ay, there's the rub;
For in that one slash burn what destruction may come,
When wind rises or humidity lowers,
Must give us pause: there's the respect
That makes calamity of our fine theories;
For who would chance the risk of a dry spring,
A late fall, or a summer's withering humidity,
The danger of escaped fire, the smoldering embers,
The blazing snags and the brands
That from East wind great holocausts make
When nature herself a far better way takes
With a green cover. Who would matches bear
To singe, to scorch and dry the woods,
But for the dread of some new idea
The way of nature without burning, from which method
No danger arises, puzzles the will
And makes us follow those customs we have
Than fly to others, good though they may be.
Thus custom does make arsonists of us all
And thus the native hue of forest green
Is sicklied o'er with the lurid pall of eternal smoke
And young forests of great promise and value
With annual fires their seeding comes to an end
And lose the power of reforestation.

FIG. 3.1. *The text of "To Burn or Not to Burn," presented by Frank Lamb at the 1925 Pacific Logging Congress.*

The Eastside Story

In the drier east Cascades of Oregon and Washington, and in the Sierra Nevada of California, fires had historically been frequent and generally of much lower intensity than on the westside. To early foresters, the continued use of such low intensity fires (called light burning during this controversy and known as underburning today) in merchantable stands of timber was seen as a continuation of Native American burning practices. "Piute forestry," a derogatory reference to the practice of light burning, was perceived as a threat to forest management (Pyne 1982) and a "challenge to the whole system of efficient fire protection" (Graves 1920a, p. 35). The controversy was largely debated and fought in California, but its results were applicable for similar forest types throughout the West.

In 1910 the debate over the practicability of light burning in pine forests began with an article promoting the use of underburning in Sierra Nevada pine forests (Hoxie 1910). Foresters replied by showing the detrimental impact of fire on seedlings and saplings, even though residual stands were well stocked (Pratt 1911). They contrasted "promiscuous" light burning with slash burning of the Pacific Northwest, which was "never allowed to run at random; it is systematically set out, and controlled absolutely" (Boerker 1912, p. 193). At the time, however, slashing fires were still a major cause of wildfires in the Northwest (Elliott 1911).

Early proponents of light burning included industrial forestry companies, including the Southern Pacific Company in the central Sierra Nevada and the Red River Lumber Company in Shasta County to the north. The Forest Service brought its growing research capacity into the fray, assigning Stuart B. Show and Edward I. Kotok from the California Region and Julius Hofmann, detailed down from the Wind River Experiment Station, to evaluate the controversy over light burning. Hofmann (1917b) argued that brushy areas on the Crater National Forest (now the Rogue River National Forest) were a direct result of burning, and stated that these extensive areas could be retimbered only with effective fire protection. Haefner (1917) claimed that southern Oregon had not experienced forest fires before its settlement by European Americans, and that late nineteenth-century burning had eliminated virgin stands in many areas. Both overlooked an extensive underburning history in the same area (Agee 1991b) and concentrated on intensely burned areas—clearly a part, but only a part, of a complex fire history in southern Oregon.

At the time, fire control in the pine forests where light burning was applicable was relatively easy; forest rangers tied branches onto their horses' tails and walked them through the forest, scattering pine needles from the path of the oncoming low flames (Munger 1917). The controversy over light burning became the subject of several articles in *Sunset Magazine* in 1920 (Graves 1920b, Show 1920, White 1920, Redington 1920), which encouraged the formation that same year of a commission to study light burning, called the California Forestry Committee. The committee, composed of federal, state, industrial, and university forestry professionals, noted that practicality and not theory was the issue. After three years of apparently unsuccessful attempts to do large-scale burning, the committee concluded that burns damaged reproduction as well as mature timber and that fire suppression was relatively easy, so that full fire protection appeared to be more practical and economical than light burning (Bruce 1923).

The classic bulletin against light burning was published a year later by Show and Kotok (1924), reinforcing the conclusions of the California Forestry Committee. The authors recognized the extensive history of light burning in Sierra Nevada forests back to the 1500s, but they lumped effects of summer wildfires with lighter spring or fall burning. By 1928 the controversy over light burning had been resolved in favor of fire exclusion (Clar 1969b), but it resurfaced decades later. Meanwhile, total fire exclusion became public policy in the drier eastside Cascade and Sierra Nevada forests.

THE GLORY DAYS OF SLASH BURNING, 1930–50

By 1930 almost all slash in the Douglas-fir region was burned in the fall, a pattern that was followed for decades. Research on methods of burning was nonexistent, since the prevailing custom of fall burning was apparently unquestioned. Debate moved from whether burning was necessary to when it should be avoided. The

Forest Service began to take a leading role in slash-fire research and concluded that Douglas-fir regenerated best with partial shade; without debris left after logging, Douglas-fir did best on unburned surfaces (Isaac 1930). McArdle (1930) found higher fuel consumption in old slash and higher proportions of small fuel removed by fires. Shepard (1931) noted high residual tree mortality, continuing for several years, after underburning partially cut stands. Isaac and Hopkins (1937, p. 278) concluded that "the harmful effects of the ordinary slash fire more than outweigh any beneficial effects it may have on the productivity of Douglas-fir forest soil," reiterating work done at Oregon State College by Fowells and Stephenson (1934).

This generally negative research focus was balanced by Munger and Matthews (1939), who concluded (again) that there is no blanket rule and listed some situations in which slash should be burned and others when it should not. They presented the first quantitative comparison of fire hazard on burned and unburned areas, noting that burning reduced rate of spread for about 6 years and resistance to control for about 17 years, after which both indices were similar and declined regardless of treatment. The prevailing interpretations of Oregon and Washington laws, however, still required almost all slash to be burned.

In the midst of the Great Depression, thousands of young men went to work in the woods through Civilian Conservation Corps programs. Those in the Pacific Northwest all had one event burned into their memories: the Tillamook Fire of 1933 in the Oregon Coast Range. A firestorm to rival that of the Yellowstone fires of 1988, the Tillamook fire consumed 90,000 ha of forest in 20 hours (Heinrichs 1983). Repeated burns of portions of the area in six-year cycles (1939, 1945, 1951) further emphasized the need for fire control and hazard reduction through slash burning. Today the former Tillamook burn is the thrifty Tillamook Forest.

During World War II, military authorities restricted slash fires west of the Cascade Range because such fires reduced visibility, and fire glow at night might have aided enemy submarines (Morris 1958). Lack of manpower and more partial cutting also resulted in less slash being burned into the mid-1940s. Nevertheless, in 1938–43, about 35,000 ha of state and private land was burned annually in western Oregon (Russell and McCulloch 1944). In Washington escaped slash fires made up one-third of total wildfire area in 1944

(Cowan 1961), suggesting that control of such fires, even with fall burning, was still a problem. There was some recognition that slash burning was not always necessary due to better access (truck rather than railroad logging), smaller unit sizes, and better utilization (Russell and McCullough 1944). Slower burning after fall rains was proposed in 1948 (Aufderheide and Morris 1948).

In the low severity fire regimes of the eastern Cascades researchers began to provide evidence that a blanket rule forbidding fire use in these areas had contributed to increased insect problems, increases in fuel hazards, and undesirable changes in species composition (Weaver 1943). Weaver's ideas were so controversial at the time that he was forced by his superiors at the Bureau of Indian Affairs to place a disclaimer as a footnote to the article: "This article represents the author's views and is not to be regarded as an official expression of the attitude of the Bureau of Indian Affairs on the subject discussed" (1943, p. 7). These footnotes continued through the 1950s, until the bureau realized such footnotes were embarrassing the institution (Biswell 1989).

POLICY REEVALUATIONS, 1950–70

In the 1950s foresters began to doubt the need for compulsory slash burning on westside cutover lands. Some foresters became known as burners and some as nonburners (Morris 1958), labels that are still heard today at fire conferences. Although the states defended the practice as a necessary evil, the WFFA concluded that if a practice was evil, it was unnecessary (Cowan 1961). Hazard abatement still meant slash burning (Heacox 1951), even on unburned but restocked lands. Hagenstein (1951) compared this practice to a man committing suicide because he was afraid he was going to suffer an accident. He hinted that some of the state reluctance to waive slash burning was because less burning might result in more wildfire and more financial responsibility placed on the state. Munger (1951) concluded that only in rare instances was burning desirable silviculturally, and that increased protection should be weighed against the costs of slash burning. Site quality (soils work) and regeneration effects were the major research topics of the decade.

The controversy over light burning reemerged during the 1950s, with Weaver's continued publications (1951a, 1955, 1957, 1959). Weaver's work encouraged Harold Biswell, a professor at the University of California, to begin research in underburning California ponderosa pine forests. Like Weaver, Biswell encountered resistance. He has summarized his experiences in a recent book (1989). His research expanded (e.g., Biswell et al. 1955) at about the same time that wildfire effects in these fire regimes of historically low severity began to consistently mimic those of high severity fire regimes, since all but the most severe fires were being contained soon after ignition.

The cooperative efforts of federal, state, and private interests were successfully reducing wildfire acreage in the Pacific Northwest, if not wildfire effects on those acres burned. As a result, by 1958, after fifty years of fighting the enemy fire, the WFFA evolved into the Washington Forest Protection Association with a much broader mandate than fire prevention.

The 1960s saw some of the first changes in slash-burning policy since the 1920s. An industry spokesman reopened the debate over whether "to burn or not to burn" by suggesting that slash should never be burned except when justified (the reverse was still current policy) (Prater 1962). Isaac (1963) borrowed the 1925 rhetoric of both Allen and Lamb when he suggested that fire was a tool, not a blanket rule, in Douglas-fir management. With apologies to Shakespeare, he listed well-founded reasons both to burn and not to burn.

Light burning (now called prescribed burning) received official governmental recognition when the Department of the Interior accepted the Leopold Report (Leopold et al. 1963), a wildlife commission report that recognized the important role such fires played in national park ecosystems. The emergence of other groups in society advocating use of fire as a tool (environmental preservationists, hunting clubs, gravel mining interests) helped to institutionalize the use of fire as a tool (Lee 1977). A new fire policy was adopted by the National Park Service in 1968, allowing natural fires to burn in some areas and permitting the use of manager-ignited fire in ecosystem restoration (USDI 1968). The first parks to use fire under the new policy were Sequoia–Kings Canyon and Yosemite (Kilgore and Briggs 1972) in California. Northwest parks did not implement new fire management plans until the mid-1970s.

At the same time, air quality legislation began to force changes in the fire liability laws that required slash burning (Larson 1966). In 1968 Oregon's liability law was altered, which immediately resulted in about a one-third reduction in burning in southern Oregon (Smith 1968) and a 25–30 percent reduction in Forest Service burning (Maul 1968). Wildfire acreage from debris burning in Oregon and Washington averaged 2,178 ha each year from 1963 to 1970 (USDA Forest Service 1964–71). Much of this occurred on the westside of the Cascades; in comparison, this two-state annual average was just slightly over the annual average for one county (Grays Harbor, Washington) during 1922–27 (WFCA 1929).

In 1969 Oregon implemented a plan to avoid excessive smoke from slash fires, concentrating on the populated Willamette Valley corridor (Cramer and Graham 1971). Washington adopted a similar smoke management plan in 1970 (Hedin 1979). Both plans focused on avoiding populated areas and incorporated more sophisticated meteorological forecasting than had been attempted before. The decade ended with a recommendation that the season for slash burning in the region be extended into spring and summer for better smoke dispersal (Dell 1969).

A reevaluation of Munger and Matthews's (1939) work in Douglas-fir by Morris (1970) concluded that unburned plots maintained higher potential rate-of-spread rates for 14 years instead of 6, and that resistance-to-control indices on unburned plots did not decline as rapidly as had earlier been assumed. These data appear to have had little impact on slash-burning decisions.

ENVIRONMENTAL CONCERNS, 1970–90

Smoke from slash burning was a dominant issue in the 1970s, eclipsing earlier concerns over fire hazard, soils, or silviculture. The brown-and-burn technique, in which desiccant is applied to a unit that weeks later is mass-ignited, came into vogue as a means to burn under wetter conditions and to burn quickly (Finnis 1970, Hurley and Taylor 1974). The flying drip torch was introduced in the mid-1970s (Orr 1977). The prescription tools of the 1960s, such as fuel moisture sticks (Morris 1966), were expanded with more detailed weather forecasts to predict smoke dispersion (Cramer and Graham 1971). Complex manuals were developed to predict environmental effects of forest residue reduction (Cramer 1974).

The average annual area of slash fire in western Washington increased from about 15,000 ha per year to about 22,000 ha per year; at the same time, tonnage burned apparently decreased from about 1.63 million tons per year to about 0.90 million tons per year (WDNR Smoke Management Plan Reports 1973–81). The tonnage decreases were attributed to better utilization, burning in wetter parts of the year, and better fuel estimates. Some of the earlier figures apparently reflected total tonnage on the site rather than the actual amount consumed (Visibility Program Subcommittee 1982). Burning was being conducted in almost every month of the year. Some of the decrease may also have been due to deliberate underestimation of fuel loads by those who felt that the probability of burn approval was enhanced when total reported tonnages were low (Fahnestock and Agee 1983).

An exceptionally large wildfire event burned in the Entiat River drainage of north-central Washington in 1970, immortalized by the movie *Wildfire*. The fire complex, which burned many tens of thousands of hectares, was caused by a series of lightning strikes in late summer. The size of the complex stimulated the formation of the National Wildfire Coordinating Group (Pyne 1982). Much of the area, burned with stand-replacing fire intensities, had previously burned with moderate to low intensity fire (J. K. Agee, unpublished data), suggesting a fuel buildup effect on fire intensity and stimulating further the use of fire in forest ecosystems.

The shift in fire policy by the National Park Service encouraged programs of underburning on national forest lands. In 1971 ponderosa pine stands were burned at Pringle Falls, Oregon, as a companion treatment to brush crushing; in 1972 underburning was used to reduce incense-cedar (*Calocedrus decurrens*) in the understory of mature ponderosa pine near Sisters, Oregon; underburning in Douglas-fir near Detroit, Oregon, was attempted the same year; and in 1975 juniper woodlands were burned in eastern Oregon (Dell 1976). However, Dell noted (p. 124) that no one had yet recognized what "the role of fire may be in wilderness management here in the Northwest."

Research was beginning on fire history in parks and wilderness.

Fahnestock (1976) presented information on the fire history of the Pasayten Wilderness of north-central Washington at the same meeting where Dell spoke, and Crater Lake National Park had commissioned a fire history for its "panhandle" area that was completed in 1976 (McNeil and Zobel 1980). The park used this information to begin small-scale underburning that year, which expanded to individual burns of roughly 1,000 ha by 1978. Although burns of that size might be justified based on the past fire history of the area, the smoke generated from the 1978 burns, which filled the Medford basin due to a wind shift, forced a decrease of an order of magnitude in the size of later prescribed fires. Crater Lake also had the region's first prescribed natural fire in 1978, the 200 ha Goodbye fire.

The Clean Air Act amendments of 1977 had the largest impact on policy of the decade, although the effect was not felt until the 1980s when the smoke management plans for Oregon and Washington were revised. Park and wilderness areas were designated as visibility-sensitive areas for slash-fire smoke because of potential visitor impact. At the same time, plans were being developed for many of these areas to allow natural fires to burn within the boundary. The position of most of these areas along the crest of the Cascades and in the Olympic Mountains further complicated slash-fire management plans.

Wildfire area from debris burns in Oregon and Washington averaged 2,100 ha annually during 1971–80, comparable to the figures of the previous decade. At the same time wildfire area from other sources declined, however, so that the percentage of total wildfire area due to debris burning increased fairly steadily from 1967 to 1980 (USDA Forest Service 1968–81).

The decade of the 1980s saw significant reevaluation and shift in fire-use policy. Regulation of slash burning increased rapidly, while natural-fire policies at the national level attracted so much attention following the Yellowstone fires of 1988 that they were regionally suspended into the 1990s.

Smoke intrusions from slash burning, although much reduced from levels of past decades, became headline news in the 1980s. Public perception replaced problems of regeneration or soil as the key factor in whether or not to burn slash (Sandberg 1989). In the public perception, smoke from woodstoves in urban areas and urban-rural interface communities came to be linked with slash-fire smoke as a cause of respiratory problems (Koenig et al. 1988). In Washington, a goal of 35 percent reduction in a decade in slash-fire emissions, along with no summer weekend burning, was voluntarily adopted by federal, state, and private participants in the smoke management plan. Meanwhile, the Environmental Protection Agency (EPA) was tightening the regulations on particulate matter standards. The regulations focused less on total particulate concentrations than on fine particulate concentrations, which make up a high proportion of particulate emissions from slash fires (Morgan 1989). The result in western Washington was a decrease from 19,500 ha burned in 1977 to 13,300 ha burned in 1988 (Stender n.d.).

Trends in drier eastern Washington were not so strong, because of the mixed message being received by forest managers: underburn more to reduce fuels and restore natural processes, but don't produce smoke. Total area burned remained at roughly 20,000 ha per year over the same period, although emissions were estimated to drop by 50 percent (Stender n.d.). Guidelines for underburning in pine-larch stands were published by the Forest Service, which estimated that about 4,000 ha was being underburned per year in this type of forest in Oregon and Washington during the 1980s (Kilgore and Curtis 1987).

The 1980s saw the rise and fall of interest in the use of fire in management of natural areas. In the early 1980s the natural-fire program at Crater Lake was emulated by similar programs at North Cascades National Park, Alpine Lakes Wilderness, Olympic National Park, and Mount Rainier National Park. Crater Lake's program of manager-ignited prescribed fire was also adopted by other natural-area managers in the Nature Conservancy and the Washington Department of Natural Resources Natural Heritage Program.

The goals of these programs began to be criticized in the early 1980s. Criticism was primarily concerned with prescribed fire in the Sierra Nevada national parks of California (Bonnicksen and Stone 1982). The issue was whether fire was being reintroduced for fire's sake—essentially a process-related goal—or for specific objectives concerning species composition, density, and size class, which was a more structurally oriented goal (Agee and Huff 1986a). Although the debate continued through the decade (Parsons et al. 1986), the stimulus for a reevaluation of the parks' fire-management goals came in 1988, when drought, strong winds, and a combination of natural fires and human-caused wildfires burned across the Yellowstone country for much of the summer. That same summer a natural fire at Crater Lake, which had been burning gently in a red fir forest for weeks, suddenly exploded across the eastern boundary onto 600 ha of Forest Service land (Jones 1990).

Several groups were convened to evaluate management and ecological aspects of the Yellowstone fires. The official federal fire-management task force concluded that the policy was sound but had been undermined by poor execution (Philpot et al. 1989), while others (Bonnicksen 1989) hinted that the policy itself should be questioned. An ecological effects task force on the Yellowstone fires made the most eloquent plea for a reevaluation of policy (Christensen et al. 1989, p. 685):

> Knowledge of the causes and consequences of natural processes such as fire is rudimentary, but we have learned enough to know that wilderness landscapes are not predestined to achieve some particular structure or configuration if we simply remove human influences. A great variety of future "natural" landscapes is possible . . . we cannot escape the need to articulate clearly the range of landscape configurations that is acceptable within the constraints of the design and intent of particular wilderness preserves.

After the Yellowstone fires, all natural-fire programs were suspended until improved management practices were in place to implement the policy, which remained largely intact (Wakimoto 1990). By the early 1990s several parks and wilderness areas had revised plans in place, which were generally more conservative.

The chief question remains: to burn or not to burn? The debate now revolves around slash burning, natural fires, and underburning for various objectives, including wildlife habitat and protection of rural developments. The consistent refrain since the 1920s has been that the question is improperly phrased: Shakespeare is indeed owed an apology. There are times when burning is appropriate and times when it is not. Such value judgments require weighing social and resource objectives in a political arena far more polarized than in the past.

The institutional response to the fire catastrophes of the 1900–1910 era was to burn all slashings, and this simple, single-focus approach to fire hazard over the next 60 years caused land management objectives to be subsidiary to fire management goals. In the 1980s the rhetoric appeared to advocate multiple objectives. Still, the air resource clearly became the dominant—and perhaps in many cases the sole—concern, although natural-fire programs were essentially exempt. Site objectives became subsidiary to air quality goals, in large measure due to regulatory "trumping" of air quality statutes over more generalized objectives for public and private forest.

The lesson of the first 80 years of modern fire use in the Pacific Northwest is that single-focus management has resulted in success in the short term but failure over the long term in optimizing societal objectives for forests. When land management became the focus, more appropriate strategies were used to support overall land management objectives. Although these lessons were primarily learned through slash-burning experience, they apply equally well to the use of fire in natural-area management. More precise goals will be necessary, and off-site impacts will have to be integrated into fire management plans.

The trends of the last two decades suggest that slash burning will be less important in the future than it was in the past, but it will remain the tool of choice in certain situations. The use of prescribed fire will depend on the demonstrated value of fire for hazard reduction (Lee 1977, Agee 1989a). Natural fire will remain alive as a concept but will be constrained in operation. The history of fire in the Pacific Northwest suggests that entrenched practices and blanket rules are not sufficiently flexible to meet the needs of the future. Adaptive management strategies, those that rationally change over time in response to changing conditions, will be necessary in the future if fire is to survive as a forest management tool.

CHAPTER 4

METHODS FOR FIRE HISTORY

EARLY ATTEMPTS TO EXPLAIN the ecological effects of fire were hampered by a lack of both ecological data and description of the fire regime. Although experiments that varied the frequency and season of burning extend back to the early part of the century (Schiff 1962), the first ecology textbook describing fire as an ecological factor was published only in 1947 (Daubenmire 1947). One of the first major reviews of fire effects (Ahlgren and Ahlgren 1960) described effects of presence or absence of fire with little detail concerning effects of multiple events over time. Today, it is possible to describe fire regimes in more detail, but the details can be confusing and sometimes misleading.

Defining a particular fire frequency can be likened to describing a cord of wood. There is a precise volumetric definition of a cord of wood, but each person stacking a cord may use different amounts of firewood to produce the stacked volume. Sometimes a short cord results! Johnson and Van Wagner (1985) suggest that clear thinking about fire history is difficult, and that mathematical models are necessary to avoid a storytelling tradition of fire history. However, mathematical models, which usually have rigid assumptions about the nature of the system, can be just another form of storytelling if they are not carefully interpreted.

This chapter reviews the common methods of determining fire history in order to provide an understanding of what the techniques are, where they are best applied, and how they should be interpreted in an ecological context. It does not provide an exhaustive review of fire history studies. The regional studies will be summarized in later chapters dealing with specific forest types, and broader reviews are also available (Alexander 1979, 1980, Mastrogiuseppe et al. 1983). This chapter emphasizes methods to determine fire frequency and predictability, but the application of techniques will depend on past fire severity as well.

FIRE FREQUENCY DEFINITIONS

The terminology used in fire history work is gradually becoming standardized. A review of terminology at a fire history workshop (Romme 1980) reached consensus on several terms. The literature published since 1980, however, has not closely reflected that consensus in usage. The phrases defined here and used in this chapter are adapted in part from Romme (1980). Preferred phrases are followed by commonly used synonyms in parentheses:

Fire frequency: a general term referring to the recurrence of fire in a given area over time.

Fire event (fire occurrence, fire incidence): a single fire or series of fires within an area at a particular time.

Fire interval (fire-free interval or fire-return interval): the number of years between two successive fire events in a given area.

Mean fire-return interval (mean fire interval): arithmetic average of all fire intervals in a given area over a given time period.

Fire rotation (natural fire rotation): the length of time necessary for an area equal in size to the study area to burn.

Fire cycle: the average stand age of a forest whose age distribution fits a mathematical distribution (negative exponential or Weibull).

The terms *fire event* and *fire interval* are simply descriptive of the presence of one or two fires on a landscape. Their interpretation is relatively unambiguous. Other terms are either derived or interpreted in several ways that make comparison between studies quite difficult. The *mean fire-return interval* is typically applied in fire regimes of low to moderate severity, using data from multiple fires in the same stand. In practice, data are often pooled over different-sized areas, which confounds ecological interpretation unless the area factor is clearly integrated into the interpretation. *Fire rotation* and *fire cycle* are terms used in fire regimes of moderate to high severity. The extent of individual past fires or present age classes across the landscape are pooled to derive an average rotation or cycle. This value may be derived from a large area containing different fire regimes, which again may confound ecological interpretation.

CREATING A FIRE HISTORY DATABASE

The first evidence to establish in creating a fire history database is fire occurrence—whether fire has been present on the landscape. This may be established from fire scars on trees, from plants that appear to have germinated after fire, or from charcoal. The most consistent indication of fire occurrence is charcoal, since the other two indicators become less reliable over time. Charcoal may be found in the surface soil, often in contact with the roots of the current stand.

Charcoal has been recovered in lake sediment cores and used to interpret fire and vegetation history. Swain (1973) used relative increases in charcoal presence and varve (annually laminated sediment) thickness to reconstruct past fire history in Minnesota. Charcoal peaks within sediment deposits in a small lake in the North Cascades (Cwynar 1987) were used to infer the relative importance of fire over the last 12,000 years. Techniques for analyzing thin sections of lake varve continue to improve. Charcoal in thin sections of lake varve can produce a local (i.e., catchment) record of fire frequency (Clark 1988, Clark 1990).

Inferences of fire occurrence may also be made from charcoal on the bark of currently living trees. The bark of coastal Douglas-fir will retain its char for centuries, based on evidence from stands in the Olympic Mountains. Two-hundred-year-old stands have occasional residual trees with substantial charcoal on the bark left by the fire that created the stand. In young stands that burn, charred residual trees will eventually appear to have a candy-stripe pattern as the tree increases in diameter: charred strips will be separated by newer bark produced since the fire. Other species retain charcoal on the bark for shorter periods of time. Ponderosa pine, for example, may shed most of its charred bark plates within several years if the tree is vigorous, so that fire evidence must be established by finding the blackened, jigsaw-shaped bark plates on the forest floor. In Okanogan County, Washington, bark char on the boles of ponderosa pine has been observed to last at least 50 years, so its presence on the bole of ponderosa pine is not necessarily evidence of a very recent fire.

Fire Scar and Age-Class Evidence

Fire scars are an excellent source of fire frequency data, because they can be used to establish precise years of past fire events. Techniques for reconstruction of fire dates from fire scars are discussed later in this chapter. It is important to make sure that the scar was caused by fire. Many other disturbance processes can also create bole scars.

A fire scar is caused when heat from a fire persists around the tree bole long enough to allow a temperature pulse to penetrate the bark and kill part of the cambium. Mechanisms of fire scar formation are discussed in chapter 5. Each succeeding year after the fire injury, the adjacent live cambium expands slowly over the surface of the scarred area and may eventually enclose it (Figs. 4.1, 4.2). Once a scar has been created, another fire is likely to rescar the same area. This happens because often a pitch or resin deposit ignites in the vicinity of the wound, the bark is thinnest at this part of the bole circumference (even if the scar has healed over), and dead, dry

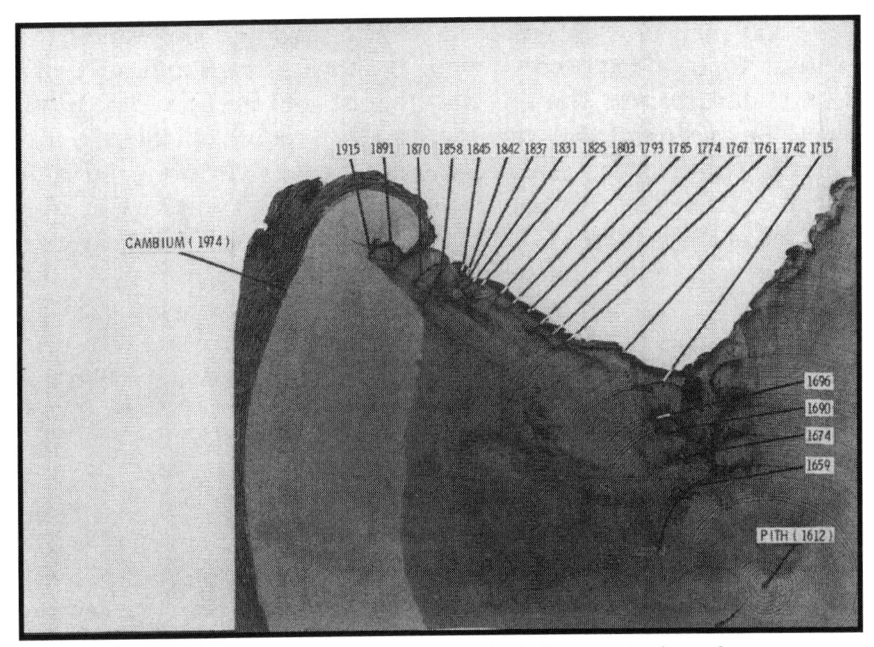

FIG. 4.1. *Cross-section of ponderosa pine with multiple fire scars in the catface.* (USDA Forest Service photo, courtesy S. F. Arno)

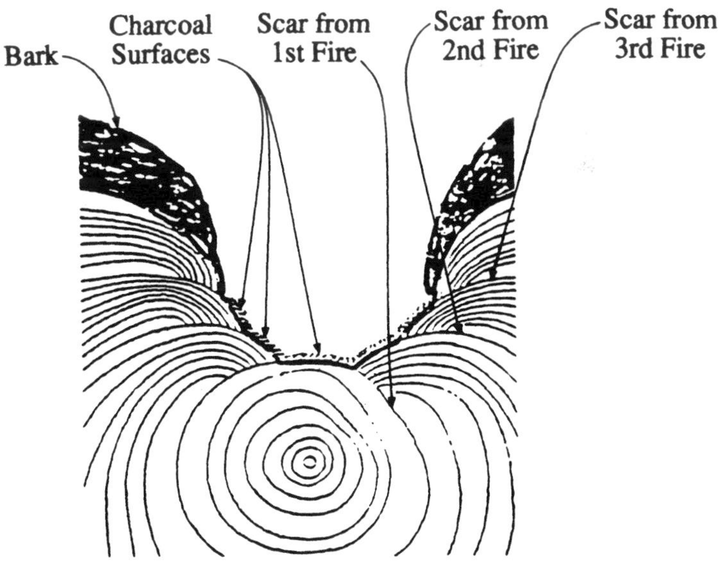

FIG. 4.2. *Characteristics of fire scars, such as the one shown in Figure 4.1.* (From Morrison and Swanson 1990)

sapwood may now be exposed on the unclosed portion of the wound. Char on exposed sapwood is almost always indicative of a second disturbance: The first disturbance (not necessarily fire) results in exposure of the sapwood, and the second, obviously a fire, chars it. Multiple small, healed-over scars may exist along the same radius of a bole cross-section (Fig. 4.3). This occurs when a scar arc is narrow, perhaps due to thicker bark shielding the cambium around other parts of its circumference, or because of a long fire-free interval after a fire, so that the scar is enclosed by later callus growth. The bark across this healed area is thinner than elsewhere around the tree and is therefore more susceptible to repeat scarring when another fire occurs.

Fire scars are usually triangular, extend to the base of the tree, occur on adjacent trees of the same age and species, date to the same year, and are associated with the presence of charcoal on exposed

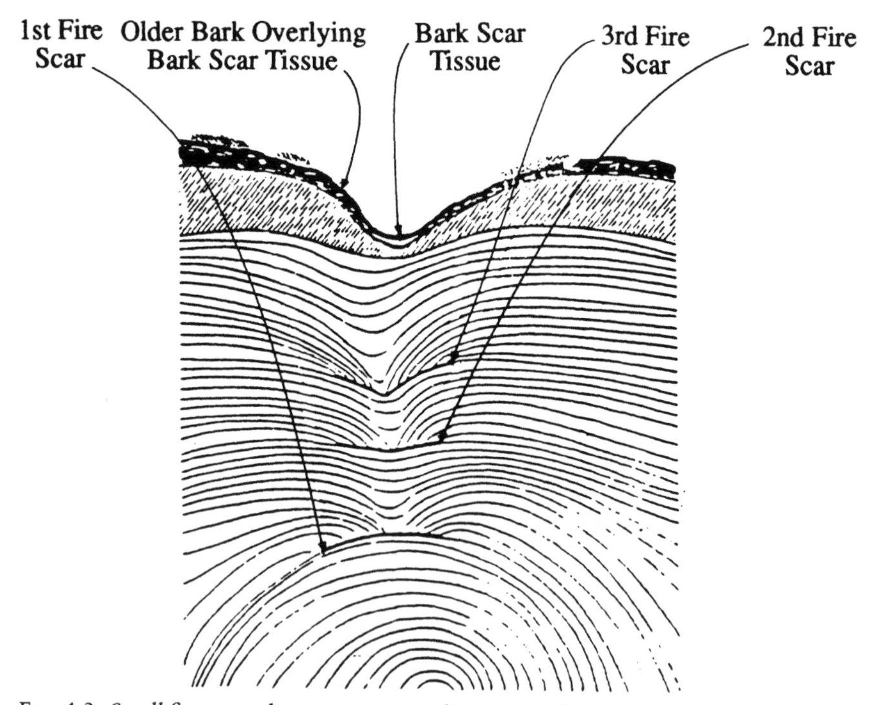

FIG. 4.3. *Small fire scars along a common radius. These often are healed over and would not be visible from the exterior of a live tree.*
(From Morrison and Swanson 1990)

sapwood or bark. Other agents may also be responsible for bole scars on trees, so that unless properly identified such scars may cause researchers to exaggerate the occurrence of fire at the site. Scars can be created by humans, other animals, physical processes, and diseases.

Humans have scarred trees for boundary or trail markers throughout the West for more than a century. Typically such scars are well above the base of the tree and may have characteristic shapes. Native Americans are known to have stripped bark off ponderosa pine to eat the live phloem tissue (Swetnam 1984). Such scars are oval to rectangular, may have tool marks on the upper and lower edge, and begin above the base of the tree and in some cases extend to a height of 3 m. Similar human-caused scars are noted on Alaska yellow-cedar in southeast Alaska (Hennon et al. 1990).

A common scarring agent in young stands is the black bear (*Ursus americanus*). Bears strip the bark off trees in spring to eat the phloem or lick the sap (Molnar and McMinn 1960). Sapwood claw marks may be present (Fritz 1951). After 30–50 years, bear scars may appear very similar to fire scars (Fig. 4.4), except that when dated

FIG. 4.4. A: *Recent bear damage on subalpine fir in Olympic Mountains.* B: *Older bear damage on Pacific silver fir in Olympic Mountains.*

they may reflect a period of 2–5 years of stripping rather than a common date of a fire event.

Red squirrels (*Tamiasciurus hudsonicus*) have been observed to create scars on most of the conifers of the northern Rocky Mountains (Stillinger 1944). Usually no tooth marks are evident (Sullivan and Sullivan 1982). Hares (*Lepus americanus*) and porcupines (*Erithrizon dorsatum*) also strip bark but leave tooth marks on the tree; hares will generally leave a shaggy sapwood with their tooth marks, while porcupines will leave broad, prominent vertical and diagonal incisor marks (Sullivan and Sullivan 1982). Voles (*Microtus* spp.) chew the bark on small trees right at the base, and deer (*Odocoileus* spp.) scrape their antlers on stiff young trees, creating scars about 1 m above the ground.

Partially successful insect attacks may also cause scars. The fir engraver beetle (*Scolytus ventralis*) has been observed to kill patches on the north side of small white fir trees (Furniss and Carolin 1977, Thomas and Agee 1986). The female beetle and the larvae etch a clearly discernible pattern in the sapwood, and the scar is usually not basal. Basal scars may be caused by attacks of the red turpentine beetle (*Dendroctonus valens*), which usually are not fatal. Such scars may be patchy and circular to rectangular, representing the larval gallery pattern.

A classic example of confusion of beetle scars with fire scars occurs in the lodgepole pine forests of the pumice plateau in south-central Oregon (Fig. 4.5). Stuart et al. (1983) found that beetle scars were caused by cambial kill on only one side of the bole (usually the north side) and superficially resembled fire scars (e.g., Gara et al. 1986). The beetle scars could be differentiated from fire scars by the presence of some or all of the following characteristics: pitch tubes, beetle emergence holes, blue stain, larval galleries, retained bark on the scar face, a lack of char, and an orange or red discoloration around healthy sapwood. After 50–100 years, however, the blue stain fades and beetle galleries on the exposed sapwood erode, making differentiation of the scar source more difficult.

Several physical disturbance agents may also cause bole scars. Windthrown trees occasionally slide down and scrape the boles of standing trees, removing the bark along one side of the tree. The sapwood may be damaged, and this damage is visible even decades later. Broken branches may also be evident on the same side of the

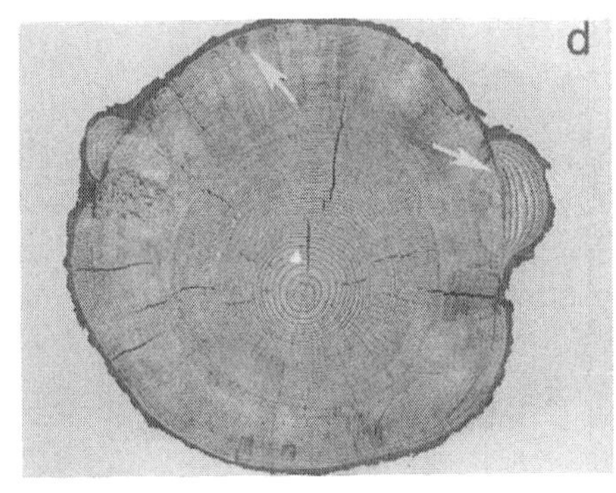

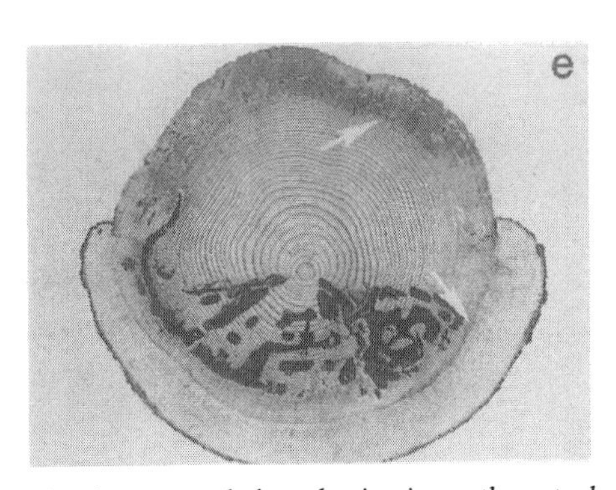

FIG. 4.5. *Characteristics of mountain pine beetle scars on lodgepole pine in south-central Oregon.* A: *Beetle scar with retained bark in scar (arrow).* B: *Pitch tubes and callus tissue along a 10-year-old beetle scar.* C: *Emergence hole of a beetle.* D: *Cross-section of a tree attacked 10 years earlier (arrows denote blue-stained sapwood).* E: *Cross-section of tree attacked in 1925 (arrow at top right denotes zone of blue stain, arrow at bottom right denotes ring of discolored wood coincident with healthy sapwood). Excavations in heartwood are from insects of Cerambycidae family.*
(From Stuart et al. 1983)

tree (Molnar and McMinn 1960), and the fallen tree often remains at the base of the scarred tree. Frost may also cause scars and create damaged annual rings in trees. Frost damage may occur at various elevated places on the bole exposed to the lowest temperatures. Another physical disturbance agent is sunscald (Fig. 4.6). If a disturbance opens a stand to greatly increased light intensity, unshaded portions of the bole on the south side of the tree may experience lethal temperatures. Sunscald scars are usually fusiform and are found at intermediate heights; lower portions of the bole have sufficently thick bark to avoid damage, and upper portions receive sufficient shading to avoid injury. Ice flow in riparian areas can

FIG. 4.6. *Sunscald damage (12 years old) to Douglas-fir created by tree removal to the south, exposing the stem.*

scrape bark off the trees on the upriver side of the tree, and can abrade the exposed sapwood to depths exceeding 5 cm (Filip et al. 1989).

Basal scars can be caused by diseases, particularly *Armillaria mellea*. In cedar-hemlock forests, *Armillaria* is found near the base of the tree, commonly on root buttresses (Arno and Davis 1980). The scarring on root buttresses can be traced to disintegrated roots, and such scars can occasionally be found as high as 1 m on the bole, although most are lower. Arrested *Armillaria* infections were found on 18–73-year-old western white pine and Douglas-fir (Molnar and McMinn 1960). Pole blight is a disease of western white pine on dry sites, which can cause stem lesions (Molnar and McMinn 1960) that can be mistaken for thin, extended fire scars. The blight is usually associated with pines on dry sites (Bega 1978).

Fire scars are a good means of establishing fire history, but they are not commonly present in all fire regimes. In high severity fire regimes, the only evidences of fire besides charcoal may be different age classes of forest across the landscape. Occasional residual trees may occur in protected areas, such as riparian zones or fuel-limited talus slopes. If the age classes can accurately establish the dates of past fires, then they can be used to determine some form of area frequency. Stand establishment dates are estimated by selecting the largest specimens of early seral tree species in the stand and aging as many of these to the pith near ground line as possible.

If logging operations have occurred in the vicinity, field counts of annual rings on stumps or samples for laboratory analysis can be easily obtained. In nature preserves, cutting of large specimens may not be possible, and age must then be established on the basis of increment borer samples. For age analysis, a core taken near the ground along an upslope-downslope centerline is most likely to hit the pith of the tree (Agee and Huff 1986b). If the pith is missed, a geometrically based correction can be applied to estimate pith age of the sample (Liu 1986). Additional years must be added to this pith date to correct for the number of years required for the tree to grow to the height at which it was bored, usually 20–40 cm height. A list of correction factors used for several species and sites is presented in Table 4.1. Such estimates usually have a possible error of at least two years, and unless a good site match is found, a site-specific set of estimates developed as part of the study is the best alternative.

TABLE 4.1. AVERAGE AGE CORRECTION FACTORS FOR CORE HEIGHT TO GROUND LEVEL FOR SEVERAL PACIFIC NORTHWEST CONIFERS

		YEARS OF AGE TO ADD AT THESE CORE HEIGHTS (CM ABOVE GROUND)						
Species	*Site Quality/ Plant Association*	*10*	*20*	*30*	*50*	*100*	*150*	*Source*
Abies grandis	Abgr/Pamy						12	Cobb (1988)
Abies lasiocarpa	Abla/Xete(10)[b]		5		9	13		Agee et al. (1986)
	Abla/Xete(50)		17		25	38		
Larix occidentalis	Abgr/Pamy					6		Cobb (1988)
Pinus contorta	Abgr/Pamy					5		Cobb (1988)
	Abgr/Hodi	1	2		4	5		Agee et al. (1986)
	"high elev."	2	3	4	6			
Pseudotsuga menziesii	Abgr/Hodi	5		8	10	12		Agee and Dunwiddie (1984)
	Abgr/Pamy						9	Cobb (1988)
	Site I			3		5	7	McArdle and Meyer (1930)
	Site III			3		7	9	
	Site V			5		9	11	
Tsuga mertensiana	"dry"	6	8	11	16			Agee et al. (1986)

[a] Abgr/Pamy = *Abies grandis/Pachistima myrsinites.*
Abgr/Hodi = *Abies grandis/Holodiscus discolor.*
Abla/Xete = *Abies lasiocarpa/Xerophyllum tenax.*
Site = average height of dominant and codominant trees at age 100 (Site I = 200; Site III = 140; Site V = 80).
[b] At core height, 10 or 50 annual rings in first cm from pith.

There is no foolproof way to determine how soon after the disturbance the oldest sampled trees became established. Relying on tree establishment dates produces a minimum rather than a true age of the event. Graphing the estimated tree germination dates over time will show the oldest sampled trees and may show a pronounced establishment pulse for certain early seral species (Fig. 4.7; Larson and Oliver 1981). These pulses may reflect delayed tree establishment. Major pulses of seral subalpine fir establishment after fire have occurred 50 years after the fire (Agee and Smith 1984). In drier subalpine ecosystems, establishment of seral lodgepole pine

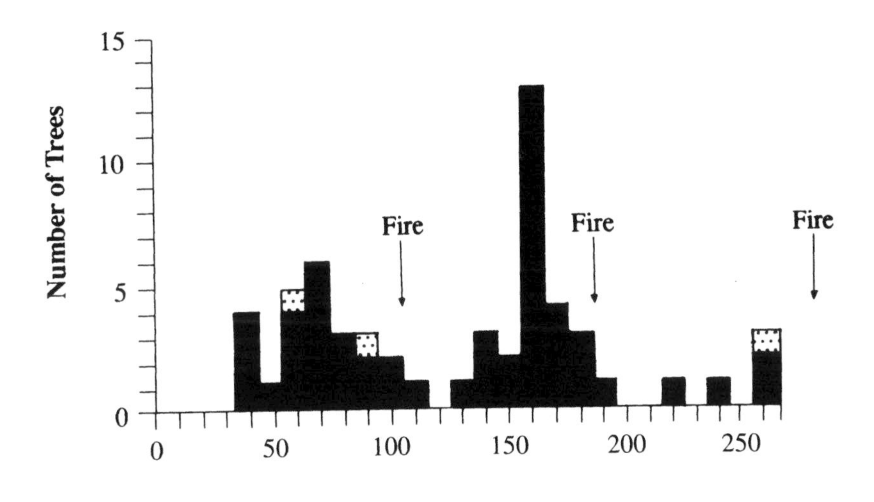

FIG. 4.7. *Pulsed establishment of Douglas-fir in the Stehekin Valley, Washington, after forest fires about 90 years apart. Disturbance opens growing space for new regeneration.*
(From Larson and Oliver 1981)

can usually be used as a disturbance "marker"; in southwestern Oregon, seral knobcone pine is a good indicator, and in moist western hemlock forests, seral Douglas-fir is a good indicator species for disturbance studies.

POINT AND AREA FREQUENCY

The various definitions of fire frequency attempt to describe one of two kinds of frequency: *point* or *area*. A point frequency is the mean fire-return interval at a single point on the landscape. An ecological interpretation derived from such data assumes that fires will burn across this point on the landscape at that mean frequency, usually with considerable variation around the mean. Technically, a point frequency should be derived solely from point data, such as a series of fire scars on a single tree. However, a point estimate is usually better derived using data from several "nearby" points, because a single tree is not likely to have recorded every fire if these events were frequent, or it may have had some scars burned out by later fires. If these points are not nearby, they become more representative of an area frequency and should be interpreted as such.

An area frequency is a mean fire-return interval for a unit area of landscape. It is commonly used in fire regimes where the only botanical evidence in most stands is of the last fire, so that evidence must be compiled from many stands across the landscape. Fire scars may be used, but stand ages are a more common data source. From the stand age data, the extent of past fires may be reconstructed, or the present stand age data may be fit to a mathematical model, from which fire frequency characteristics can be derived.

Selection of Samples

The landscape evidence of fire will determine whether a point or area frequency technique is applicable. Where fire scars are abundant, fire regimes of low to moderate severity are implied, and point frequency techniques can be used. Several techniques for point frequency sampling are summarized by McBride (1983); these include sampling within reconnaissance transects, adjacent to points on a grid, or randomly. The area should be stratified so that sample units reflect homogeneous units of physiography or forest habitat types, which are likely to have similar fire histories. Sampling density for point frequencies can be 100 percent of the scarred trees or some lesser percentage determined by total area and budget. The clustering of samples may be important if combinations of samples are to be interpreted as a point frequency.

In fire regimes of moderate to high severity, sampling density depends on the complexity of the fire regime and the degree to which it is desirable to reconstruct the past disturbances. Table 4.2 (adapted from Morrison and Swanson 1990, with some additions) shows sampling schemes ranging from less than 1 to more than 50 trees per km^2. Agee (1991) suggests that in the extremely complex fire regime of the mixed-evergreen forests of southwestern Oregon,

TABLE 4.2. SAMPLING DENSITIES OF SELECTED FIRE HISTORY STUDIES

Study	*Location*	*Trees Sampled (per km²)*	*Origin and Probable Fire Scar Dates (per km²)*	*Study Area Size (km²)*
Agee (1991)	Southwestern Oregon	52	5.6	2.5
Teensma (1987)	Central Cascades	19.7	24.9	109.7
Morrison and Swanson (1990)	Central Oregon Cascades	13.8	19.8	39
Hemstrom and Franklin (1982)	Mount Rainier	1.3	1.3	770
Tande (1979)	Jasper Nat. Park, Alberta	8.1	9.6	432
Kilgore and Taylor (1979)	Southern Sierra Nevada	12.1	36.6	18
Arno (1976)	Northern Rocky Mtns.	2.3	14.5	73
Heinselman (1973)	Minnesota	0.8	0.9	3,480

SOURCE: Adapted from Morrison and Swanson 1990.

a sampling density higher than commonly used in other places was still not sufficient to address all the aspects of the fire history, even on a very small area. Similarly, Morrison and Swanson (1990) felt their high sampling density was not sufficient in the western Cascades of Oregon. They note, however, that for even-aged stands with one or two fires per site, little is gained by extremely intensive sampling. Some knowledge of the fire regime is very helpful in planning sampling density for fire history.

DETERMINING A POINT FREQUENCY

The tree is the ideal sampling locus for determining fire-return intervals, being both a repository of fire scar information and essentially a point on the landscape. Selection of trees to sample requires a thorough survey of the study area. Look for trees with multiple fire scars. Each scar will present itself as a vertical seam with the usually triangular "catface" (Fig. 4.8). The best sample trees can be marked with flagging for ease of location. On slopes, most of the fire scars will be on the upslope side of the tree, so reconnaissance is most effective working downslope from ridgetops to valleys.

Two techniques for sampling fire scars on live trees are commonly employed: cutting sections out of live trees, and using increment boring techniques for dating fire scars.

The cleanest and most accurate way to sample a scar or series of scars on an individual tree is to cut the tree down and take an entire cross-section back to the laboratory. Because this kills the tree, and creates a very bulky sample collection, an alternative sampling

FIG. 4.8. *An eastern Cascades Douglas-fir with two fire scars: the blackened strip down the center of the tree, and the whiter strip to either side.*

technique is preferred. Techniques for sampling partial cross-sections of fire scars without killing the tree are summarized by Arno and Sneck (1977). Two parallel cuts about 5 cm apart are made on one side of the catface at a height that will include the most scars (Fig. 4.9). A vertical cut is then made on the interior portion to include the scars, and finally the tip of the saw is pushed in vertically behind the parallel cuts to free the partial cross-section. Once the sample has been removed, excess portions can be trimmed off. Two problems with this approach are that the resulting cuts on the tree are unsightly and the tree will be at increased risk for windthrow. A technique for thinner, less conspicuous sections was presented by McBride and Laven (1976). They took thin wedge-shaped sections, not necessarily to the pith, thus avoiding the need to place a chainsaw-wide vertical cut behind two parallel cuts. This technique may not be feasible for deeply embedded scars. Trees with significant

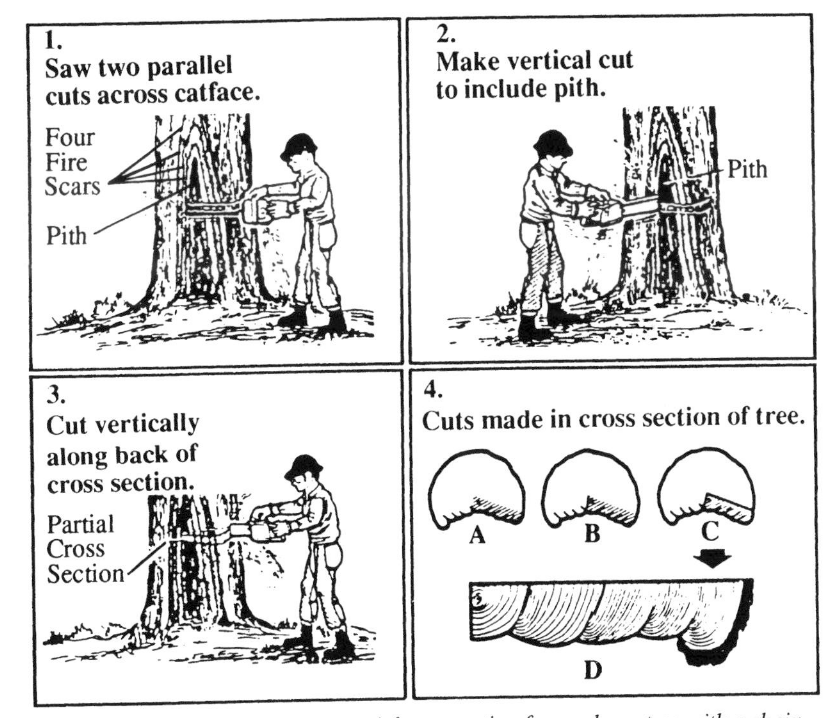

FIG. 4.9. *Method of collecting a partial cross-section from a large tree with a chain saw.*
(From Arno and Sneck 1977)

sections removed are at increased risk for windthrow from lack of ability to resist compression or tension stress (Fig. 4.10). The section can be filled with cement or a flattened rock to avoid compression stress, and metal straps can be nailed vertically across the cut to reduce tension stress. However, if the straps are later grown over, they can pose a safety hazard if the tree is cut at that height.

The use of increment borers to date scars provides an alternative to wedge sections. This technique has been in use for many years, but a clear description of the methods was published only in 1988 (Barrett and Arno 1988). For a tree with a single scar, several cores

FIG. 4.10. *Ponderosa pine windthrown in first winter after fire scar removal by method shown in Figure 4.9. Due to substantial growth since the last fire, a significant portion of the tree cross-section was necessarily removed in sampling.*

can be taken along the scar edge, combined with a "back boring" core to better represent the growth pattern of the tree (Fig. 4.11A). The cores along the scar must include one at or very near the scar tip. The back-bored core should have a growth anomaly corresponding to the year represented on the scar tip. It may be a significant growth decline, representing damage to the tree, such as crown scorch; a significant growth increase, representing little tree damage but loss of surrounding competitors; or an initial growth loss followed by a growth increase, representing some tree damage but less than that suffered by competitors (Barrett and Arno 1988). Another single-scar technique recommended by Barrett and Arno is the face-boring procedure, best used on small trees that can be completely bored through (Fig. 4.11B).

The accuracy of the increment core procedure, if cores are cross-dated, compares favorably with that from wedges (Sheppard et al. 1988). Cross-dating is a method of matching annual ring patterns of a sample with known annual ring patterns from a master chronology. The master chronology is often built from the ring patterns of 10–20 mature, undisturbed trees in the local area. Such a comparison will help identify missing or false rings in the sample and correct aging errors on the non-cross-dated sample. In Douglas-fir and western hemlock, Means (1989) found no age difference in 21 core samples compared to wedge samples, with the remaining 4 samples off by 1, 2, 3, and 4 yr. More significant errors were found if cores were not cross-dated.

Techniques for boring multiple scars are also summarized by Barrett and Arno (1988). These procedures tend to stretch the limits of the increment borer technique for determining fire history, but can be very useful in trees of diameter <60cm. Extensive experience with increment boring and with coring fire-scarred trees is necessary for successful application of increment boring techniques on multiple fire scars.

Calculating a Mean Fire-Return Interval

The simplest way to calculate a mean fire-return interval is to establish the set of fire-return intervals from a single point sample and take the arithmetic average. Typically, the interval between

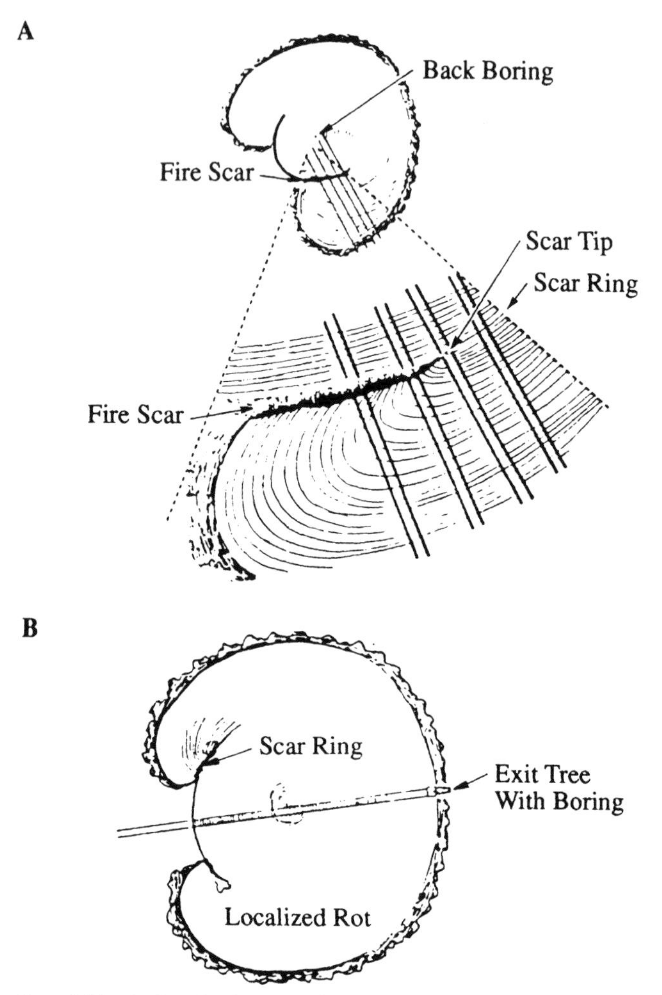

FIG. 4.11. A: *Scar-boring procedures for a tree with a single scar.* B: *Face-boring procedures for a tree with a single scar (assumes borer is long enough to pass completely through tree).* (From Barrett and Arno 1988)

scarring events is determined and then averaged within the sample (Laven et al. 1980). In some cases, the number of years from the pith to the first scar has been included as the first interval (e.g., Houston 1973), but this is not commonly done. The sample may be cross-dated (Stokes and Smiley 1968) to establish the exact year of each event—a step that may have value beyond the scope of the initial study, if regional comparisons are made later (e.g., Swetnam and Betancourt 1990). If a sample has rot, soaking the sample in water and freezing it allows a clear razor-knife-cut surface to be prepared and counted, without significant radial expansion of the specimen (Herman et al. 1972).

The problem with using a single point sample to derive a point estimate of fire frequency is that not every fire will scar every tree, and some scars will have been burned out by later fires (Kilgore and Taylor 1979). When fire does scar the tree, it may not do so across the entire existing catface (Dieterich and Swetnam 1984). Therefore, the mean fire-return interval derived from any single tree can overestimate the true mean fire-return interval. The average of a set of single mean fire-return intervals does not correct the problem. In Table 4.3, the mean fire-return interval for individual point samples ranges from 20.4 yr to 25.6 yr. An arithmetic average of these over the 50 ha study area is 23.4 yr. Yet it is probable that not every fire

TABLE 4.3 HYPOTHETICAL FIRE DATES DETERMINED ON SINGLE TREES AND THEIR USE IN DETERMINING MEAN FIRE-RETURN INTERVALS

<table>
<tr><th>Tree 1</th><th>Tree 2</th><th>Tree 3</th><th>Tree 4</th><th>Tree 5</th></tr>
<tr><td>Point</td><td></td><td></td><td></td><td></td></tr>
<tr><td colspan="2">---------- 0.25 ha ----------</td><td></td><td></td><td></td></tr>
<tr><td colspan="3">-------------------- 1 ha --------------------</td><td></td><td></td></tr>
<tr><td colspan="4">------------------------------ 10 ha ------------------------------</td><td></td></tr>
<tr><td colspan="5">-- 50 ha --</td></tr>
<tr><td>1910</td><td>1902</td><td>1910</td><td>1920</td><td>1902</td></tr>
<tr><td>1902</td><td>1884</td><td>1884</td><td>1902</td><td>1890</td></tr>
<tr><td>1875</td><td>1840</td><td>1864</td><td>1864</td><td>1870</td></tr>
<tr><td>1850</td><td>1825</td><td>1825</td><td>1835</td><td>1842</td></tr>
<tr><td>1840</td><td></td><td>1810</td><td>1825</td><td>1805</td></tr>
<tr><td>1800</td><td></td><td></td><td>1800</td><td>1800</td></tr>
<tr><td colspan="5">Mean Fire Intervals</td></tr>
<tr><td>22</td><td>25.6</td><td>25</td><td>24</td><td>20.4</td></tr>
<tr><td colspan="2">------------- 13.8 -------------</td><td></td><td></td><td></td></tr>
<tr><td colspan="3">----------------------- 11.0 -----------------------</td><td></td><td></td></tr>
<tr><td colspan="4">-------------------------------- 10.0 --------------------------------</td><td></td></tr>
<tr><td colspan="5">-- 7.5 --</td></tr>
</table>

passing a point leaves a fire scar record, so that the fire scar records of multiple trees in the vicinity may provide a better estimate of the mean fire-return interval.

The integration of fire scar records from multiple trees is called the composite fire interval. It allows separate records to be combined in a single estimate of mean fire-return interval. If the records of trees 1 and 2 (Table 4.3) are assumed accurate and combined, the composite mean fire-return interval is roughly half the individual mean fire-return intervals. In this example, increasing the area within which records are combined continues to decrease the mean fire-return interval. At some area size, the mean fire-return interval ceases to function as a point estimate of fire frequency and becomes instead an area estimate.

One of the key assumptions in developing a composite mean fire interval is that the date of each event at each point is accurate. Inaccurate dates can lead to significant errors in calculating mean fire-return intervals. Sheppard et al. (1988) present a simple example. Assume that four single-scarred trees, all adjacent, are tentatively fire-dated to 1810, 1860, 1863, and 1910. The mean fire-return interval is 33 yr. If the 1860 and 1863 events are assumed to be the same fire event, and if the record is adjusted accordingly, the mean fire-return interval is 50 yr.

There are two methods to produce a database (a master fire chronology) for developing a composite mean fire-return interval. The first (Arno and Sneck 1977) is less accurate but may be sufficient for some purposes. It requires the plotting of all dates from all trees on a chart with tree number along the top (arranged in some geographical order) and year along the side. The dates for each tree are then reconsidered based on the records of nearby trees. Trees with poor quality sections are usually adjusted to coincide with records of nearby trees. Preference is given to adding years to a given scar date, since missing rings are more common than false rings (Soeriaatmadja 1966, Madany et al. 1982). When a given date is moved back in time, all preceding dates for that tree are similarly moved in the same direction.

The second method involves some degree of cross-dating specimens based on the overall ring-width patterns of the sample (Madany et al. 1982). The patterns on the fire-scarred samples are compared to composite skeleton plots of rings from unscarred trees or from regional master chronologies (Stokes and Smiley 1968, Fritts 1976). Missing or false rings are accounted for, and the scars are then dated according to the apparent rings in which they occurred. Whether or not to cross-date samples depends on the purposes of the study and the fire regime involved (Madany et al. 1982). In ponderosa pine and other forest types with historically high fire occurrence, the cross-dating approach should be employed, although it can greatly increase the time required for analysis. In ecosystems with longer fire-return intervals, the first method may suffice (Madany et al. 1982). Means (1989) showed that for western hemlock and Douglas-fir, the absence of cross-dating resulted in negligible dating error on 20 of 22 single-scar specimens, with 1 specimen having a 3 yr error, and one having a 22 yr error due to missing rings.

Interpretation of a composite fire interval requires knowledge of how it was produced, how large an area was included in the estimate, and what time interval was considered. The decrease in mean fire-return interval with size of area, hypothetically shown in Table 4.3, is common. Kilgore and Taylor (1979), in giant sequoia–mixed conifer forests of California, showed a decrease in mean fire-return interval from 13–22 yr to 2 yr as area considered increased from single-tree estimates to about 1,000 ha. Data regraphed from Arno and Peterson (1983) show the same trend (Fig. 4.12). In what must be considered the "record" low mean fire-return interval for forests, Dieterich (1980a) found the mean fire-return interval on a 40 ha site in Arizona between 1790 and 1900 to be 1.8 yr, with one point sample having a mean fire-return interval of 4.1 yr. As the area increases in size, one might eventually produce a composite mean fire-return interval of less than one year, so the area bias inherent in composite methods must be taken into account in interpreting the ecological significance of such intervals.

Two other examples indicate the need for careful interpretation of composite fire intervals. The first is an evaluation of Native American burning in Montana (Barrett and Arno 1982). Stands ranging in size from 80 to 230 ha were sampled, and fire-return intervals for various periods were calculated for stands with little Native American use and those with heavy use. Unless one assumes that each fire event covered all of each area, there is an area bias of unknown magnitude associated with a threefold variation in stand size.

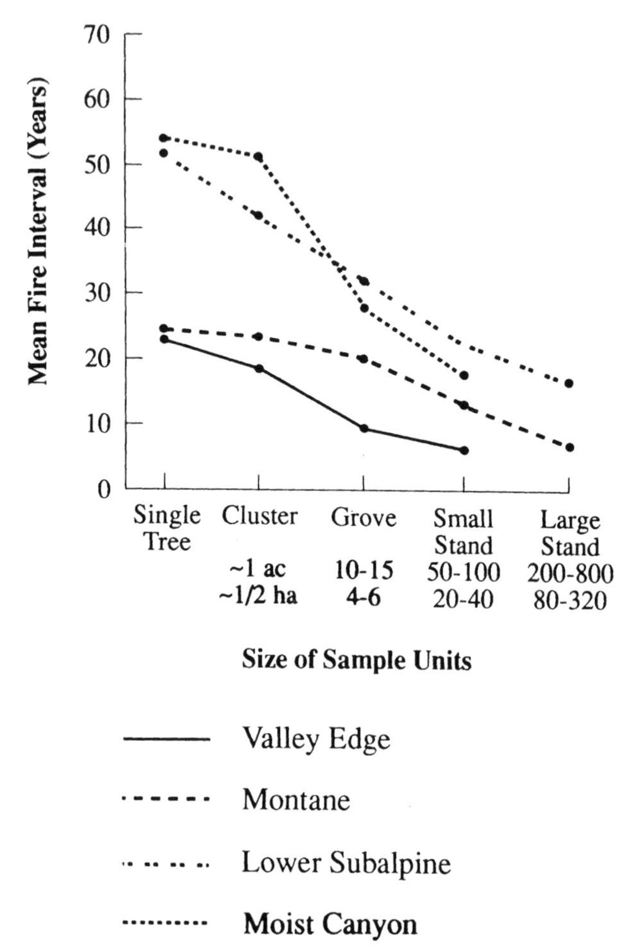

FIG. 4.12. *Shrinking fire-return intervals as the area considered increases from a point to a large stand.* (From Arno and Peterson 1983)

The second example comes from data of Weaver (1959) and Bork (1985). Weaver's estimates of fire frequency were based on individual point samples, and he found on several north-central Oregon *Pinus ponderosa/Purshia tridentata* sites mean fire-return intervals of 11–16 yr. Bork, working in similar vegetation about 80 km to the south at Pringle Butte, found a mean site fire-return interval of 4 yr. Did Weaver miscount his samples, or is the fire frequency so much higher to the south? In fact, the authors are probably expressing similar data in different ways (Fig. 4.13). Weaver presented only single-tree point estimates, while Bork had similar point estimates but also developed a variety of area estimates. Although Bork's composite fire-return intervals are shorter, the mean fire-return intervals for individual trees show a wider variation.

In most Pacific Northwest studies of fire history, fire-return intervals tend to lengthen in periods earlier than about 1650, due to the lack of available sample trees and associated problems in developing good composite fire chronologies. Comparisons between sites must be interpreted very conservatively before the mid-seventeenth century if based on fire-scarred trees.

Determining an Area Frequency

Area fire frequencies are mean fire-return intervals meant to represent landscape rather than point fire frequencies. Sometimes mean fire-return intervals calculated solely from fire scars represent area frequencies rather than point frequencies. More commonly, area frequencies are applied to landscapes where fire scars are less common and forest stand ages are a primary source of fire evidence.

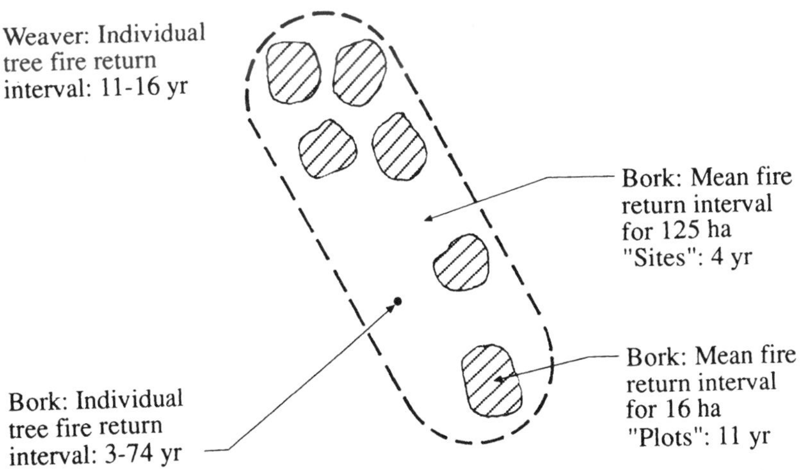

FIG. 4.13. *Comparison of fire-return-interval data from Weaver (1959) and Bork (1985). Weaver's data would be represented as a point frequency, while Bork's data in a similar ponderosa pine/bitterbrush community represents varying levels of area frequencies.*

Mean fire-return intervals represent a landscape average of fire frequency.

The first method of determining area frequency is called the natural fire rotation method. It was first proposed by Heinselman (1973) in his classic paper on fire in the Boundary Waters Canoe Area of Minnesota. Using forest age class and fire scar evidence, and interpretations of natural fire barriers such as swamps, lakes, and streams, he reconstructed historic fires that had burned across the area back to A.D. 1595. He expressed measures of different fire intervals over time by percent area burned per century, and then proposed a new term, *natural fire rotation* (NFR): the average number of years required in nature to burn over and reproduce an area equal to the total area under consideration. In simple mathematical terms, the calculation becomes:

$$\text{NFR (yr)} = \frac{\text{Total time period}}{\text{Proportion of area burned in period}}$$

For example, if an area of 10,000 ha has a total of 40,000 ha burned in 200 yr, the NFR is calculated as 200/[40,000/10,000] = 200/4 = 50 yr. Over such a long time period, some points on the landscape will burn four or five times and some not at all. Variation or periodicity is not directly considered in the calculation, which as Heinselman (1973) recognized masked the variability in fire between and within community types. He described the NFR as a valid measure of the role of fire in the total system, which needed interpretation at the community level.

Natural fire rotation was the technique used by Hemstrom and Franklin (1982) in their study of fire history at Mount Rainier National Park, Washington. They provided a more specific set of assumptions used in reconstructing disturbances:

1. *Uniformity.* The nature of vegetation responses to fire has not changed during the period covered by this study.
2. *Age continuity.* Trees in stands of similar age, separated by younger stands but not by significant vegetative or topographic firebreaks, probably originated after the same fire episode. Although fires may burn across them, large expanses of subalpine parkland, rocky ridges, and distances of more than 4–5 km were considered significant firebreaks.

3. *Fire behavior.* Fires tend to burn upslope to some topographic or fuel barrier.
4. *Topographic consistency.* This assumption recognizes correlations between particular types of disturbances, topographic location, and charcoal deposits. A surface layer of charcoal indicates past fire regardless of topographic location. Other disturbances occur in characteristic topographic locations and do not leave a charcoal layer at the mineral soil surface. Lahars (mudflows) typically affect stands on the valley bottom and lower slopes (Crandell 1971). Snow avalanches commonly operate on steep, gulleyed slopes at relatively high elevations (Luckman 1978).
5. *Regeneration span.* The age distribution of the early seral cohort on the western slopes of the Cascade Range frequently spans more than 75 yr (Franklin et al. 1979, Hemstrom 1979).
6. *Conservative limits.* Mapping reconstructed episodes of disturbance emphasizes conservative boundaries as limited by assumptions 1–5. Even if there were no physical barriers, disturbances were not extended into areas where indicative survivor trees were lacking, except as allowed by assumptions 2 and 3. Reconstructed burn areas may, therefore, be considered the minimum areas probably burned.

An example of how a natural fire rotation may be calculated from a stand-age map is shown in Figure 4.14. The stand-age map is developed from a set of assumptions, such as those of Hemstrom and Franklin (1982), using fire scars, stand ages, and an interpretation of fire barriers. On smaller areas, the broad kinds of fire barriers used in these studies may not be present. In southwestern Oregon, natural fire rotations were calculated on a 200 ha area by using tree age and fire scar evidence to establish disturbance points, and then by drawing circles of 250 m radius around each point and connecting areas of closely spaced circles (Fig. 4.15; Agee 1991). In their advice to refrain from "storytelling" Johnson and Van Wagner (1985) are partly criticizing the potential bias introduced by this type of reconstruction. Objective techniques perhaps biased in unknown ways due to lack of point evidence of fire can result in biased natural fire rotations. The erasure of earlier fire evidence by later

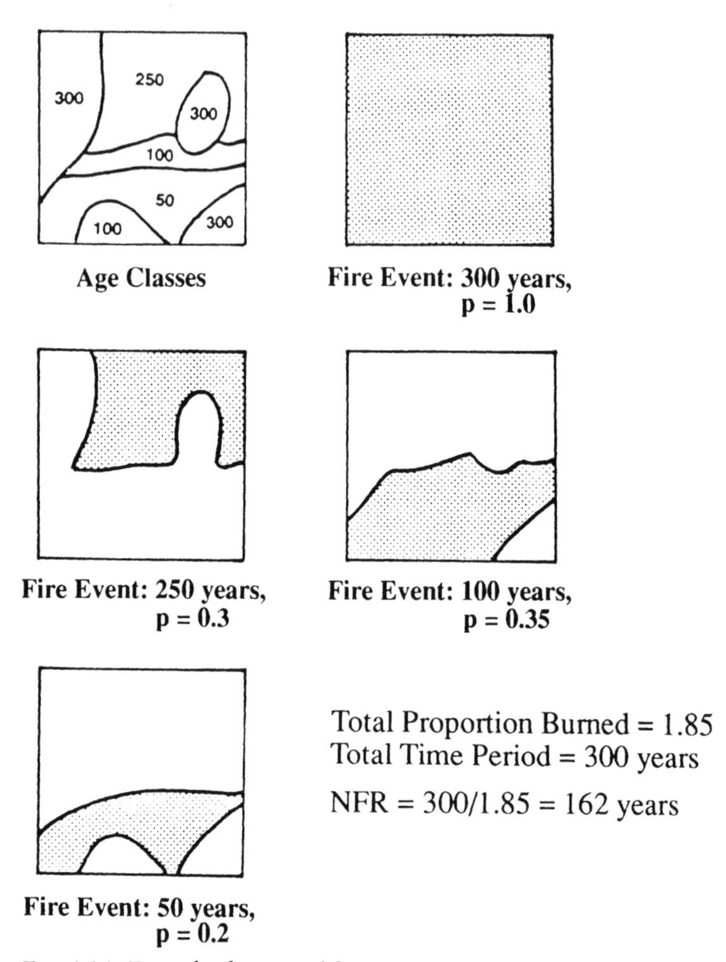

FIG. 4.14. *Example of a natural fire rotation (NFR) calculation.*

fires results in limited records of early periods, and perhaps longer natural fire rotations.

Variation in natural fire rotation for a specific area can be defined in several ways. Hemstrom and Franklin (1982) divided their long temporal period into three segments: a modern fire suppression era (NFR = 2,583 yr), a settlement period (NFR = 226 yr), and a 650 yr presettlement era (NFR = 465 yr). At Desolation Peak in the North Cascades, natural fire rotations were calculated by century, by aspect, and by community type (Agee et al. 1990), addressing Heinselman's suggestion to more specifically interpret the natural fire

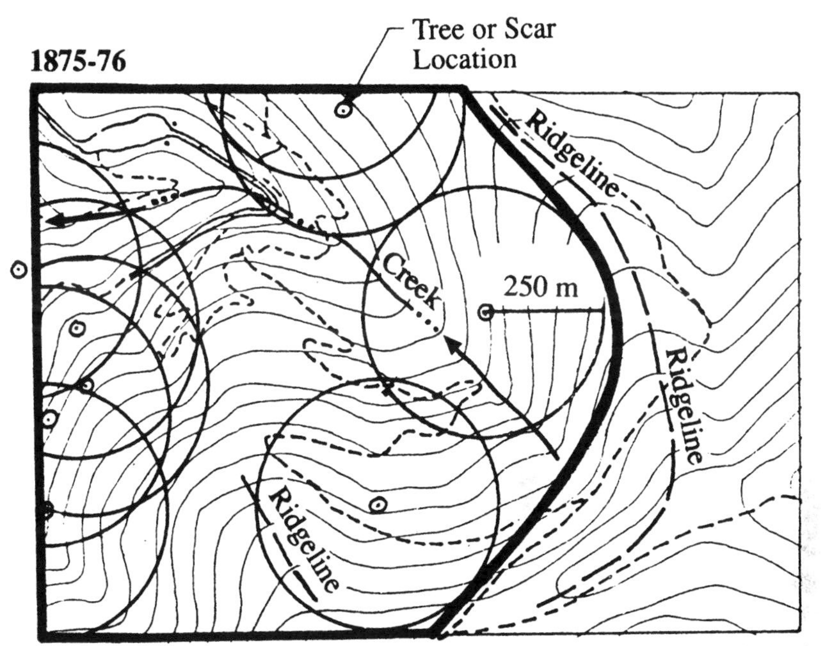

FIG. 4.15. *Reconstruction of individual fire events based on multiple point evidence of fire (fire scar or age class). In this case, circles of 250 m radius were drawn around each point at each time period. For 1875–76, shown here, roughly two-thirds of the area is included in the reconstructed burn.*
(From J. K. Agee, "Fire History along an Elevational Gradient in the Siskiyou Mountains, Oregon," Northwest Science 65 [1991]: 197. Copyright 1991 by Washington State University Press, Pullman. All rights reserved. Reprinted by permission of the publisher.)

rotation. Variation in fire severity was presented by Morrison and Swanson (1990) and Teensma (1987), who calculated separate natural fire rotations for fires of low to moderate and high severity in the central Oregon Cascades.

A mathematical method of addressing the variability in natural fire rotation over the whole area is outlined by Sugita (S. Sugita, University of Minnesota, personal communication, 1992). Sugita divided the Boundary Waters Canoe Area into 131 grids, each about 7 km on a side. He counted the number of fire occurrences in each grid over 362 years from Heinselman (1973). He then assumed that the distribution of fire frequencies (from none in some grids to eight in others) followed a Poisson distribution:

$$P\,(X = k) = \frac{e^{-a}\,a^{k}}{k!}$$

where

X = a discrete random variable assuming the possible values 0, 1, . . . n, representing the number of fires in one cell of the grid

k = a discrete value of X

e = base of natural logarithms

a = Poisson parameter, estimated by average number of occurrences per cell.

Depending upon the grid size and how the presence of fire in any grid is counted (for example, any presence of fire counts as an occurrence, or 10 percent must be burned to count, and so forth), the Poisson parameter, divided into the time period being considered, should approximately equal the natural fire rotation. The two figures will not be exactly equal because of the transformation of a continuous area into a discrete grid. With the assumption that the fire frequency approximates a Poisson distribution, the probability of 0, 1, 2, ... n fires (Fig. 4.16), and associated natural fire rotations, in the period of interest can be calculated. For the Boundary Waters Canoe Area, Sugita found a = 3.0922, or a natural fire rotation of 117 yr, comparable to Heinselman's estimate of 100 yr for the same time period. The actual distribution of fire frequencies did not significantly differ from a Poisson distribution (Kolmogorov-Smirnov test). The variability of frequency over the landscape is apparent from such an analysis: Some areas burn repeatedly and others not at all. Inspection of the grid map will indicate the spatial distribution of this variability.

A similar analysis was applied to the Mount Rainier data of Hemstrom and Franklin (1982). Their natural fire rotation for the forested portion of the park for the entire time period was 405 yr. The average number of occurrences that burned more than 20 percent of a cell within a grid size approximately 1,700 m on a side (2.9 km^2) resulted in a = 1.9209, equivalent to a natural fire rotation of 420 yr. Compared to the Boundary Waters Canoe Area, there is higher probability of no fires within any cell and lower probability of multiple fires (Fig. 4.16). The observed distribution of fire frequency

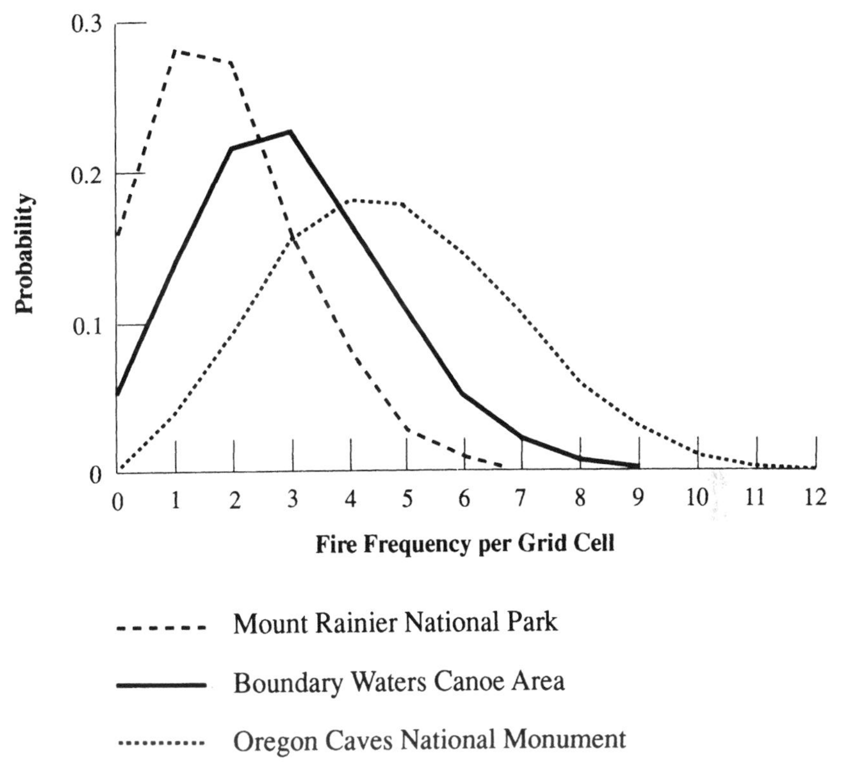

FIG. 4.16. *The Poisson distribution for fire presence in landscape grids for three natural fire rotations: Mount Rainier, Boundary Waters Canoe Area, and Oregon Caves. The cell size is different for each example shown.*

at Mount Rainier was also not significantly different from a Poisson distribution (Kolmogorov-Smirnov test). A third comparison was made with data from Oregon Caves National Monument (Agee 1991), a small preserve (200 ha) with a natural fire rotation of 73 yr since 1650. A grid size of 0.67 ha was used, and presence of any fire in the grid was counted as an occurrence. The Poisson parameter was calculated as 4.954, equivalent to a natural fire rotation of 69 yr, and the observed and expected distributions, as in the other comparisons, did not significantly differ from one another. On an area this small, the fit of a Poisson distribution is remarkable because of the expectation that in such a small area every fire would burn much or all of the area.

This technique may be quite useful in explaining the variability across landscapes subjected to a natural fire rotation analysis. On a uniform landscape with little mixing of different forest types or fire regimes, or on a small landscape where most fires might burn the entire landscape, the Poisson distribution may not fit as well as in these three examples. Observed and expected distributions should always be compared before the Poisson distribution is accepted for purposes of explaining natural fire rotation variability.

The Fire Cycle

The use of statistical distributions to model disturbance frequency was demonstrated by Everitt (1968) for floodplain dynamics of the Little Missouri River. Everitt showed that the area of floodplain forests of varying ages fits a negative exponential distribution. The greater area occupied by younger forests reflects the destruction of older forests by channel migration, but the probability of destruction is independent of stand age (for example, the channel migration might as easily undercut a 25-year-old stand as a 100-year-old stand). Stands of any age can be destroyed at one location, while new forest habitat is created at another location by sediment deposition. Average stand age (the recurrence interval of the stand-initiating event) or the proportion of stands greater than any specified age can be calculated once the shape of the curve is known. The use of this same statistical distribution to model fire disturbances was proposed by Van Wagner (1978). Its major advantage is reliance on present age classes across the landscape; reconstruction of all past fire events is not necessary. If the present age-class mosaic is not steady-state, however, significant bias may appear in a fire cycle calculation based on age structure at one point in time (Baker 1989).

Van Wagner's idealized scenario assumes a simple forest on a uniform site, composed of many equal-sized stands, all of equal flammability regardless of age. Climate is uniform, such that lightning, over the study period, causes an equal number of fires per year at random, and each fire burns only one stand. The frequency of an age-class x then follows a geometric distribution. If p is the probability of a fire in any single year, and the fire cycle exceeds about 20 yr, so that $p < 0.05$, then the distribution can be approximated by the negative exponential distribution:

$$f(x) = pe^{-px}$$

where

$f(x)$ = frequency of an age class
e = base of natural logarithms
p = probability of fire in any one year.

The negative exponential distribution has several convenient properties. The mean age of all stands, or the fire cycle (C), is equal to $1/p$. The median age is 0.693 C. The cumulative proportion of all stands up to age x is

$$\Sigma f(x) = 1 - e^{-px}$$

This expression approaches 1 as x approaches infinity. The probability of a stand burning n times during one cycle is $e^{-1}/n!$, and the probability of survival throughout a rotation is e^{-1}, or 36.8 percent.

The negative exponential distribution is best applied to large areas composed of many stands (Fig. 4.17), such that only a small proportion of the area is likely to burn at any one time. Johnson and Van Wagner (1985) and Baker (1989) note that the study area must be large relative to the size of the largest expected fire event. Hemstrom and Franklin (1982) found the method inapplicable to individual watersheds at Mount Rainier that were less than 4,000 ha in size, although it produced reasonable results for the entire park (53,000 ha). A regional forest-type application by Fahnestock and Agee (1983), using the cumulative form of the distribution and dealing with areas greater than 50,000 ha, again appeared to produce reasonable results. Van Wagner (1978) noted that the model is fairly robust against departures from admittedly narrow constraints. The assumption most open to question is that of uniform flammability with age. Without much other than empirical observations, some researchers believe that older forests are more flammable than younger forests, but objective evidence is hard to find (Van

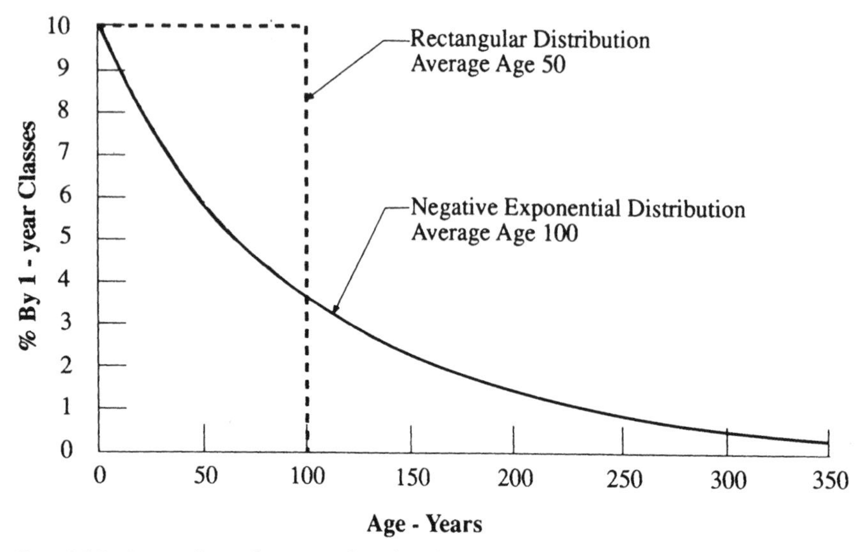

FIG. 4.17. *Comparison of two age-class distributions: the negative exponential, with average age 100, and a balanced rectangular age distribution, with average age 50.* (From Van Wagner 1978)

Wagner 1978). In Douglas-fir forest burned with crown scorch intensity, the youngest forests had the highest potential surface fire behavior, with a pronounced dip in mid-succession and an increase in late succession (Agee and Huff 1987). Given the variability from stand to stand, the assumption of equal flammability may not be terribly flawed.

The negative exponential method was applied to the Boundary Waters Canoe Area data of Heinselman (1973). Although the natural fire rotation method of reconstructed fire events resulted in a fire rotation of 100 yr, the fire cycle calculated from present stand ages was 50 yr. Van Wagner (1978) suggested that since the negative exponential distribution did not rely on event reconstruction but on Heinselman's more precise age-class maps, it produced an estimate closer to reality. Heinselman (1973) noted at the time that his estimates were conservative, and he later agreed (Heinselman 1981) that the negative exponential calculations from his data appeared to better describe fire frequency in the Boundary Waters Canoe Area. Another analysis of the same data (Baker 1989) suggested that the area could be divided into three fire regions, and that the negative exponential model appeared inappropriate to describe any of the three. The negative exponential distribution, applied to the Mount Rainier data, produced a fire cycle of 306 yr, considerably less than the natural fire rotation method, but with a poor goodness-of-fit, generally <0.60 (Hemstrom 1979).

The negative exponential model, with a constant flammability regardless of age, is in fact a special case of a more flexible but less easily calculated model, the Weibull distribution. The Weibull requires a similar set of assumptions to the negative exponential but its shape parameter, c (equivalent to flammability), can vary (Johnson and Van Wagner 1985). The Weibull distribution can be expressed either as a fire interval distribution (Fig. 4.18A) or as a time-since-fire distribution (Fig. 4.18B). The cumulative fire interval distribution is:

$$F(t) = 1 - \exp[-(t/b)^c]$$

where

b = a scaling parameter (annual percent burned, with 1/b the fire cycle)

exp = e (base of natural logarithms)

c = shape parameter (>0) interpreted as a flammability index for fire studies ($c < 1$: decreasing flammability with age; $c = 1$: equal flammability with age; $c > 1$: increasing flammability with age).

In the southern Canadian Rockies, Johnson and Larsen (1991) used the time-since-fire distribution to partition two time components of a 380 yr fire record. The fire cycle (C), or 1/b, was calculated from the Weibull distribution for the periods 1600–1730 ($C = 60$ yr), and 1730–1980 ($C = 90$ yr). The time periods were separated on the basis of a graphical method to determine the break in slope when data on stand age were plotted on semilog paper (Fig. 4.19). This graphical technique should be cautiously applied, because omitted from the real data is the theoretical tail of the curve, so that semilog plots will appear to descend with increasingly negative slope over the range of age classes (M. Finney, in prep.).

The probability-density fire interval function is the frequency or probability of having fires with intervals of age t:

$$f(t) = [ct^{c-1}/b^c]\exp[-(t/b)^c]$$

Weibull Fire Interval Distribution

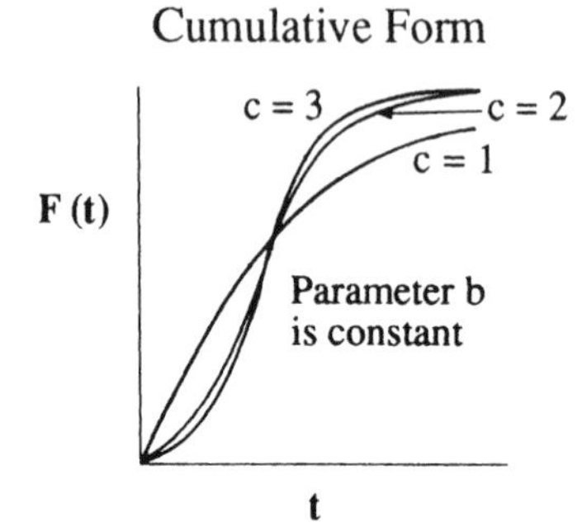

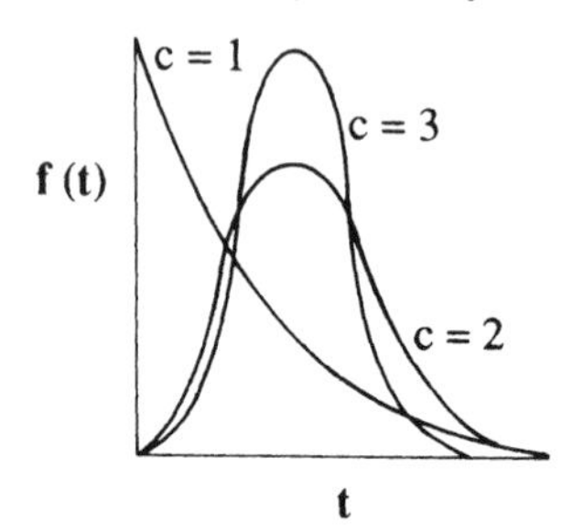

Weibull Time-Since-Fire Distribution

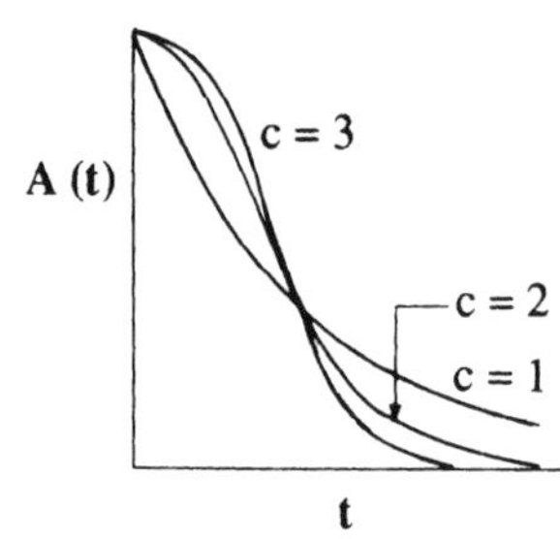

FIG. 4.18. *The different distributions for the Weibull fire history model. Graphs show distributions with different shape parameter (c) settings and constant fire recurrence or scale parameter (b).*
(From Johnson and Van Wagner 1985)

Note that when $c = 1$ and $1/b = p$, the equation reduces to the probability density function of the negative exponential distribution. Johnson (1979) applied this form of the Weibull model to fire recurrence in the subarctic. He collected data in several recently burned stands and observed directly the fire interval data. Although the Weibull function can be used on point frequency data, describing fire recurrence at a point, it is also used to describe fire frequency on an area basis, combining fire recurrence data over many stands.

The cumulative time-since-fire distribution is:

$$A(t) = 1 - F(t) = \exp[-(t/b)^c]$$

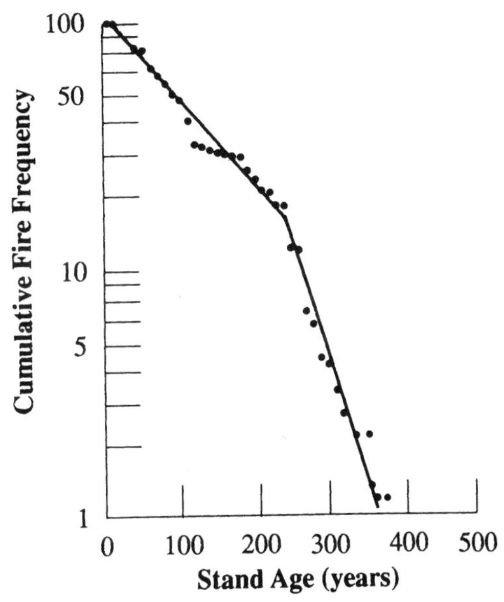

FIG. 4.19. *A mixed time-since-fire distribution from the southern Canadian Rockies. The dots represent 10 yr age classes. Once segregated, the data are individually rescaled to 100, and separate estimates of fire cycles are then calculated.*
(From E. A. Johnson and C.P.S. Larsen, "Climatically Induced Change in Fire Frequency in the Southern Canadian Rockies," Ecology 72 [1991]: 194–201. Copyright 1991 by Ecological Society of America. Reprinted by permission.)

In this equation, Johnson and Van Wagner (1985) note that t is interpreted not as a fire interval but as the time that has passed since the last fire, the "running time." Age-class distributions can be calculated from A(t) when it is multiplied by another factor (see Johnson and Van Wagner 1985 for details).

The negative exponential model is essentially a random selection model, which burns equal numbers of stands chosen randomly from all stand ages. The greater number of younger stands (Fig. 4.17) is due to the annual depletion across all cohorts, which then begin as new stands. The tail of the distribution, composed of old stands, exists because of chance, not because of decreased flammability with age. The Weibull model is an age-selection model, ranging from decreasing flammability with age ($c < 1$), to no selection ($c = 1$; the negative exponential) to increasing flammability with age ($c > 1$). The tendency toward a bell-shaped distribution when $c > 1$ is due to low probability of selection of a young stand, many intermediate intervals, and (due to high probability of burning) few older stands.

CHAPTER 5

FIRE EFFECTS ON VEGETATION

FIRE HAS VARIABLE BUT PREDICTABLE EFFECTS on individual plants. Translating these from the physical to the physiological helps us understand how fire affects vegetation not only at the level of the individual plant but also at the levels of the plant community and the landscape. Further, vegetation provides the fuel that makes fire possible, so we can view fire effects on vegetation as an interaction rather than just a unidirectional effect.

This chapter first describes mechanisms of fire damage to individual plant parts. The differential response of plants may be due to fire variation or to specific adaptations of plants to survive as individuals. Some species with clear adaptations to fire may appear to increase the flammability of the community within which they grow, but the issue of fire-dependent ecosystems is a controversial topic on which debate is likely to continue. Also controversial, at least historically, has been the role of fire in plant succession theory. In recent years the emphasis has moved away from the development of all-encompassing theories of vegetation development to the construction of process-oriented ecological models. These dynamic models can incorporate fire as an ecological process, and several models applicable to the Pacific Northwest are conceptually described in this chapter. Such models have value in linking physiological processes of individual plants to interactions at the community and landscape levels.

FIRE AND THE INDIVIDUAL PLANT

Trees can be injured by fire in several ways. Foliage and buds can be killed in the crown of the tree, the bole can be heated to a point where part or all of the cambium is killed, or the roots can be heated and killed. Each of these types of damage depends upon lethal heat contacting live tissue. Although fire temperatures usually exceed 600°C in the flaming zone, plant tissue may be protected from these temperatures by insulation provided by the soil, by tree bark, or by location in the crown of the tree far above the flames and lethal heat.

There is no single critical temperature for tissue death, although duration of 60°C for 1 minute is often noted by researchers as a "standard" lethal time-temperature combination (Kayll 1968, Methven 1971). Others (Lorenz 1939, Seidel 1986), using water baths or dry heat, have shown limited survival of seedlings up to 65°C for 1 minute. Wright and Bailey (1982) reported that *Stipa comata* culms survived exposure to >60°C for 1 minute regardless of the moisture content. For most purposes the 60°C/1 min combination is an acceptable lethal heat criterion for breakdown of cell protoplasm. Heat experiments have primarily been made with seedlings, and when fire burns across a site, even if it is a low intensity underburn, it usually consumes the seedlings in its path. The use of seedlings in laboratory experiments, however, has been helpful in predicting effects on plant tissue. The results are applicable to mature plants, which may be subjected to lethal heat over only a portion of their leaves, stems, or roots.

Crown Damage

Above every fire is a zone within which living tissue will be killed by the hot gases. The maximum height of the zone is related to fireline intensity (the rate of energy release per unit length of fireline; see chapter 1), the ambient temperature, and the windspeed (Van Wagner 1973). Increasing fireline intensity will increase scorch height as the energy release rate increases. As ambient temperature increases, less heat is needed from the fire to reach lethal temperatures. Increasing wind tends to level the convection column, reducing scorch height for a given fireline intensity. If wind is increased with other factors constant, this can cause fireline intensity to increase and scorch height to increase.

The no wind/25°C relation between fireline intensity (I, in kW m^{-1}) and scorch height (h_s, in m) is

$$h_s = 0.148\ I^{2/3}$$

(see Fig. 5.1). Corrections for different ambient temperature and windspeed conditions (Fig. 5.2A,B) can be graphically applied to the scorch height calculated from this equation (Van Wagner 1973; see Albini 1976 for English units).

In the derivation of this relationship, Van Wagner (1973) recognized that slightly higher temperatures might be required for dormant tissue to be killed. Variable sensitivity to temperature in the tree crown due to the dormancy effect and bud size was incorporated into a crown damage model for northern Rocky Mountain conifers (Peterson and Ryan 1986). Peterson and Ryan defined three levels of temperature considered lethal after one minute: 60°C, 65°C, and 70°C. The 60° level was applied to trees with small buds, such as Douglas-fir, during the growing season. The 65° level was applied to small-bud species during the dormant season and to large-bud species like ponderosa pine during the growing season. The 70° level was applied to large-bud species during the dormant season.

Actual crown mortality requires the integration of crown scorch height with crown dimensions of an individual tree. (The ability of crowns to resprout from axillary or adventitious buds after being scorched or burned will be discussed later.) For completely scorched

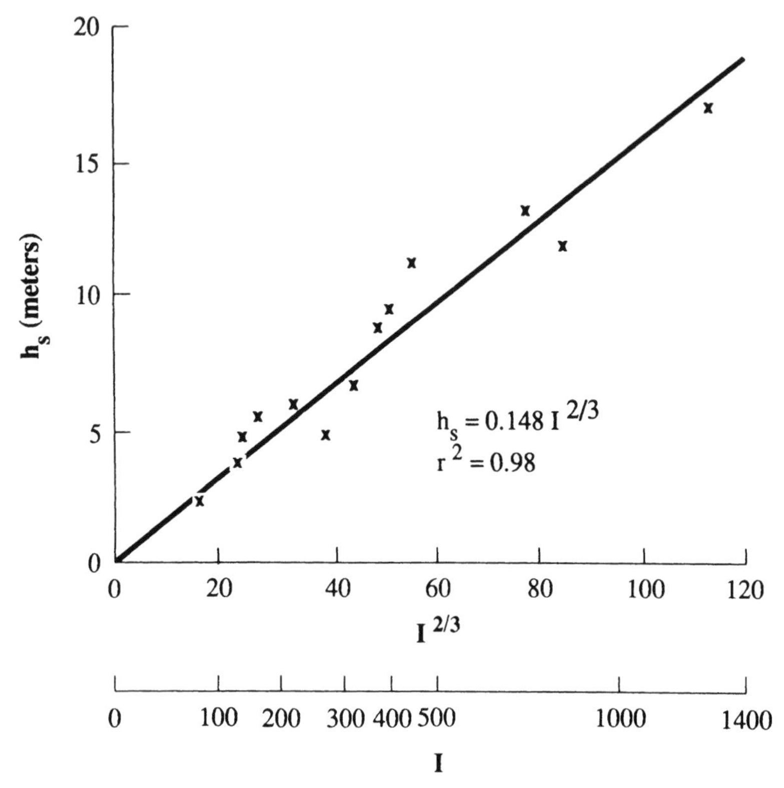

FIG. 5.1. *Experimental relation between scorch height, h_s, and fire intensity, I, in kW m^{-1}.*
(From Van Wagner 1973)

trees, crown morphology may not be relevant, as the entire crown may be killed. For taller trees, the base of the live crown, the shape of the crown, and the height of the tree will influence the proportion of the tree crown that is likely to be killed. Proportion of crown volume killed is a better predictor of postfire tree condition than scorch height (Peterson 1985), because the former is more highly associated with injury to the tree than the latter.

Geometric solids have been used to describe crown shape (Table 5.1), and crown volume can be calculated based on crown diameter and height. Specified solids may not accurately represent crown shapes, particularly if one crown shape is assumed to represent all individuals of one or more species. Scorch height may not be the same all around the tree, particularly on slopes, where scorch

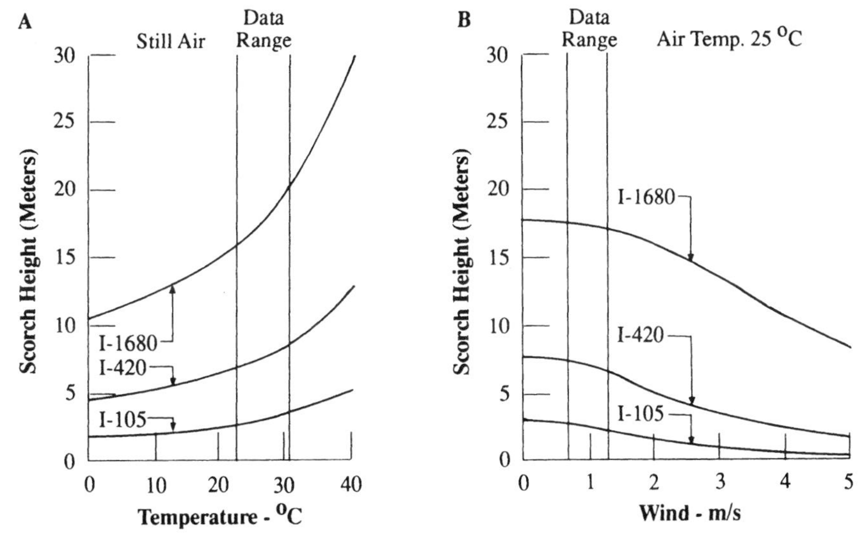

FIG. 5.2. A: *Theoretical relation between scorch height and air temperature for three fireline intensity levels.* B: *Theoretical relation between scorch height and wind at 25°C for three fireline intensity levels. Actual data range shown by vertical lines.* (From Van Wagner 1973)

height will be higher on the upslope side of the tree (Peterson 1985). An average scorch height can be applied to estimate affected crown volume in this case (Fig. 5.3).

It is often desirable to reconstruct fireline intensity in postfire monitoring to compare fire effects in different fires or different areas of a single fire. Because scorch height depends on windspeed and ambient temperature as well as fireline intensity, some estimates of

TABLE 5.1. CROWN SHAPES OF SEVERAL PACIFIC NORTHWEST TREE SPECIES (PERCENTAGES)

Crown Shape	*Douglas-Fir*	*Lodgepole Pine*	*Subalpine Fir*	*Western Redcedar*
Paraboloid	78.0	41.1	23.8	57.1
Cone	16.0	33.0	48.8	18.2
Ellipsoid	5.0	22.3	22.7	24.7
Cylinder	1.0	3.6	4.7	0

SOURCE: Peterson 1985.

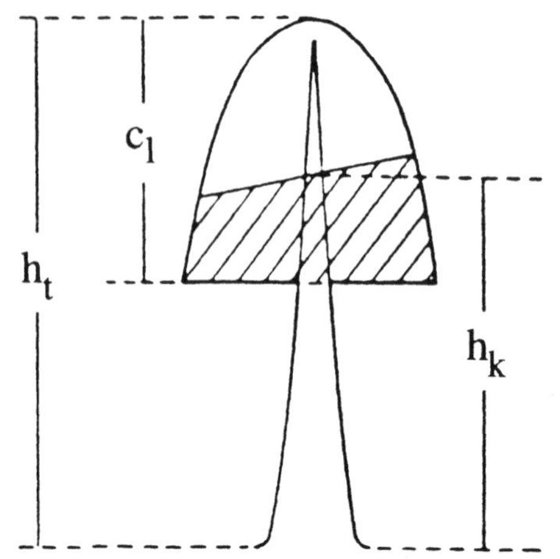

FIG. 5.3. *Fraction of crown volume killed calculated by height of crown kill* (h_k)*, tree height* (h_t)*, and crown length* (c_l)*. Shaded area indicates crown kill.*
(From Peterson and Ryan 1986)

the windspeed and temperature are essential if a fireline intensity estimate within a narrow range is desired. In one experiment on loblolly and shortleaf pines (*Pinus taeda* and *P. echinata*), fireline intensity of underburns was underestimated by a factor of three when reconstructed from scorch height (Cain 1984).

BOLE DAMAGE

Tree stems may also be damaged by fire, even though they often have a thick bark that insulates the cambium against the heat from forest fires. When a fire approaches a tree bole, it rarely produces an even distribution of heat around the stem. This uneven distribution of heat will sometimes result in the death of the cambium on one side of the tree, creating a fire scar; in other cases, enough heat penetrates the bark to kill the cambium around its entire circumference.

The uneven distribution of heat around a tree bole as a fire passes is caused by the presence of a cylinder (the bole) in three-dimensional air space. The cylinder effect was demonstrated by Gill (1974), who positioned metal rods of varying diameters at various distances from a stationary flame source. Maximum flame heights were observed on the leeward side of the rods. The diameter of the rods also affected flame height, but only on the lee side. These results suggest that the lee side of the tree is more likely to receive higher heat loads at the bark surface, and that larger diameter trees may experience more heat than smaller diameter trees. This might help to explain a commonly observed pattern of fire scarring in ponderosa pine forests, where frequent, low intensity fires historically occurred. Some trees remain unscarred until they are perhaps 30–50 cm in diameter, and then become scarred and rescarred by most later fires at frequent intervals. The smaller trees, even with thinner bark, may not create enough of an eddy effect on their lee side to cause fire scars to form.

The flame experiment results of Gill (1974), which were collected in a laboratory setting, are consistent with temperature data collected around the bases of trees during wildland fires. Synthetic trees placed in a grassland with fuel 70 cm tall recorded highest temperatures on lee aspects and at a height of 40 cm (Tunstall et al. 1976). On trees surrounded by pine litter, Fahnestock and Hare (1964) found lee aspects to have the highest temperatures (Fig. 5.4). The highest temperature recorded (846°C) was from a headfire on the lee side of a tree at about 1 m above the ground.

Bark plates had higher temperatures than bark fissures, and more heat was applied by heading than backing fires to the lee side of the tree up to 1 m above the ground. In simulated fires at 0.3 m above a lighted wick around the tree bole, Hare (1965) found lee aspect temperatures of 538°C compared to 260°C on the windward aspect. Fire scars are more likely to occur on the upslope sides of tree boles because of the eddy effect from headfires, and also because headfires on a slope are usually of higher intensity than backing fires due to preheating of fuels ahead of the fire by radiation and convection. The upslope side of the tree may also act as a catchment for fuels rolling downslope between fire events, possibly increasing fire intensity on that side of the tree, but such fuel accretion is not necessary for accelerated fire behavior and fire scarring.

The effect of bole heating on the cambium depends on how well it is insulated from the heat applied to the exterior of the bark. Heat transfer through bark is a complex process, but with several simplifying assumptions it can be adequately modeled to predict passage of a temperature pulse through bark. Thermal diffusivity, by

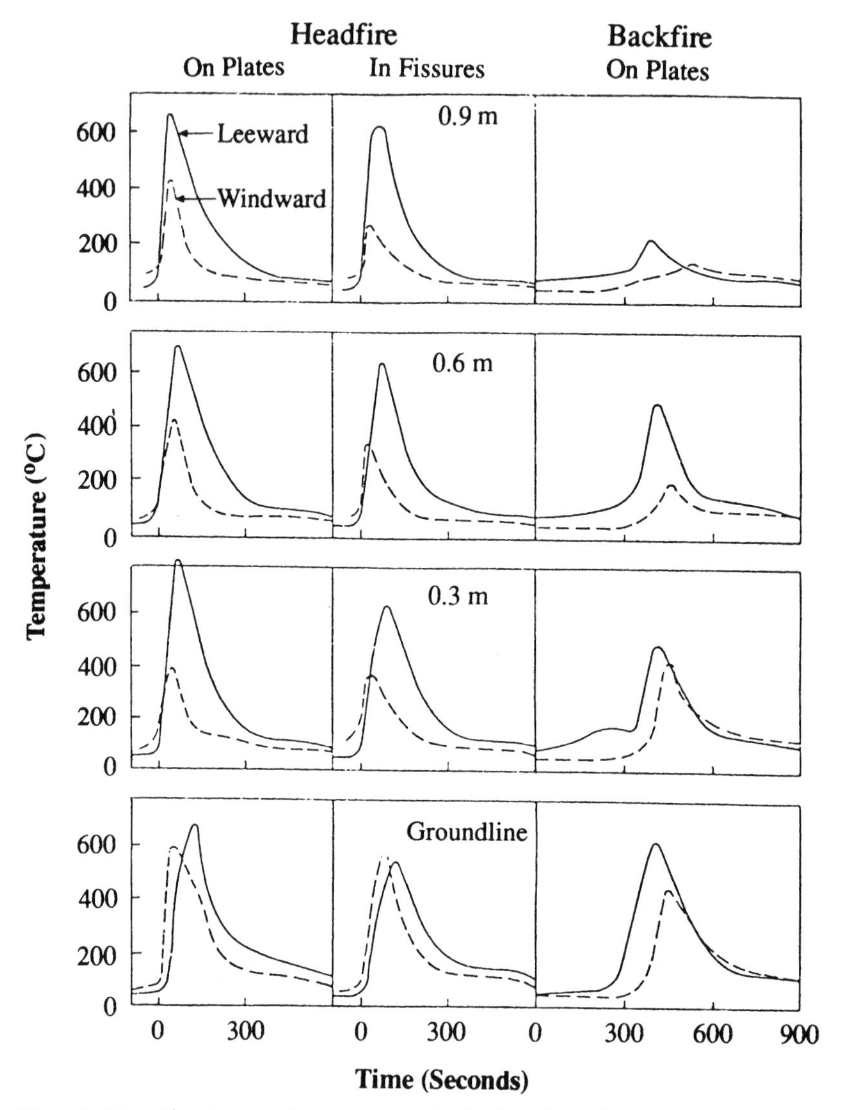

FIG. 5.4. *Mean time-temperature regimes at the bark surface of three trees in headfires and three trees in backfires. Top graphs are at roughly 1 m above ground line, and bottom graphs are at ground level.*
(From Fahnestock and Hare 1964)

Fourier's equation for heat conduction (Brown and Marco (1958, is the rate at which a temperature pulse moves through a material:

$$a = \frac{k}{c\,p}$$

where

a = thermal diffusivity ($cm^2\ sec^{-1}$)
k = thermal conductivity ($cal\ cm^{-1}\ sec^{-1}\ deg^{-1}$)
c = heat capacity ($cal\ gm^{-1}\ deg^{-1}$)
p = density of material ($g\ cm^{-3}$)

Thermal diffusivity increases as conductivity increases, but decreases as density or heat capacity increases. Because diffusivity is the ratio of several quantities, its dimensions are not readily interpretable in a physical sense (Spalt and Reifsnyder 1962). A good insulating material, such as asbestos, has a thermal diffusivity of $2.5 \times 10^{-3}\ cm^2\ sec^{-1}$. An average thermal diffusivity for bark is about $1.1–1.3 \times 10^{-3}\ cm^2\ sec^{-1}$ (Martin 1963, Reifsnyder et al. 1967), which makes bark more efficient than asbestos as an insulator. Bark, however, is a composite of several different materials: the bark solids, entrapped cells of air, and a seasonally varying moisture content. Thermal diffusivity might vary considerably based on bark structure, surface texture, and moisture content, particularly with the high thermal conductivity and high heat capacity of water.

As moisture is added to bark, thermal diffusivity does indeed decrease, primarily due to the high heat capacity of water (Fig. 5.5). Reifsnyder et al. (1967) suggested that the actual moisture content of the outer bark of red pine (*Pinus resinosa*) does not seasonally vary as widely as the limits in Figure 5.5. Therefore, differences in cambial heating resulting from thermal diffusivity changes due to bark moisture content or species-specific bark structure are expected to vary by a factor of 1.5–2 at most. There appears to be a much more important factor involved in cambial protection: bark thickness.

The thickness of the material through which the temperature pulse is moving has a large effect on the temperature experienced at the inner edge of the material. Bark thickness is the most important bark characteristic in assessing cambial protection from fire (Martin

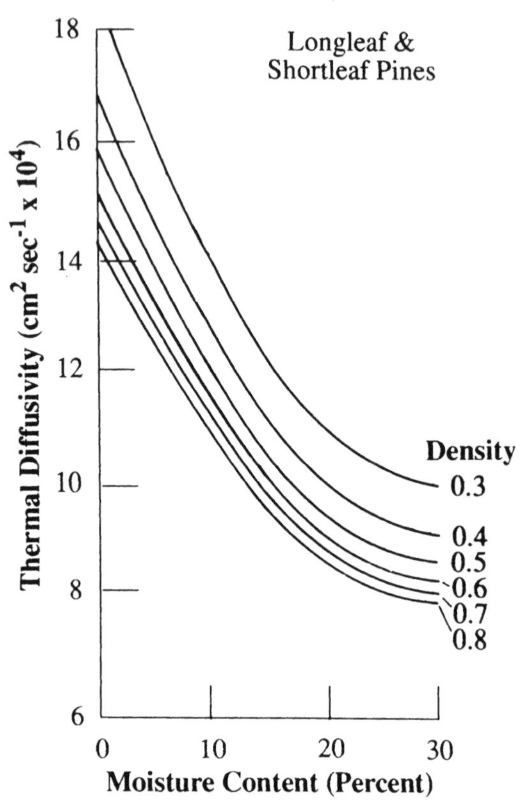

FIG. 5.5. *Thermal diffusivity of bark as a function of moisture content and dry density.*
(From Reifsnyder et al. 1967)

1963, Reifsnyder et al. 1967, Uhl and Kauffman 1990). Protection increases with the square of the bark thickness. Ryan and Reinhardt (1988) found a 40-fold difference in computed bark thickness for seven Pacific Northwest conifers, implying differential susceptibility to injury from fire. Bark thickness can be predicted from diameter of outside bark (Table 5.2), so that cambial damage from a given fireline intensity can be predicted from stand-level information.

A critical time for cambial kill can be derived as a function of tree diameter (Fig. 5.6). This relationship is based on bark thickness as a function of species and diameter. Peterson and Ryan (1986), using the work of Hare (1965) and Spalt and Reifsnyder (1962) and assuming a fire temperature of 500°C, developed the equation:

$$t_c = 2.9\ x^2$$

where

t_c = critical time to cambial kill (minutes)
x = bark thickness (cm)

For example, a species with a bark thickness of 0.6 cm could survive for about 1 minute, while a tree with bark thickness of 2.6 cm could survive for about 20 minutes, independent of other deleterious effects of the fire.

Bark thickness is not uniform around the tree; instead, it is thicker along some radii and thinner along others. The so-called bark fissures, comprising younger bark, will experience lower temperatures in a fire (Fahnestock and Hare 1964), but often have a higher probability of cambial kill due to the thinner bark. On thick-barked species, such as Douglas-fir, the fissures are likely places for repeated cambial kill along the same bole radius (Fig. 5.7). Species with rough-textured bark may experience lower average temperatures at the bark surface than smooth-barked species (Uhl and Kauffman 1990).

If the heat pulse is sufficient to kill the cambium but of limited spatial extent on the bole, fluid movement within the tree's inner bark or xylem may translocate heat sufficiently to prevent cambial kill (Vines 1968). It was not clear from Vines's experiment whether this reaction was a physiological response on the part of the tree or a

TABLE 5.2. SINGLE BARK THICKNESS (BT) AS A FUNCTION OF DIAMETER OUTSIDE BARK (DOB) FOR SEVEN WESTERN NORTH AMERICAN CONIFER SPECIES

Species	*Equation (cm)*	*Source*
Douglas-fir	BT = 0.065 DOB	Monserud 1979
Western larch	BT = − 0.1143 + 0.0629 DOB	Faurot 1977
Engelmann spruce	BT = 0.189 + 0.022 DOB	Smith and Kozak 1967
Lodgepole pine	BT = 0.0688 + 0.0143 DOB	Faurot 1977
Subalpine fir	BT = 0.015 DOB	Finch 1948
Western redcedar	BT = 0.386 + 0.021 DOB	Smith and Kozak 1967
Western hemlock	BT = 0.056 + 0.043 DOB	Smith and Kozak 1967

SOURCE: Ryan and Reinhardt 1988.

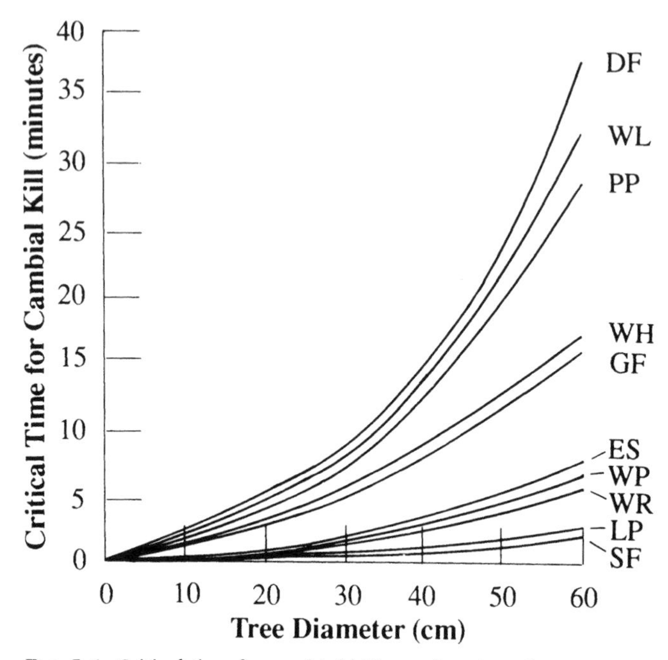

FIG. 5.6. *Critical time for cambial kill as a function of tree diameter and species. Species are Douglas-fir (DF), western larch (WL), ponderosa pine (PP), western hemlock (WH), grand fir (GF), Engelmann spruce (ES), western white pine (WP), western redcedar (WR), lodgepole pine (LP), and subalpine fir (SF).* (From Peterson and Ryan 1986)

physical effect of heating the bark. A factor that can have the opposite effect is bark flammability. If the bark catches fire, it will produce heat that can affect the cambium at the same time as its thickness—and insulating ability—is being reduced (Gill and Ashton 1968, Vines 1968).

Height of bark char has been used to estimate flame height (not length) and fireline intensity (McNab 1977). An experiment in loblolly and shortleaf pine (Cain 1984) demonstrated that height of bark char underestimated flame length for both heading and backing fires. If the bark is flammable or has heavy lichen cover, the height of bark char may exceed flame height on the landscape, resulting in overestimates of fireline intensity.

FIG. 5.7. *Douglas-fir showing two grown-over fire scars (arrows). Note the thinner bark at the intersection of the tree circumference with a radius drawn through the two scars.*

ROOT DAMAGE

Damage to roots from prescribed or wild fires has not been studied intensively. Small-diameter (fine) roots are more susceptible to damage than larger roots because of thinner bark. Root damage usually accompanies bole damage, so its effects may be confounded with heating of the basal portion of the tree stem. However, roots can be killed at some distance from the stem while the stem is unaffected by heat.

Early reports of fine-root kill to 2.5 cm soil depth in southern pine (Heyward 1934) concluded that little harm was done, because many of these roots die annually anyway. More recent reports have documented fine-root mortality from low intensity underburns and have linked the root effects to growth reductions in young ponderosa pine stands (Grier 1989) and to tree mortality in low vigor, old-growth ponderosa pine stands (Swezy and Agee 1991). The impact on the tree from fine-root kill may also depend on season of burning; trees may be less dependent on fine roots in autumn, a time of natural root turnover, than in spring while entering the season of drought. Both ponderosa pine studies measured fine-root impact from spring burning; surveys of old-growth pine burned in autumn months found no increased levels of tree mortality over a 10 year postfire period (Swezy and Agee 1991).

Fire Adaptations

Methods of categorizing tree response to fire were developed early in the history of American forestry. The U.S. Geological Survey published the first table of relative fire resistance for species in the Cascade Mountains (Gorman 1899). A more detailed description of fire resistance of common species of the Pacific Northwest was available by 1925 (Flint 1925). This system was slightly expanded by Starker (1934) in a more national survey (Table 5.3). These systems were based on the relative resistance of mature individuals to fire.

The first treatment of fire adaptations in an ecology textbook came in 1947 with Daubenmire's *Plants and Environment.* Fire was recognized as an environmental "factor" equivalent to light and water "factors," and distinct from other effects of the "temperature factor." Daubenmire (1947) defined seven adaptational features that facilitated woody plant persistence under repeated burning. Although Daubenmire's work marked an important beginning to the classification of plant adaptation to fire, these seven adaptations are not all exclusively related to fire, and at least one may have little to do with fire.

1) *Germination.* Some shrubs possess hard-coated seeds that lie dormant in the soil until a fire passes. Density of snowbrush (*Ceanothus velutinus*) in the soil may exceed 12 million ha^{-1} (Weatherspoon 1988). Some seeds are consumed by fire, but others are scarified, which later allows them to imbibe water and germinate in a nutrient-rich and competition-free environment. Such seeds are common in *Ceanothus* (Fig. 5.8), *Arctostaphylos,* and *Rhus* (Daubenmire 1959), and these seeds are thought to remain viable in the soil for several decades (Quick 1977). Kauffman and Martin (1991) reported optimum scarification temperatures for deerbrush (*Ceanothus integerrimus*) of 75–100°C, with wet heat more effective than dry heat at 4–8 min exposure. Exposure above 120°C for 4–8 min caused significant mortality, suggesting a heat threshold above which survival is low to absent. This is why such seedlings will not be found within the perimeter of small burn piles in mixed-conifer forest but will germinate in profusion around the edge.

2) *Rapid growth and development.* Some woody species have life history characteristics that enable them to complete a life cycle quickly. This enables them to provide seed in the event of two closely spaced fires. Bishop pine (*Pinus muricata*) along the California coast can produce viable seed at a young age (Daubenmire 1959). Longleaf pine (*Pinus palustris*) in the Southeast undergoes a "grass stage" where it resembles a perennial grass for several years while developing a deep root system (Fig. 5.9). The terminal bud remains in the center of this cluster of foliage, protected from the typical frequent, low intensity fires that burned through the region. The shoot then begins to grow rapidly, carrying the terminal bud above the zone of lethal heat injury. A similar adaptation is present in the Mexican pines *Pinus montezumae* and *Pinus michoacana* (Perry 1991). An example of a developmental adaptation to fire is an increase in flowering (Biswell and Lemon 1943). In the Pacific Northwest, Thurber's needlegrass (*Stipa thurberiana*) and Great Basin wildrye (*Elymus cinereus*) show increased flowering after burning (Wright and Klemmedson 1965, Kauffman 1990).

3) *Fire-resistant foliage.* Daubenmire defined this adaptation on the basis of absence of high resin or oil content, so that normal foliar moisture content could stop a fire encountering the plant.

Table 5.3. Relative Fire Resistance of the Most Important Trees of Oregon and Washington in Order of Greatest Resistance

Species	*Bark Thickness of Old Trees*	*Root Habit*	*Branch Habit*	*Canopy Cover*	*Lichen Growth and Color*	*Foliage Inflammability*	*Most Common Way of Killing*
Western larch	very thick	deep	high & open	open	light/black	low	most resistant
Douglas-fir	very thick	deep	high & dense	dense	none–heavy/ gray	high	crown fires
Ponderosa pine	thick	deep	moderately high & open	open	light/black	low	crown fires
White/grand fir	moderately thick	shallow	low & dense	dense	none–heavy/ gray	medium	root char, crown fire
Western redcedar	thin	shallow	low & dense	dense	none–moderate/ gray	high	root char, crown fire, burn down
Mountain hemlock	medium	medium low	low & dense	dense	none–moderate/ gray	high	root char, crown fire
Noble fir	moderately thick	medium thick	high & dense	dense	medium–heavy/ gray	high	foliar scorch or crowning, core burn
White pine	medium	medium	high & moderate	dense	moderate/ gray & heavy	medium	scorching cambium or crowning

Lodgepole pine	very thin	deep	moderately low & open	open	moderate–heavy/ gray, black	medium low	scorching cambium or crowning
Western hemlock	medium	shallow	moderately low & dense	dense	none–heavy/ gray	high	root char, crown fire, core burn
Engelmann spruce	very thin	shallow	low & dense	dense	none–heavy/ gray, black	very high	root char, scorching cambium, crowning
Sitka spruce	very thin	very shallow	moderately high & dense	dense	none–heavy/ gray, yellow	high	root char, occasional crowning

SOURCE: Starker 1929.

FIG. 5.8. *Snowbrush* (Ceanothus velutinus) *plant that has germinated at the edge of a circle of burned debris.*

FIG. 5.9. *Longleaf pine* (Pinus palustris) *of the southeastern United States in the grass stage.*

In fact, this is not an adaptation to fire at all. Considerable debate has accompanied a hypothesis that some fire-dependent communities may have evolved flammable characteristics in response to natural selection (Mutch 1970), but to propose that absence of such traits constitutes a fire adaptation is without basis. Some plants (e.g., *Atriplex* spp.) possess higher than average levels of chemicals (particularly silica-free ash content and the sum of calcium plus phosphorus content) that inhibit the combustion process (Philpot 1970, Nord and Green 1977), but alternate explanations for such compounds may be more easily defended than the proposition that these are fire-derived traits. High moisture content of foliage clearly reduces the probability of foliar ignition, but this trait appears to be more a function of site and leaf age than an adaptation to fire.

4) *Fire-resistant bark.* Bark thickness, as noted earlier, can be a critical factor in determining plant survival. Western larch, Douglas-fir, and ponderosa pine all have bark thicker than associated species (Fig. 5.6), and are more likely to survive fires of low to moderate severity. Even thinner bark on mature trees, such as Oregon white oak (*Quercus garryana*), is sufficient to withstand cambial kill in savanna environments, where fires tend to be flashy and of low duration.

5) *Adventitious or latent axillary buds.* Many shrub species and a wide variety of tree species can regenerate crowns after fire by sprouting of buds along the stem, at the stem base, or from lateral roots (Fig. 5.10A–C). Epicormic sprouting from aerial buds depends on whether the crown was scorched or consumed. In crown fires, buds behind thin bark on fine branches will be killed, and epicormic sprouting will occur only along branches and stems exceeding a certain diameter, representing the lower limit of bark thickness that withstood the lethal temperatures. Such "fire columns" are common in *Eucalyptus* spp. and pitch pine (*Pinus rigida*) in the eastern United States. Coast redwood (*Sequoia sempervirens*), bigcone Douglas-fir (*Pseudotsuga macrocarpa*), and the oaks (*Quercus* spp.) commonly sprout new crowns after crown scorch or consumption. Basal sprouting may occur (*Betula, Arbutus, Lithocarpus, Quercus, Ceanothus* spp.) if aerial foliage and buds, or the cambium of the

FIG. 5.10. *Three types of sprouting response to fire.* A: *Sprouting along the stem in oak* (Quercus *spp.*). B: *Root suckering in aspen* (Populus tremuloides). C: *Sprouting from the root crown in chamise* (Adenostoma fasciculatum).
(Aspen photo courtesy Dr. L. Brubaker)

main stem of the tree, are severely injured or killed. Size and age are likely to influence the ability of a plant to sprout after fire (Kauffman 1990). Species like quaking aspen (*Populus tremuloides*) or Oregon white oak can produce sprouts from lateral roots after the death of the main stem. Such clonal development can be a primary regeneration mechanism for some species.

6) *Lignotubers.* A lignotuber is a basal swelling at the interface between root and shoot that contains buds and food reserves (James 1984). When the shoots are killed by fire, dormant buds insulated by the soil remain alive and are stimulated to sprout, using the stored food reserves. Lignotubers in Australian eucalyptus have exceeded several meters in diameter, but commonly are much smaller in the western United States. The issue of when a normal root crown with buds becomes a lignotuber has been debated since Jepson (1916) first described these root crown "burls." From local observation, species with lignotubers (*Arctostaphylos, Adenostoma*) typically have a spherical mass of tissue from which roots emerge on one side and shoots on the other (Fig. 5.11), in contrast to nonlignotuberous species that may have individual shoot- or root-related basal swellings (e.g., bitterbrush [*Purshia tridentata*]). Lignotubers may maintain dominance of the individual in the commmunity over several generations, until the dormant buds at the surface of the lignotuber are pushed from the soil by growth of the lignotuber or exposed by erosion of soil around the base; a succeeding fire can then kill the buds and the plant will die. I have observed this in chamise (*Adenostoma fasciculatum*) in coastal California chaparral.

7) *Serotinous cones. Serotinous* means "late opening," and this adaptation refers to cones that retain seeds in the tree canopy for a long time (Fig. 5.12). Viable seed has been removed from lodgepole pine (*Pinus contorta* var. *latifolia*) cones 75 years after cone maturation (Clements 1910). Not all cones within a tree or species are typically serotinous, so that the species may respond to other disturbances, such as insect attack. For example, much of the lodgepole pine in the pumice region of south-central Oregon and the Cascades and Sierra Nevada (*P. contorta* var. *murrayana*) is nonserotinous (Stuart et al. 1989). Fires burning through the crown melt the resin

FIG. 5.11. *Manzanita* (Arctostaphylos *spp.*) *with lignotuber as an aid in sprouting response. The chamise plant of Figure 5.10C has a lignotuber too.*

seal on serotinous cones, and they open soon after. Perry and Lotan (1977) found 100 percent of heavily serotinous cones of lodgepole pine opening after treatment in a water bath at 60°C for two minutes, but many cones from less serotinous trees opened at room temperature. The California closed-cone pines, such as knobcone pine (*Pinus attentuata*), bishop pine, and Monterey pine (*Pinus radiata*); the Mexican closed-cone pines; Baker cypress (*Cupressus bakeri*) in the Siskiyou and Klamath Mountains; and semi-serotinous black spruce (*Picea mariana*) and jack pine (*Pinus banksiana*) of the boreal forest are other examples of species with serotinous cones.

A broad biological classification of plant response to disturbance was developed by Rowe (1983). He defined five categories of plant response based on life-history characteristics of species. Some of the categories may not be exclusively related to fire, but characteristics of fire regime are incorporated into this system (e.g., *resister* category):

FIG. 5.12. *Serotinous cones of lodgepole pine opening the day after a wildfire. Seeds blow through the air and become buried in fresh ash from the fire.*

1. *Invaders.* Highly dispersive, pioneering fugitives with short-lived disseminules. Plants such as fireweed (*Epilobium angustifolium*), Scouler's willow (*Salix scouleriana*), and cottonwood (*Populus* spp.) are typical *invaders,* generally needing disturbance to occupy a site.
2. *Evaders.* This category includes species with relatively long-lived propagules that are stored in the soil or canopy. The species thus evades elimination from the site. Daubenmire's "germination" and "serotinous cone" adaptations both fit the *evader* category.
3. *Avoiders.* These are generally shade-tolerant, late successional species that slowly reinvade burned areas and have essentially no adaptation to fire. Hemlocks (*Tsuga* spp.), western juniper (*Juniperus occidentalis*), and subalpine fir (*Abies lasiocarpa*) are good examples of *avoider* species. Herbaceous species with reproductive parts in the litter layer are likely to be killed even by low intensity fires (Flinn and Wein 1977), and would also be classified as *avoiders.*
4. *Resisters.* These are species that can survive low intensity fires

relatively unscathed. Thick-barked species, such as Douglas-fir, ponderosa pine, and western larch, are *resisters*.
5. *Endurers*. These species have the ability to resprout from the root crown, lateral roots, or the aerial crown. Oaks, Pacific madrone (*Arbutus menziesii*), and various shrub species are among the many species classed as *endurers* in the Pacific Northwest.

This system is a useful way to broadly classify the species on a site, and it may be used to develop generalized responses to fire regimes: a low severity fire will favor *resisters*, while a high severity fire will favor *invaders*, *evaders*, and *endurers*. Fire suppression will generally favor *avoiders* by allowing late successional species to eventually occupy the site.

Some species may fit more than one category because of multiple adaptations or changes in morphology or physiology over time. For example, lodgepole pine may be an *invader* in terms of its rapid colonization of sites after fire, but it can also be an *evader* due to its serotinous cones. Douglas-fir is an *avoider* when young but a *resister* when mature. *Ceanothus* has an evader strategy with refractory seeds, and an endurer strategy because it sprouts after fire. Researchers should evaluate the possibility of dual classification when considering each species on a site.

Another universal life-form classification was developed in Australia by Gill (1980), but it has the disadvantage of assuming 100 percent leaf death before entering the classification key (Table 5.4). Species in fire regimes of moderate to low severity (particularly *resisters*) are not well classified by this system, and although it may fit the Australian experience, it is less universal than the system developed by Rowe.

FIRE AND PLANT COMMUNITIES

The adaptations discussed so far focus on characters of individual species that allow them to respond to disturbance. Observations that communities with many fire-adapted species are also more "flammable" because of chemical or physical properties of fuels led Mutch (1970, p. 1047) to propose the following hypothesis of community-level adaptation to fire: "Fire-dependent communities burn more readily than non-fire-dependent communities because natural selection has favored development of characteristics that make them more flammable."

The *resister* ponderosa pine, for example, relies on periodic disturbance for continued dominance in most forest types and tends to have fuel characteristics that encourage burning: resinous needles and well-aerated litterbeds. The *endurer/evader* chamise (*Adenostoma fasciculatum*), a chaparral species, exhibits decreasing fuel moisture and increasing extractive content of the live needles during the fire season (Philpot 1977). The "Mutch hypothesis," although still frequently discussed, has never been tested. An attempt to do so by Buckley (1983) was criticized by Snyder (1984, "Mutch Ado about Nothing") as focusing more on reproductive success with or without fire than on flammability. Christensen (1985) noted that characters thought to be facilitated by fire, such as chemical composition of needles, made little difference in most field situations, and stated that the hypothesis implied selection for disturbance facilitation on an improbably wide scale.

A more fundamental flaw in Mutch's hypothesis as stated is that communities are aggregations of individual species that have migrated at unique rates over time (Davis 1981, Brubaker 1986). They

TABLE 5.4. CLASSIFICATION OF WOODY PLANT SPECIES IN RELATION TO INTENSE FIRE

Category	Class
A. Plants in the reproductive phase just subject to 100% leaf scorch by fire die (Nonsprouters):	
(a) seed storage on plant	I
(b) seed storage in soil	II
(c) no seed storage in burn area	III
B. Plants in the reproductive phase just subject to 100% leaf scorch by fire recover (Sprouters):	
(a) subterranean regenerative buds	
(i) roots suckers, horizontal rhizomes	IV
(ii) basal stem sprouts, vertical rhizomes	V
(b) aerial regenerative buds	
(i) epicormic buds grow out	VI
(ii) continued outgrowth of active aerial prefire buds (tree ferns, monocots)	VII

SOURCE: Gill 1980.

have likely changed associated species faster than they could evolve fire-related characters in response to selective pressures. Therefore, any hypothesis proposed at the community level is improbable. Still, if restated at a species level, the hypothesis has some intriguing aspects that would be worth testing. There is no consensus on whether fire-dependent ecosystems are more flammable due to natural selection. But flammable features do influence fire regimes and resulting successional trajectories (Christensen 1988).

Fire and Successional Theory

The concept of species composition change over time has been a central theme in North American plant ecology during the twentieth century (Christensen 1988). Many of the current views were stated (perhaps in amended form) at the beginning of this century-long debate, but remained subordinate to other ideas for most of the period. Today's views of succession more appropriately incorporate disturbance, and they eschew general theory in favor of mechanistic explanations of plant community change.

The dominant plant succession paradigm of the century was developed by Clements (1916, 1936): the monoclimax theory. It is ironic that Clements, who earlier had completed one of the first fire ecology studies (on lodgepole pine in the Rocky Mountains), inadequately incorporated disturbance into his model. The primary assumptions of the theory are (1) that regional macroclimate determines the endpoint of succession (the climatic climax); (2) that only one "climatic climax" community, capable of self-perpetuation through regeneration and composed of only one dominant life-form, would eventually dominate the site (hence, monoclimax); and (3) that all other communities now present were in the process of altering their environments, making them more suitable for the next community "stop" on the successional "railroad" (called autogenic succession), with the last stop being the climatic climax. Associated with this replacement sequence, later called "relay floristics" (Egler 1954) because of the discrete passing of one community into another, was the concept of the plant community as a superorganism, with emergent properties similar to developmental stages of an organism.

Although the monoclimax theory was the dominant paradigm in plant ecology for over half a century, it was one of several approaches to plant succession during that time. European phytosociologists recognized the plant community as the basic unit of vegetation, although they criticized the superorganism concept (Braun-Blanquet 1913). The "individualistic" concept proposed by Gleason (1917) denied the existence of the plant community. The individualistic theory directed attention to the individual plant as it disperses propagules and is selected by the environment on the basis of its physiological attributes. Disturbance was assumed to operate at the level of the individual, so that vegetation change in the presence of fire, for example, was the sum of fire effects on individuals. From a vegetation management perspective, Gleason's theory was much less satisfactory than Clements's, because it provided little opportunity to order vegetation; however, it may be much closer to the truth.

The polyclimax theory was a third successional theory developed in the same period (Tansley 1924); it bridged the monoclimax and individualistic theories of succession. The polyclimax approach did not assume eventual community convergence, and climate was not the sole controlling factor of vegetation change. Tansley defined multiple (poly-) climaxes, recognizing fire climaxes, biotic climaxes (meadows maintained by grazing), and edaphic climaxes (serpentine soils), among others. Succession led toward a mosaic of different climax communities, reflecting the mosaic of habitats rather than a single climax community (Kimmins 1987).

Tansley's significant contribution was the differentiation of autogenic succession (which drove the monoclimax "train") and allogenic succession (caused by external factors, such as fire), although the definition of disturbance as solely allogenic began to blur as early as 1926 (Cooper 1926) and essentially disappeared by the 1970s (e.g., White 1979).

In the 1950s two new adaptations of these successional theories were developed in the Pacific Northwest. The "climax pattern" hypothesis of Whittaker (1953) was applied to the vegetation of the Siskiyou Mountains (Whittaker 1960) in the formulation of the "prevailing climax." Whittaker integrated elements of the individualistic and polyclimax approaches by recognizing that vegetation varied continuously across the landscape, but that the "climax" could be shown as a mosaic of populations representing positions along complex environmental gradients. Disturbance gradients, such as fire, were considered equivalent to climate, soil, and other gradients. Whittaker defined the "prevailing climax" as the vegetation mix covering the widest area of a region. He attempted to bridge the gap between systems defining discrete communities, which were favored in management applications, and those recognizing continuous patterning on the landscape. The attempt was sensible, but it was difficult to apply the "prevailing climax" to management scenarios. Whittaker's approach was complicated by his choice of the Siskiyou Mountains, the most complex and diverse forest region in the Pacific Northwest. His "prevailing climax" was never widely used or adapted to other areas.

The "habitat type" concept of Daubenmire (1952) was a more straightforward application of polyclimax theory. Plant associations were defined on the basis of potential overstory and understory dominants in the absence of disturbance. Although climate was still assumed to be the primary environmental factor affecting succession, other factors could be used to explain local variation within a climatic region. Daubenmire (1968b) used the example of pure stands of ponderosa pine that are climax being always drier than those in which Douglas-fir is the climax dominant. Both types occur widely throughout the West. Should stands be grouped on the basis of similarity of undergrowth—in which case many ponderosa pine stands would be grouped with locally present Douglas-fir stands—or with other ponderosa pine stands that had no understory species in common? Daubenmire defended the latter approach, which "permits emphasis on climate as a master factor" (1968b, p. 261). He placed all ponderosa pine climaxes, or the ponderosa pine *series*, in an ecologically meaningful relation to the Douglas-fir series. A *Pinus ponderosa/Festuca idahoensis* ("Pipo/Feid," named by combining the first two letters of the generic and specific latin names of the dominant overstory and understory species) habitat type, for example, represents the dominant species found at the successional endpoint on relatively dry ponderosa pine sites. A variety of earlier successional communities, or current cover types, may occur on such sites. A *Pseudotsuga menziesii/Symphoricarpos mollis* (Psme/Symo) habitat type represents the successional endpoint on moister, cooler sites and may also have a variety of earlier successional communities currently dominant. Both types of sites may currently be dominated by ponderosa pine, but the different series designation allows an interpretation that the overstory dominants, without disturbance, will eventually change to Douglas-fir in the Psme/Symo habitat type.

The habitat type concept, with its incorporation of the concept of discrete communities, is a system easily integrated with management. The theoretical trajectory toward a climax community (the plant association, defined on the basis of overstory and understory dominants) is a useful model in which to apply disturbance scenarios. The system, however, does not define mechanisms, which leaves considerable latitude for incorporation of other successional models within this classification framework. The habitat type concept has been widely applied in the Pacific Northwest on public lands managed for multiple uses (Henderson et al. 1989, Williams and Lillybridge 1983, Cooper et al. 1991) and is the basis for identification of natural areas to be preserved (State of Washington 1991).

During the 1960s the concept of the monoclimax community was expanded to an ecosystem level by Odum (1969). Systems ecologists argued that ecosystems have emergent properties: not those of the individual parts but those resulting from the organization of the ecosystem (McIntosh 1981). Odum (1969, p. 265) listed numerous emergent properties—which he called "trends to be expected in succession"—such as increasing stability, production, and diversity. Later criticisms suggested that many of the trends were based on untested or inadequate assumptions, data, or logic (Drury and Nisbet 1973). For example, late successional communities in the Pacific Northwest may be neither the most stable (Agee 1981a) nor the most diverse (Huff 1984).

A now traditional view of the development of scientific paradigms was presented by Kuhn (1970), who proposed that one paradigm replaces another over time as knowledge accumulates, and that the paradigm shift is often rather abrupt. This theory of paradigm development is analogous to a "relay floristics" approach. Within plant ecology, paradigm development has instead been more like an "initial floristics" model: Most of the paradigms were proposed at the start of the process in the first quarter of the twentieth century, with one being more dominant at first and others emerging much later. As with the initial or relay floristics models, the successional endpoint of the continued debate about population- versus ecosystem-based views is not yet clear. As Christensen (1988) notes, a grand unified theory of community succession may not be possible or even desirable.

Recent Succession Models

In recent years plant ecologists have focused attention on mechanisms of vegetation change rather than on a unified theory, and such mechanisms can be successfully incorporated into the habitat type concept. A polyclimax approach using habitat types is useful in defining a classification system, within which various successional pathways may be defined. In a sense, this integrates the individualistic and polyclimax approaches, something first suggested by Cooper (1926). Such models have been applied with the habitat type framework to produce alternative successional pathways (e.g., Arno et al. 1985).

Recent models of mechanisms of vegetation succession have adopted an individualistic approach. Even now, such mechanisms are not fully understood. Forests are long-lived, so that hypothesized changes are difficult to observe in a single career; interactions between species and environment are difficult to measure; and field experiments are unwieldy (Kimmins 1987). Connell and Slayter (1977) proposed a three-pathway model adapted from the concept of initial or relay floristics proposed by Egler (1954): *facilitation, tolerance,* or *inhibition.*

A further development of the three-pathway model was proposed by Noble and Slayter (1980). This multiple-pathway model was a more specific individualistic approach to mechanisms of succession, proposed by Mediterranean-climate ecologists as a better means of incorporating disturbance into successional models. The multiple-pathway model incorporated species adaptations to disturbance into the system, recognizing that most ecosystems will be disturbed one or more times on the successional "railroad," and that the multiple "tracks" of plant community composition and structure commonly observed in nature are the result of individual species' responses to the permutations of the disturbance regimes.

The multiple-pathway system (Connell and Slayter 1977) incorporates three types of information into its graphical representations of species responses: method of persistence, conditions for establishment, and critical life history events. *Method of persistence* defines recovery of the individual (equivalent to an *endurer* life history strategy) or reproductive strategy (*evader, invader, avoider*) after disturbance. *Conditions for establishment* define the species capable of responding to facilitation, tolerance, or inhibition pathways in the three-pathway model. *Critical life history events* define the time necessary to reach critical stages in life history: reproductive age, thick bark for fire resistance, longevity, and so forth. Three to four subcategories of response are defined for each of these factors. Capital or lower-case letters are assigned to each subcategory, and species response diagrams can be developed for the associates of each community and integrated into community development pathways. The system, although workable, is cumbersome to use, and the assignment of letters to species is not an absolute one: In communities where one species is relatively more shade-tolerant than in others, letter reclassification will be required.

A simpler conceptual example of a multiple-pathway model can be applied within a habitat type classification (Table 5.5). Five species are identified (A–E) in two plant associations. Plant association C/x is represented by a no-disturbance regime and a frequent, low severity fire regime (Fig. 5.13), while plant association E/y is represented by a no-disturbance regime after wildfire and an infrequent, variable severity fire regime (Fig. 5.14). The species characteristics in Table 5.5 define individualistic responses relative to the responses of associated species. In plant association C/x, named for overstory species C and shrub species x, which would eventually dominate in the absence of disturbance, the no-disturbance regime favors species C in the tree understory because of its shade tolerance and eventually allows it to dominate the overstory (Fig. 5.13A). In the presence of frequent, low intensity disturbance, species A and B, early successional species, are able to survive in the tree understory because the frequent fires maintain open growing space for them and they are more tolerant of fire (Fig. 5.13B). These species can continue to codominate the overstory over time.

Plant association E/y also contains species C but has two new species, D and E. Species E, being relatively most shade-tolerant, captures growing space in the understory in the absence of major

TABLE 5.5. CHARACTERISTICS OF SPECIES USED IN A PRACTICAL APPLICATION OF THE MULTIPLE-PATHWAY MODE

SPECIES	RELATIVE LEVEL OF:				
	Longevity	*Seed Dispersal*	*Early Growth*	*Shade Tolerance*	*Fire Tolerance (Sapling/Mature)*
Habitat Type C/x					
A	high	low	high	low	high/high
B	high	high	high	low	high/high
C	high	high	low	high	low/high
Habitat Type E/y					
C	high	high	high	low-moderate	high/high
D	low	high	high	low	low/low
E	moderate	low	low	high	low/low

NOTE: The relative position of one species to another is important in defining the levels from high to low. Species C has a different set in habitat C/x than in type E/y. Results of successional patterns in two disturbance regimes and with no disturbance are shown in Figures 5.13 and 5.14.

disturbance (Fig. 5.14A), and would eventually dominate the overstory. In the presence of periodic disturbance, it shares dominance with the other species (Fig. 5.14B). Species D, a short-lived but fast growing species, may dominate the site as long as disturbance continues. Species C assumes the role of an early successional species in this plant association, being favored by periodic but not exceptionally frequent disturbances. Different permutations of disturbance regimes (type, frequency, intensity, periodicity, extent, and so forth) would result in different graphical outputs for a given plant association. This easily conceptualized system enables a multiple-pathway set of successional mechanisms to be applied within a habitat type classification framework.

MODELING FIRE RESPONSE IN ECOSYSTEMS

Successional diagrams in the multiple-pathway model are a graphical way of understanding successional mechanisms in ecosystems subject to periodic disturbance. Microcomputers have opened an

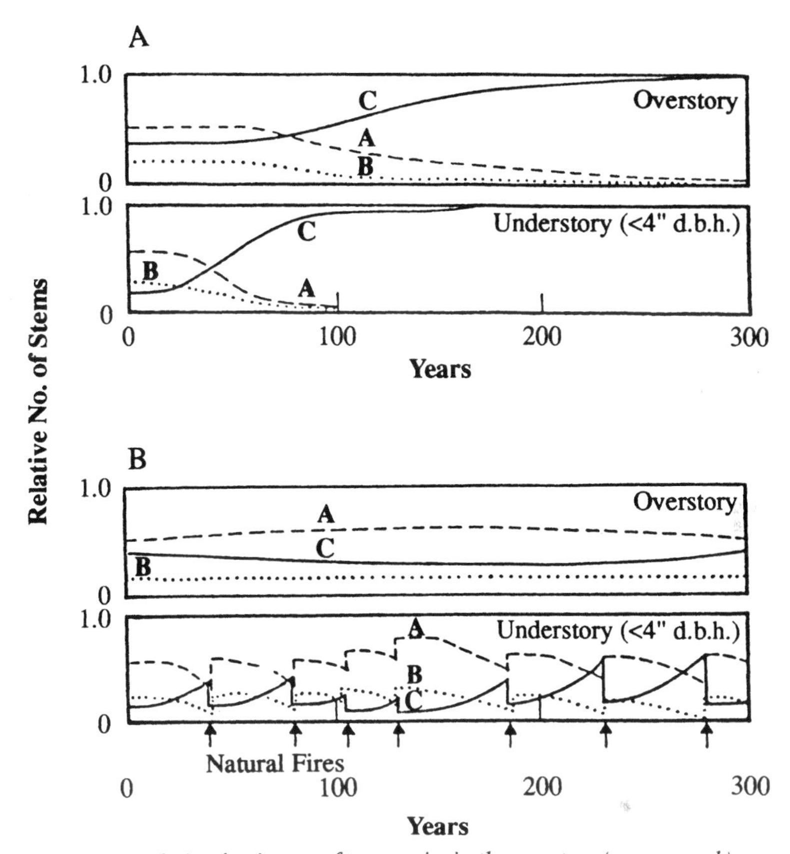

FIG. 5.13. *Relative dominance of tree species in the overstory (upper graph) and understory (lower graph) for the C/x plant association.* A: *No disturbance results in dominance by the shade-tolerant species C.* B: *Periodic low intensity fire maintains the dominance of species A. See Table 5.5 for characteristics of species used in these diagrams.*

array of simulation techniques to mimic successional processes in wildland ecosystems with or without disturbance. These models, which can incorporate autogenic and allogenic change through deterministic or stochastic processes, are an exciting and fruitful way to evaluate mechanisms of succession in the presence of periodic disturbance. They need not be burdened with the early "emergent property" claims of systems ecologists, mentioned above.

Forest successional models have emerged as our understanding of stand development patterns and autecological responses to fire has

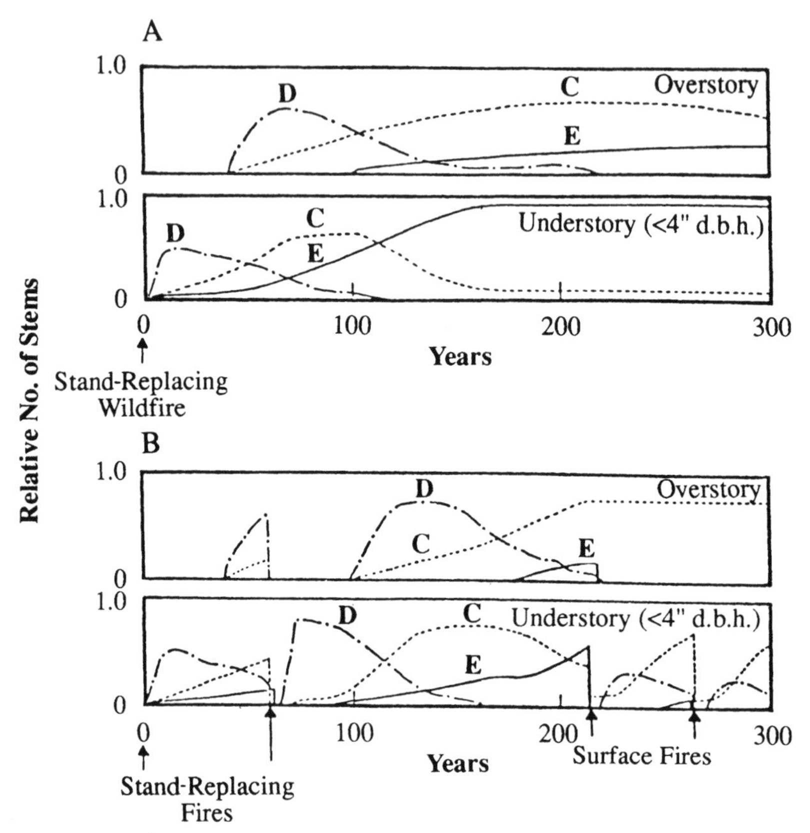

FIG. 5.14. *Relative dominance of tree species in the overstory (upper graph) and understory (lower graph) for the E/y plant association.* A: *No disturbance results in dominance by the shade-tolerant species E.* B: *Periodic, variable severity fire allows species C and D to dominate the overstory. See Table 5.5 for characteristics of species used in these diagrams.*

increased, along with our computing power (Table 5.6). Most are gap-replacement models, based on the JABOWA models developed in the eastern United States (Botkin et al. 1972). Individual trees in a small area are grown deterministically using climatic inputs and an annual time step. The competitive interactions between species are incorporated into growth and establishment subroutines over a limited spatial extent commonly known as a gap (Shugart and West 1980). Models designed to incorporate disturbance by fire adapt this

TABLE 5.6. DEVELOPMENT OF DYNAMIC ECOLOGICAL MODELS INCORPORATING DISTURBANCE BY FIRE AND APPLICABLE TO THE PACIFIC NORTHWEST

Model Name	*Source*	*Species*
FYRCYCL	van Wagtendonk (1972)	ponderosa pine, white fir, incense-cedar, sugar pine
SILVA	Kercher and Axelrod (1984)	those above, plus black oak, Jeffrey pine
CLIMACS	Dale and Hemstrom (1984)	21 PNW conifers, tested on smaller subset
NONAME	Peterson and Ryan (1986)	9 inland NW species, incl. western hemlock, western redcedar
FIRESUM	Keane, Arno, and Brown (1989)	10 inland NW species, incl. subalpine larch, whitebark pine

protocol but add disturbance to the model, either stochastically or deterministically.

One of the earliest working fire models, FYRCYCL, was developed by van Wagtendonk (1972) for mixed-conifer forests of the Sierra Nevada. An impressive incorporation of simulated weather inputs based on actual records drove the stochastic ignition and spread subroutines of the model. Fire-return intervals were produced from these data, instead of the usual method of trying to stochastically reproduce fire scar records. Fuel consumption and energy release from fire were the primary initial outputs of the model. Later versions have incorporated stand dynamics of four tree species (van Wagtendonk 1983, 1985).

The SILVA model, designed by Kercher and Axelrod (1984), integrated many elements of van Wagtendonk's model with a JABOWA-like forest stand model and more species to produce a more sophisticated mixed-conifer forest model. Certain events, as in FYRCYCL, are probabilistic, and the model also runs on a yearly time increment. Each year, each tree grows an increment based on its leaf area, response to light, respiration, and site factors. New trees are added based on stochastically generated seed crops adjusted to the number of trees expected to survive to 6–8 years of age (which often will be zero under a mature canopy). Tree death can occur as random "ecological death," growth-suppression death, or death by fire. Death by fire is a function of crown scorch height and tree diameter (a dummy variable representing both bark thickness and tree height).

Another forest successional model, CLIMACS, was developed for western Cascade forests in the early 1980s (Dale and Hemstrom 1984). It is based on a second-generation JABOWA model known as FORET. This model was not developed primarily to be a disturbance model, although it contains a disturbance subroutine. Although verification was completed only for the western hemlock/Douglas-fir type, the model is capable of adaptation to a wider range of Pacific Northwest species. Disturbance by fire or wind is provided in the model by equations based on size and species and weighted by a user-defined intensity scale from 0 (no impact) to 1 (most severe fire or wind). Species are grouped into one of three classes of general fire tolerance, with the probability of dying a function of diameter (size) and a scaled fire intensity level from 0 (low intensity) to 1 (high intensity) defined by the modeler. Dale and Hemstrom (1984) note that the model was designed primarily for stand-replacement fire events and is inappropriate for forest types with low severity fire regimes.

Peterson and Ryan's (1986) NONAME model recognized that conifer mortality was a function of both crown and bole damage. The model used adaptations of Van Wagner's (1973) crown scorch and Hare's (1965) bark thickness algorithms to integrate the effects of crown and bole damage into a single equation (Fig. 5.15):

$$P_m = c_k \exp(t_c/t_l - 0.5)$$

where

P_m = probability of tree mortality (0–1)
c_k = fraction of volumetric crown kill (0-1)
t_c = calculated critical time for cambial kill (min)
t_l = calculated duration of lethal heat (min)

The term t_c is based on bark thickness, and t_l is based on the burning time of three fuel size classes, with a maximum of about eight minutes. Peterson and Ryan recognized the limitation of omit-

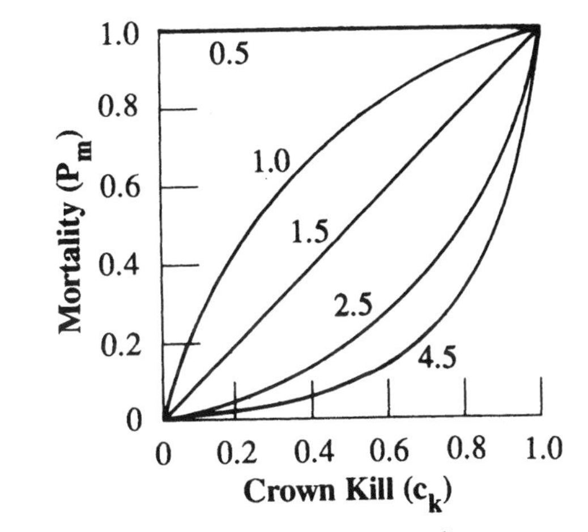

FIG. 5.15. *Probability of tree mortality as a function of fractional volume of crown killed (c_k), critical time for cambial kill (t_c), and duration of lethal heat (t_l). Each curve represents different values of t_c/t_l, such as 0.5, 1.0, and so on.*
(From Peterson and Ryan 1986)

ting consideration of smoldering logs and duff in the model due to the current lack of models of combustion in these fuels. Where t_c is equal to or less than $0.5t_l$, the equation implies that the tree is girdled, as P_m exceeds 1. Conversely, when $t_c > 4.5t_l$, P_m is determined entirely on the basis of crown damage. As applied to an "average" stand in the northern Rocky Mountains under moderate fire behavior (Table 5.7), the model predicted no impact on the basal area of typically large ponderosa pine or western larch, a 75 percent probability of reduction to basal area of medium-sized Douglas-fir, and complete mortality of the other species of moderate to small size.

A SILVA-type model called FIRESUM, for some of the same species and many new ones, was developed in the late 1980s (Keane et al. 1989). It includes the form of the Ryan and Reinhardt (1988) mortality function:

$$P_m = 1/[1 + \exp(-1.466 + 1.910\ BT - 0.1775\ BT^2 - 0.000541\ C_k^2)]$$

where

P_m = probability of mortality (0–1)
C_k = fraction of volumetric crown kill (0–1)
BT = bark thickness (cm)

It is applicable to the inland Northwest, including 10 species covering fire groups 1–7 (see Table 1.5). Additional disturbance subroutines due to insects and white pine blister rust are available. Simulation of crown fires is a submodel still in development. FIRESUM is capable of simulating fuel loadings and tree basal area over time for the ponderosa pine/Douglas-fir (see chapter 10) and whitebark pine/subalpine fir forests of the inland Northwest and northern Rocky Mountains.

These models produce output similar to that shown in Figures 5.13 and 5.14, and have significant potential to evaluate ecological effects of alternative fire management regimes. Like all models, they have limits and are often valuable in highlighting information gaps. Such models are expected to be refined in the near future as researchers continue to focus on mechanistic models of succession.

TABLE 5.7. PROBABILITY OF MORTALITY FOR EACH SPECIES IN A DOUGLAS-FIR STAND

SPECIES	BASAL AREA				MORTALITY	
	Diameter (cm)	*Fraction of Stand*	t_c/t_1	c_k	P_m	*Fraction of Stand*
Douglas-fir	23.9	0.80	0.78	0.38	0.75	0.60
Subalpine fir	8.6	0.02	0.02	1.00	1.00	0.02
Grand fir	8.6	0.01	0.06	1.00	1.00	0.01
Lodgepole pine	18.2	0.10	0.05	0.32	1.00	0.10
Ponderosa pine	54.1	0.04	4.38	0	0	0
Engelmann spruce	13.2	0.01	0.20	0.94	1.00	0.01
Western larch	34.0	0.01	1.90	0	0	0
Western redcedar	32.0	0.01	0.56	0.41	0.95	0.01
Western hemlock	23.0	<0.01	0.51	0.52	0.99	<0.01
Total mortality						0.75

SOURCE: Peterson and Ryan 1986.

NOTE: An "average stand" was developed from inventory data from the Northern Rocky Mountain Region of the USDA Forest Service. Fire characteristics include scorch height of 10 m, fuel model 10, and moderate fuel moisture.

CHAPTER 6

ENVIRONMENTAL EFFECTS OF FIRE

FIRE HAS DIRECT AND INDIRECT INFLUENCES on almost every component of the ecosystem. Direct impacts on vegetation were summarized in chapter 5; this chapter focuses on other ecosystem components, including physical remnants of earlier cultures. The interactions among components of any ecosystem are complex enough without adding consideration of the disturbance factor of fire, which is itself complex and variable. The effects of fire have often been confounded with effects of logging, chemical use, or other management actions. Prediction of ecological impact from fire is difficult, although general trends can be identified. This chapter describes fire impacts on major ecosystem components and processes using data from a wide variety of fires and ecosystems, including but not limited to the Pacific Northwest. The soil resource is a central concern because of its intimate link with vegetation, but effects on wildlife, air, and cultural resources are also summarized.

THE SOIL COMPONENT

Soil is a fundamental resource, and most sustainable management objectives include maintenance of soil productivity as a key factor. Fire interactions with soil are significant because most fires spread by combustion of organic matter in contact with or part of the soil. Fire creates physical, chemical, and biological changes that may be either desirable or detrimental in the context of long-term soil productivity.

Soil Heating

Heat created by combustion is important in understanding soil impacts from fire. Temperature, as a measure of heat, is usually described as a function of soil depth in terms of maximum temperature and duration of elevated temperature experienced at that depth. Temperatures directly above the soil can exceed 700°C (Daubenmire 1968), and recorded litter surface temperatures have exceeded 800°C in intense fires (DeBano et al. 1979). At the soil surface and below, maximum temperatures are usually less than 700°C (Wells et al. 1979). The duration of elevated temperature depends primarily on the duration of combustion. In a grass fire with limited fuel, flaming and glowing combustion at many points are complete in a minute or less, with little or no soil heating. In forests with thick forest floors and large fuel pieces, smoldering combustion can continue for many hours, with associated heat input and elevated temperatures in the soil (Fig. 6.1).

Temperature at depth in the soil is usually much less than at the surface. In intense chaparral fires, maximum temperature at 2.5 cm depth did not exceed 200°C (DeBano et al. 1979), but under piled debris it reached 275°C (Beadle 1940). Maximum temperatures under burning lodgepole pine logs in south-central Oregon reached about 60°C at 5 cm depth (Gara et al. 1985).

Moist soil generally experiences less temperature rise than dry soil, given the same heat input. In the field, when soils and fuels are moist, less fuel may be consumed and less heat applied to the soil

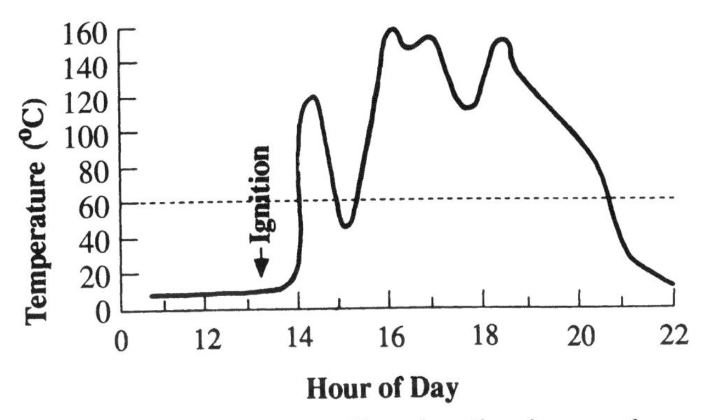

FIG. 6.1. *Time-temperature profile at the soil surface near the base of a large ponderosa pine. Broken line is lethal temperature after one minute; lethal temperatures were exceeded for over five hours.*
(From Swezy and Agee 1991)

surface. Because water has a high specific heat, it absorbs considerable energy (4.18 joules) for each degree (°C) of temperature rise. Water also has a high heat of vaporization, absorbing 2,450 joules g^{-1} vaporized, carried in latent form. Fires burning over moist soil (Fig. 6.2) may not increase soil temperature above the boiling point of water (100°C) until all water has been vaporized.

Although burning over moist soil is often recommended, it is by no means always desirable; for example, burning in moist spring conditions may have undesirable physiological effects on plants (Parker 1987, Grier 1989). Furthermore, although high temperature "spikes" may be avoided, increased temperature at depth can occur in moist soils due to the high heat conductivity of water (Fig. 6.3). An analogous situation is the "hot pad effect." If one picks up a hot object with an insulated pad, the surface temperature of the side of the pad touching the object will be high while the side touching the hand is cool. If the same pad is saturated with water, the side touching the hot object will usually be less than 100°C, but the side touching the hand may be so hot as to scald the hand.

This soil heating effect is a function of changing thermal diffusivity of soil as moisture content increases (Rose 1966). Generally, thermal diffusivity increases from air-dry soil moisture to about 10–15 percent moisture content (Baver 1940), as water is replacing air, which has very low heat conductivity. Thermal diffusivity then

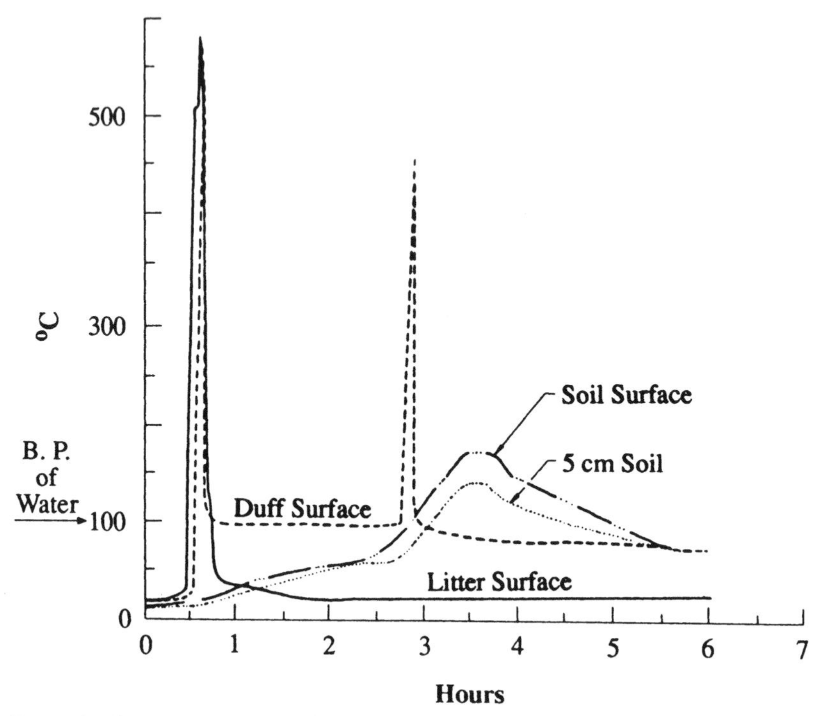

FIG. 6.2. *Time-temperature profile in mixed-conifer forest under moist springtime conditions. After an initial pulse, duff temperature declines to the boiling point of water. After all water is vaporized, soil temperatures begin to rise.*
(From Agee 1973)

slowly decreases as soil becomes wetter, due to the high specific heat of water. Solids in the soil can markedly affect heat dynamics. Quartz has a thermal diffusivity about three times that of clay minerals, and organic matter that does not combust has even lower thermal diffusivity (Wells et al. 1979).

Soil temperature in postfire years may be significantly altered, although after low intensity underburning the effects are often insignificant. The changes are due to loss of overstory canopy, removal of forest floor, and blackened residual organic matter. The loss of overstory canopy and forest floor results in increased insolation reaching the soil, while the blackened soil more efficiently absorbs radiant energy. Daytime soil temperature often increases. The lack of canopy and the blackened forest floor also allow increased radiant energy loss at night, so nighttime minimum tem-

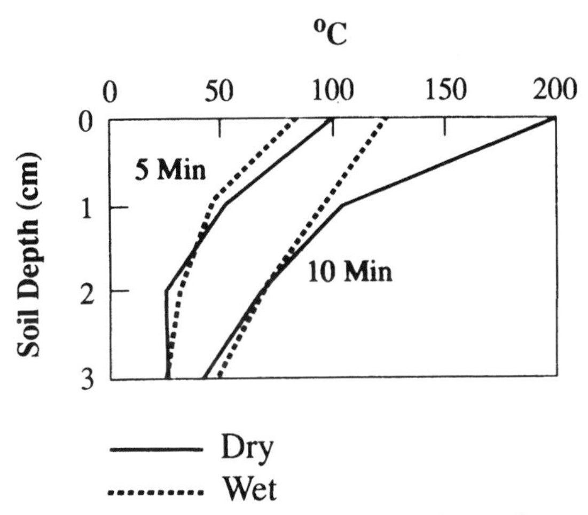

FIG. 6.3. *Temperature at various soil depths 5 and 10 minutes after similar heat application to dry and wet soil. Notice higher temperatures of dry soil at surface and higher temperatures of wet soil at depth.*
(From DeBano et al. 1976)

perature is typically lower after fire. In the central Washington Cascades, Fowler and Helvey (1978) documented this effect in a burned lodgepole pine stand (Fig. 6.4).

The most dramatic alteration of soil temperature regime by burning occurs in the permafrost regions of the taiga forest (Fig. 6.5). Postfire effects include a longer permafrost-free season and increased depth of annual thaw (Viereck 1973). This effect slowly disappears over several decades as an insulating moss-lichen mat recolonizes the site.

PHYSICAL PROPERTIES OF SOIL

Physical properties of soil may be less affected than chemical properties, but the properties that are affected may recover more slowly. Soil texture is rarely changed, even by repeated burning (Isaac and Hopkins 1937, Burns 1952, Ahlgren and Ahlgren 1960). Severe heating may fuse clay particles into stable sand-sized particles (Donaghey 1969, Dyrness and Youngberg 1957) due to a loss of adsorbed water or hydroxyl ions. Temperatures above 200°C can drive

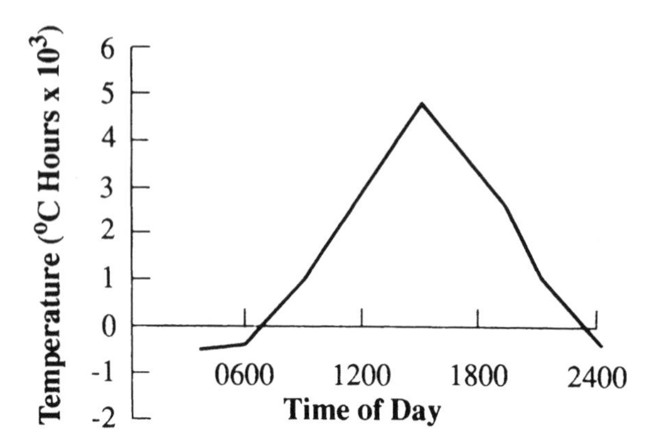

FIG. 6.4. *Departures from normalized (unburned surface) temperatures after burning. If the burned area were the same as the unburned area, departures would be zero. Burned sites are warmer during the day and cooler at night.* (From data in Fowler and Helvey 1978)

off water adsorbed on montmorillonite and illite clays, and temperatures above 500°C can drive off hydroxyl ions from these and kaolin clays (Kohnke 1968). The effect is rarely observed below 2–3 cm in the soil because high temperatures do not reach great depth (Beadle 1940).

Soil bulk density usually increases after repeated or intense burns (Hussain et al. 1969, Tarrant 1956a, Beaton 1959). Since soil porosity is inversely related to bulk density, porosity may decrease after repeated or intense burns. Total porosity and macropore space were found to decrease after severe burning in the Pacific Northwest (Tarrant 1956a, Beaton 1959). Destruction of insects and other macroorganisms that tunnel in the soil may be associated with decreased porosity (Kittredge 1938). Where single or repeated low intensity fires were studied, less significant effects on bulk density and porosity were found (Heyward 1937, Tarrant 1956b, Metz et al. 1961, Agee 1973). Water-holding capacity of the soil may be affected in severe fires in which textural and structural properties are changed; effects are usually restricted to the top few centimeters of soil (Austin and Baisinger 1955).

Soil structure, or the arrangement of particles in the soil, has usually been measured qualitatively. Less favorable structure due to moderate repeated burning (Heyward 1937) or intense burning

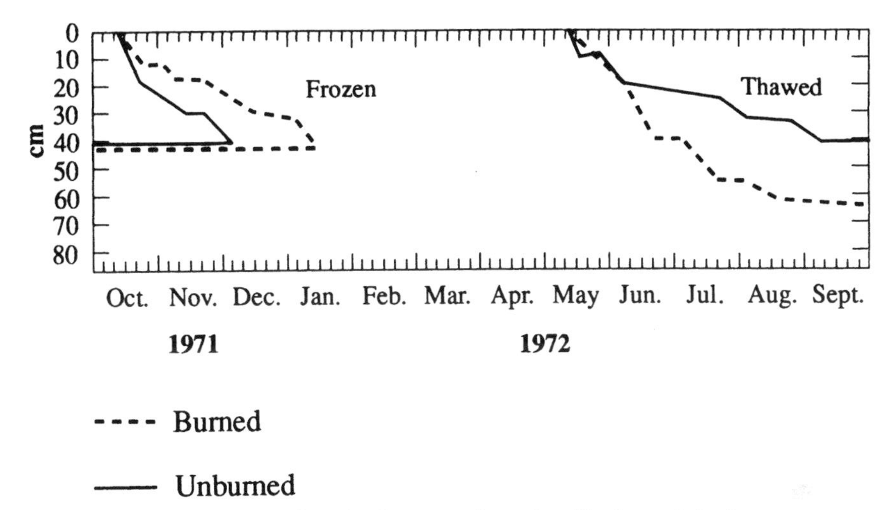

FIG. 6.5. *Changes in permafrost the first year after a fire. The burned site freezes more slowly in the autumn, and permafrost level is 50 percent deeper the following year.* (From Viereck 1973)

(Isaac and Hopkins 1937) has been reported. Reduction in soil organisms has been identified as one reason for the decline (Heyward 1937, Fritz 1932). Increase in particle aggregation may occur after intense fire (Scott and Burgy 1956), but if heat-fused aggregates are disregarded, remaining aggregated particles are often less stable (Dyrness and Youngberg 1957).

Wettability of soil has received much attention in the last 25 years because of concern over flooding and erosion. If a water droplet is placed on water-repellent soil, it will "ball up" and be absorbed very slowly into the soil (DeBano et al. 1979). Fire can create a water-repellent surface or subsurface soil layer. Soils most subject to development of water repellency are those with resinous litter, coarse-textured soils, and soils in areas with potential for intense fires (DeBano et al. 1970). Water repellency is caused by the coating of soil particles by hydrophobic organic substances from plant decomposition or fungi. A water-repellent condition is usually in place before a fire burns the site. Resinous vegetation, such as chaparral or coniferous forest, is the source of much of the hydrophobic material. Coarse-textured soils are most vulnerable because they have low internal surface area, and are thus more easily coated by a given amount of hydrophobic material than the higher surface areas of loam to clay soils.

Wettability can be significantly reduced by fire. As fire burns above the soil, the hydrophobic materials are vaporized. Some drift off as smoke, but others diffuse downward into the soil, where they recondense on cooler soil particles at depth. Water repellency can be intensified at temperatures above 175°C (DeBano and Krammes 1966). Many of the hydrophobic materials are completely vaporized above 370°C for 15 minutes. A very hot fire, then, can destroy water repellency at the surface while leaving a subsurface water-repellent layer intact (Fig. 6.6). The water repellency is maximized when soils are dry, so that water infiltration is most restricted in the first rains of the wet season. Once the soil has been slowly wetted, the water-repellent effect is less pronounced (DeBano et al. 1976).

Fire-induced water repellency in chaparral ecosystems has been intensively studied (Wells et al. 1979). However, Harr et al. (1975),

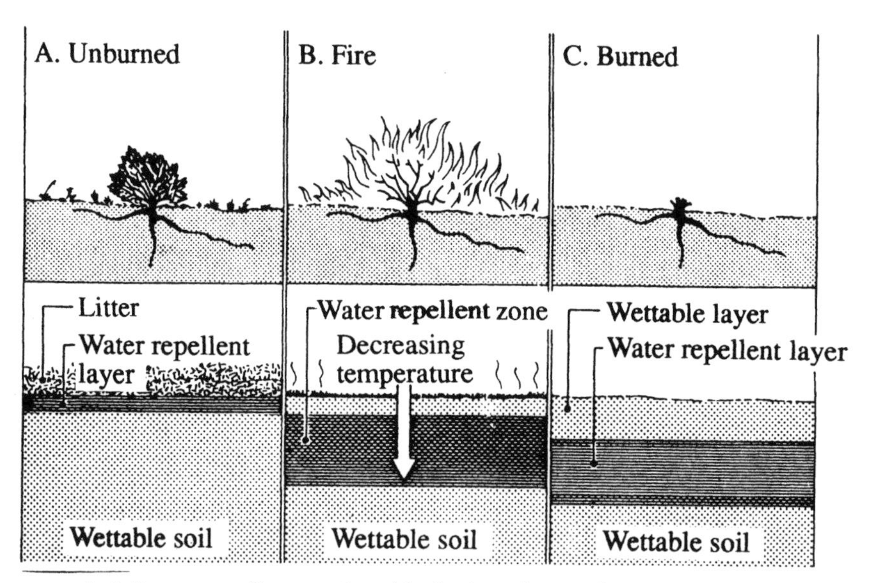

FIG. 6.6. *Soil water repellency as altered by fire in a chaparral ecosystem with intense fire.* A: *Before fire, hydrophobic substances accumulate in the litter layer and mineral soil immediately beneath.* B: *Fire burns the vegetation and litter layer, causing hydrophobic substances to move downward along temperature gradients.* C: *After fire, a water-repellent layer is present below and parallel to the soil surface on the burned area.*
(From DeBano et al. 1979)

working in four Pacific Northwest watersheds, and DeByle (1973), working in medium-textured forest soils in Montana, found little water-repellent effect as a result of fire. Low intensity fires in mixed-conifer forest were associated with minor water-repellent effects (Fig. 6.7; Agee 1973). Sagebrush ecosystems in the Pacific Northwest are not substantially affected (Salih et al. 1973), probably because of the low duration of high temperature in fires in this vegetation type. In the pumice soils of south-central Oregon, however, Dyrness (1976) found a water-repellent layer 2.5–23 cm thick, which lasted for up to five years after a wildfire in lodgepole pine forest. Similar water-repellent layers can be found under single logs that burn on pumice soil.

Water infiltration is the process of water entry into the soil, a function of many of the soil properties already discussed. Where

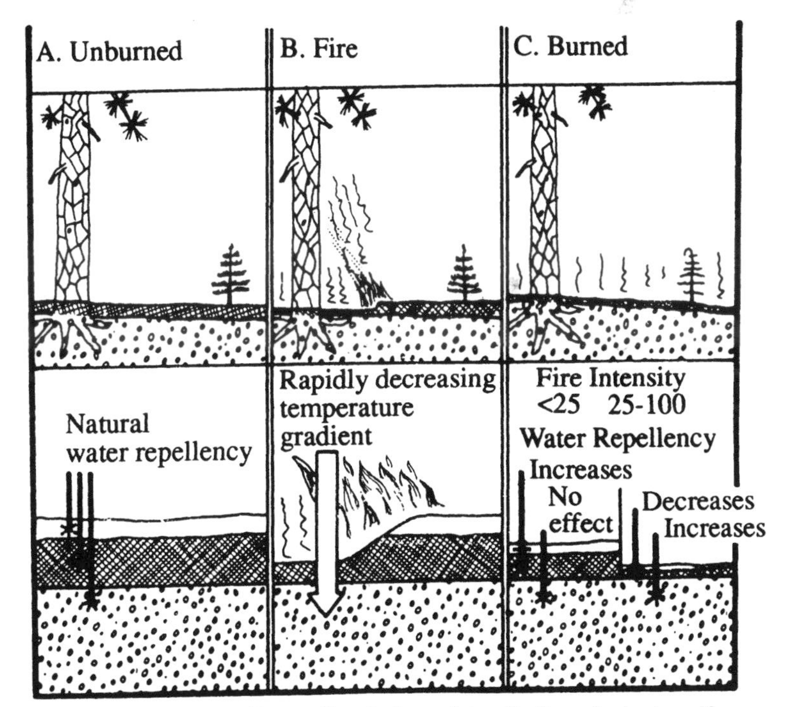

FIG. 6.7. *Water-repellency effects in lower intensity fires of mixed-conifer forest. If residual duff is present, it may become more water-repellent after fire. If all duff is consumed, soil water repellency may increase, but generally not as much as in chaparral ecosystems.*
(From data in Agee 1973)

fires are of high intensity or of repeated moderate intensity, infiltration is usually decreased (Kittredge 1938, Arend 1941, Sampson 1944, Burns 1952, Beaton 1959, Hussain et al. 1969). Some studies have shown no effect on infiltration from intense burns (Veihmeyer and Johnson 1944, Scott and Burgy 1956, Scott 1956). Tackle (1962) reported an initial decrease in infiltration followed by an increase relative to control plots after four years. He attributed this to favorable plant growth, and Roe et al. (1971) stressed the importance of soil cover to prevent raindrop splash and sealing of surface pores. A decrease in infiltration after low intensity fires was attributed to reduction in forest floor depth (Agee 1973), but in some areas such fires have resulted in no change in infiltration (Gilmour and Cheney 1968).

Overland flow occurs when precipitation or snowmelt exceeds the infiltration rate of water into the soil. After fire temporary reservoirs for surface flow, such as surface vegetation or litter, may be reduced, and in some cases water repellency may be increased. Overland flow is unusual in undisturbed forested watersheds (Rice 1973), but it can be increased after fire until vegetation cover, and its associated litter cover, is replaced on the site (DeByle and Packer 1972).

Chemical Properties of Soil

One of the most misunderstood effects of fire is on chemical properties of soil. Much of the confusion has come from studies reporting available rather than total amounts of nutrients, or reporting nutrient changes in one "storage reservoir," such as forest floor or vegetation, while ignoring others. Long-term effects depend on the amount of various nutrients mobilized by fire, the ability of other ecosystem "reservoirs" to capture the nutrients before they are lost from the system, and the ability of the system to regenerate stores of nutrients by such mechanisms as fixation of nitrogen from the atmosphere.

Fire has the capability of volatilizing nutrients, which may then be lost from the system. Such loss is a function of the amount of biomass consumed and the temperature of the fire. A simple way of looking at such losses is the soil nutrient thermometer (Fig. 6.8). Carbon begins to be lost at temperatures above 100°C, and those hydrocarbons begin to create water repellency at temperatures

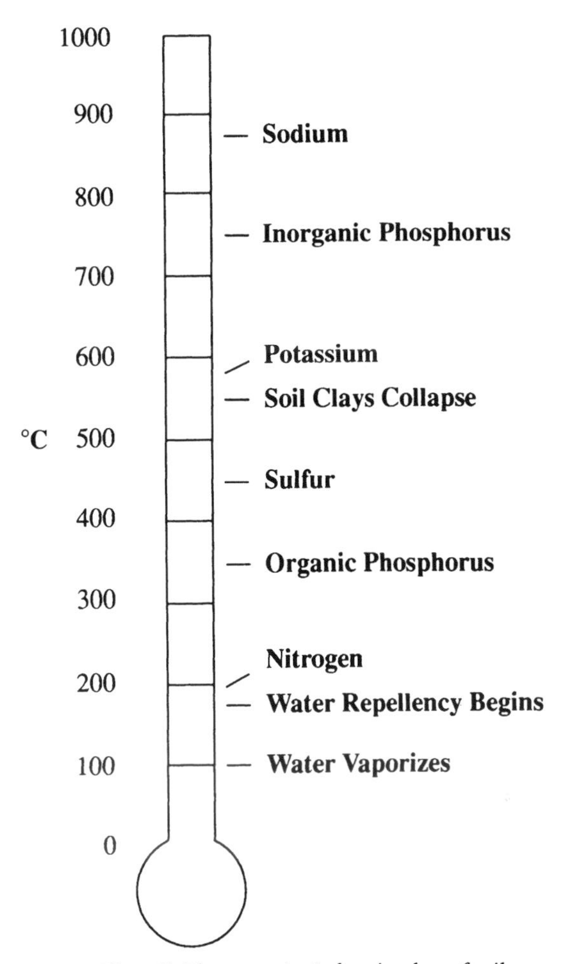

FIG. 6.8. *The soil "thermometer," showing loss of soil nutrients to volatilization with increasing temperature. Nitrogen and sulfur are the two most susceptible nutrients.*

above 175°C. Nitrogen is easily volatilized, with losses beginning at 175°C–200°C (DeBell and Ralston 1970, White et al. 1973). Sulfur begins to be lost in significant amounts above 375°C (Tiedemann 1987). Potassium can be lost in significant amounts above 550°C (Jackson 1958). Inorganic phosphorus is lost at 770°C (Weast 1982). Because fire temperatures rarely exceed 750°C, calcium, magnesium, and sodium usually remain on site and are major constituents of ash. Immediately after a fire, organic matter is reduced, since it was the fuel for the fire. In material totally consumed, nitrogen will be volatilized, and depending on combustion temperature, some to all of the sulfur and potassium is lost to the atmosphere. In residual organic material that is blackened, some nitrogen and sulfur generally remain, and other nutrients are usually present in altered, more mobile form.

The availability of nutrients may follow different trends than those for total site nutrients. The availability of nutrients depends in part on pH, the measure of soil acidity, which is often changed during burning. Soil acidity usually decreases after burning (e.g., a pH increase), as various oxides in the ash react with hydrogen ions (Ahlgren and Ahlgren 1960). Where fires consume much organic material, as in slash burns, the copious ash may increase pH in surface soil from 5.0 to 6.2, and some effect may be noted to at least 15 cm depth (Isaac and Hopkins 1937). This change may persist (Neal et al. 1965) or return to prefire levels after a year (Tarrant 1956c). Periodic low intensity burns may result in persistently higher soil pH levels (Heyward and Barnette 1934). Such pH increases may affect nutrient availability independent of amounts left in ash; in one study, Fe^{3+} was found to be less available on slash-burn sites with pH 6.16 than on undisturbed sites with pH 4.68 (Perry et al. 1984).

Exchangeable bases in the soil almost always show an increase after burning due to their release from aboveground biomass and subsequent leaching into the soil. The amount of the increase is proportional to the amount of ash created. Austin and Baisinger (1955) reported an 830 percent increase in calcium and a 166 percent increase in potassium after intense burning. Grier (1975) reported losses from volatilization and leaching of 855 kg ha^{-1} nitrogen, 75 kg ha^{-1} calcium, 33 kg ha^{-1} magnesium, 282 kg ha^{-1} potassium, and 698 kg ha^{-1} sodium. Because potassium and sodium have lower ion exchange potential as monovalent cations, they are more likely than calcium and magnesium to be leached through the soil profile (Viro 1969). Relatively infertile soils, with low cation exchange capacity, experience more nutrient loss through leaching than soils of more fertile sites for a given amount of biomass consumption. Excess anions, such as nitrate, sulfate, bicarbonate, and organic acids in solution, may displace cations from soil colloids and result in leaching of some cations (McColl and Cole 1968). Residual or colonizing shrubs, trees, and grasses may capture these nutrients before they are leached below the rooting zone (Lay 1956, Daubenmire 1968b, Stark 1973). If not, they are are lost from the ecosystem and are likely to find their way into streamwater.

Nitrogen transformations are very complex, which in part explains the mix of increases (Heyward and Barnette 1934, Vlamis et al. 1955), decreases (Wagle and Kitchen 1972, DeBano and Conrad 1978), and no effects (Ahlgren and Ahlgren 1960, Wells 1971) reported for available soil nitrogen after burning. There is clearly an immediate loss in total nitrogen. During intense burns, this may total as much as 0.9 t ha^{-1} (Grier 1975). But in residual fuel, ammonium nitrogen, a form available to plants, may be increased over prefire levels (Christensen and Muller 1975, DeBano et al. 1979). Later oxidation of ammonium to nitrate can increase nitrate levels in the soil, although it will be subject to leaching.

An additional complexity in the nitrogen dynamics after fire is a possible increase in plants supporting nitrogen-fixing symbionts. These plants form symbiotic relationships with nitrogen-fixing bacteria associated with nodules on the plant roots: The plant provides a carbon source for energy, and the bacteria are able to fix atmospheric nitrogen into forms available for absorption by the plant. After disturbance, plant genera capable of forming such symbiotic relationships—such as *Alnus, Ceanothus, Lotus,* and *Lupinus*—may increase in abundance or even completely dominate the site. *Ceanothus velutinus,* which has an *evader* strategy in its hard-coated seeds, was reported to fix 0.71–1.08 t ha^{-1} over a 10 year period (Youngberg and Wollum 1976), or an equivalent 0.1 t ha^{-1} over 1 year (McNabb and Cromack 1983). Such internal mechanisms could replace the nitrogen lost from an intense fire in as little as a decade if a pure stand were present. Others have reported lower levels of nitrogen fixation, which would require several times as long to replace the volatilized nitrogen (Zavitkowski and Newton 1968). This variation may be due to the degree of nodulation, which is often less where *Ceanothus* plants have been absent from the site for a long time (Wollum et al. 1968). This mechanism of nutrient retention has been referred to as an emergent property of ecosystems (Woodmansee and Wallach 1981). A less controversial point is that suppositions about long-term productivity based on nutrient loss from fire must also consider acceleration of nutrient accrual through such mechanisms as symbiotic nitrogen fixation.

Biotic Properties of Soil

We know less about soil biota than about any other aspect of the soil, so it is not surprising that our state of knowledge about fire interactions with soil biota is still rudimentary. Soil microflora, including bacteria, fungi, algae, and bryophytes, are important in the decomposition process and in various symbiotic relationships with higher plants. These symbioses may be nutrient-related or may involve a defense against other, deleterious microorganisms. Soil microfauna may also be important for decomposition and may be important factors in maintenance of physical soil structure.

Nonvascular plants may exist in vegetative states but are usually also represented by inactive spores, which are more resistant to stresses such as fire (Warcup 1981). Hot fires have more significant and long-lasting effects than cooler fires (Wells et al. 1979). Annual burning in southern pine forests at low intensity for 50 years had little effect on soil microorganisms (Berry 1970).

Burning in moist soils can have more severe effects on microflora than burning in dry soils, apparently due to more efficient heat conductivity. Although bacteria tend to be more resistant to heat than fungi, both groups suffer significant reductions under lower temperatures in wet soil conditions. Lethal temperatures for bacteria were found to be 210°C in dry soil and 110°C in wet soil (Wells et al. 1979). For *Nitrosomonas*, the bacterial group that oxidizes ammonium (NH_4^+) to nitrite (NO_2^-), lethal soil temperatures may be as low as 140°C in dry soil and 75°C in wet soil, while fungi may tolerate temperatures of only 155°C in dry soil and 100°C in moist soil (Dunn and DeBano 1977).

The initial impacts on soil microflora are soon overwhelmed by responses of vegetative residuals and spores. With increased pH, bacteria are usually favored over fungi. Both typically show an immediate postfire decline followed by an increase, with bacteria being favored (Warcup 1981).

The reduction in fungi may have indirect effects on nutrient status. Perry and Rose (1983) showed that low intensity slash burns had higher ratios of bacteria to fungi. The affected fungi were the primary producers of hydroxymate siderophores (HS), which chelate iron and make it available to plants. The dual effect of a reduction in HS and increase in pH in the burned soils led Perry et al. (1984) to suggest possible iron deficiencies for seedling growth on these sites.

Mycorrhizal fungi, which form symbiotic associations with tree roots and increase the water- and nutrient-absorption capabilities of roots, generally show decreases in numbers and species after slash burning (Wright and Tarrant 1957, Perry et al. 1984). This effect appears to be partly due to direct soil heating, and it is most significant on poor sites where mycorrhizae may be critical to tree survival and growth (Borchers and Perry 1990). Most mycorrhizal activity is above or within the soil surface in logs or humus, so that log consumption by fire or heat conducted into logs or soil can directly kill the fungi (Harvey et al. 1986). Wright and Tarrant (1959) found that Douglas-fir seedlings on unburned areas had 65 and 100 percent mycorrhizal associations at age one and two years, but those figures dropped to 40 and 79 percent on sites where slash was burned. Of seedlings that had ectomycorrhizal associations, those growing in unburned sites had about twice as many ectomycorrhizal root tips. Many of the studies evaluating effects of fire on mycorrhizae have taken place on clearcuts where the host trees have been removed. Cutting of trees is often associated with temporary declines in mycorrhizal populations, but slash burning, particularly if very hot, can lead to further declines (Borchers and Perry 1990). On some sites reduction in humus and decayed wood has been associated with declines in mycorrhizal activity (Harvey et al. 1981), while in other locations no effect was measured (Schoenberger and Perry 1982). Borchers and Perry (1990) note that enumeration of mycorrhizae without regard to diversity of the organisms may tell only part of the story.

The reaction of moss to burning has not been well studied. In Tasmania, Cremer and Mount (1965) found a succession from *Funaria* spp. and *Ceratodon* spp. followed by *Polytrichum juniperinum*. They noted that *Polytrichum* was not as fire-dependent as the first colonizers and grew more slowly; they thought this succession might be common to many temperate climates. Another species common to new burns in Australian forest is *Leptobryum pyriforme* (Warcup 1981), also identified as a postfire oak woodland colonizer in Washington (Agee, unpublished data). Ahlgren and Ahlgren

(1960) identified the lichen genera *Biatora, Baeomyces,* and *Cladonia* as fire-colonizing species.

Important plant diseases can be affected by fire. The fungus *Rhizina undulata* tends to increase after hot burns in many places around the world. It infects conifer roots in acid soils and kills the plant. In the Pacific Northwest it has been identified as a seedling disease of Douglas-fir after regeneration on hot slash burns. A survey of Oregon and Washington forestland found it to decrease from Washington to Oregon, with average loss in northwestern Washington of 1 percent of seedlings (Thies et al. 1977). Some individual units had higher loss figures, but except on unusual sites *Rhizina undulata* appears to have little significant effect in the Pacific Northwest.

Other root diseases, such as *Phellinus weirii,* may have been little affected by fire west of the Cascades because of the infrequency of fire. When fire did occur, the disease was normally not removed from the site, because of the low probability of root consumption west of the Cascades. In this area it remains an endemic disease that is treated by planting *Phellinus*-resistant species rather than by direct control. In the eastern Cascades, higher fire frequencies and drier conditions may have resulted in stumps and root corridors being burned out and sterilized, reducing spread of the disease. Fire suppression has likely allowed *Phellinus* to spread more widely through the forest east of the Cascades. Under fire exclusion policies, shifts in species composition to shade-tolerant, susceptible species, such as eastside Douglas-fir and grand fir, will eventually increase the importance of *Phellinus.*

Spore germination in *Heterobasidion annosus* (annosus root rot) has been inhibited by passage of smoke across colonizing surfaces (Froehlich et al. 1978, Parmeter and Uhrenholdt 1976). However, the management efficacy of such treatments remains to be evaluated (Thies 1990). There are surely a variety of interactions between fires and diseases that we do not yet fully understand.

A LANDSCAPE APPROACH TO SOIL AND WATER

Much of the rigorous research on soil and water impacts from fire has been conducted on the scale of the soil sample, typically measured in square centimeters, or possibly square meters. Much of the relevant geomorphic and hydrologic impact of fire, however, occurs at the scale of the subbasin or watershed, measured in hectares to square kilometers. Unfortunately, at this scale we see a variety of confounding impacts associated with weather variation, other aspects of management (logging, roads, salvage operations), and variation in fire severity. Nevertheless, some generalizations can be made about the direction of possible landscape effects of fire on soil and water.

Changes in soil and water regimes are most significant where fire has a large-scale impact on the vegetation and soil properties of the ecosystem (Swanson 1981; Fig. 6.9). Where fires are of low intensity or affect a small part of a large watershed (Bethlalmy 1974), changes in soil and water regimes are difficult to detect (Beschta 1990, McNabb and Swanson 1990). If direct or indirect effects of fire on soil and vegetation are significant, associated changes in hydrologic and geomorphic processes are likely to occur (Anderson 1954). Some changes may be subtle, such as increased seasonal soil water leading to prolonged seasonal soil creep. Some may be associated with increased disturbances of other kinds, such as snow avalanches. Others, such as soil mass movements, streambank cutting, or aggradation of channels, may have direct watershed effects. These changes are a function of storm events in the years following fire as well as of fire severity.

Annual water yield can be significantly increased after fire due to the reduction of interception loss and transpiring vegetation, compared to generally lower increases in evaporation. This effect is proportional to the amount of watershed area burned and to annual precipitation (Beschta 1990). After the 1933 Tillamook fire, annual water yield increased about 10 percent for the first decade after burning (Anderson et al. 1976). The seasonal distribution of water yield may also be affected. Peak flows may increase (Rich 1962, Helvey et al. 1976). Summer flow generally shows the highest

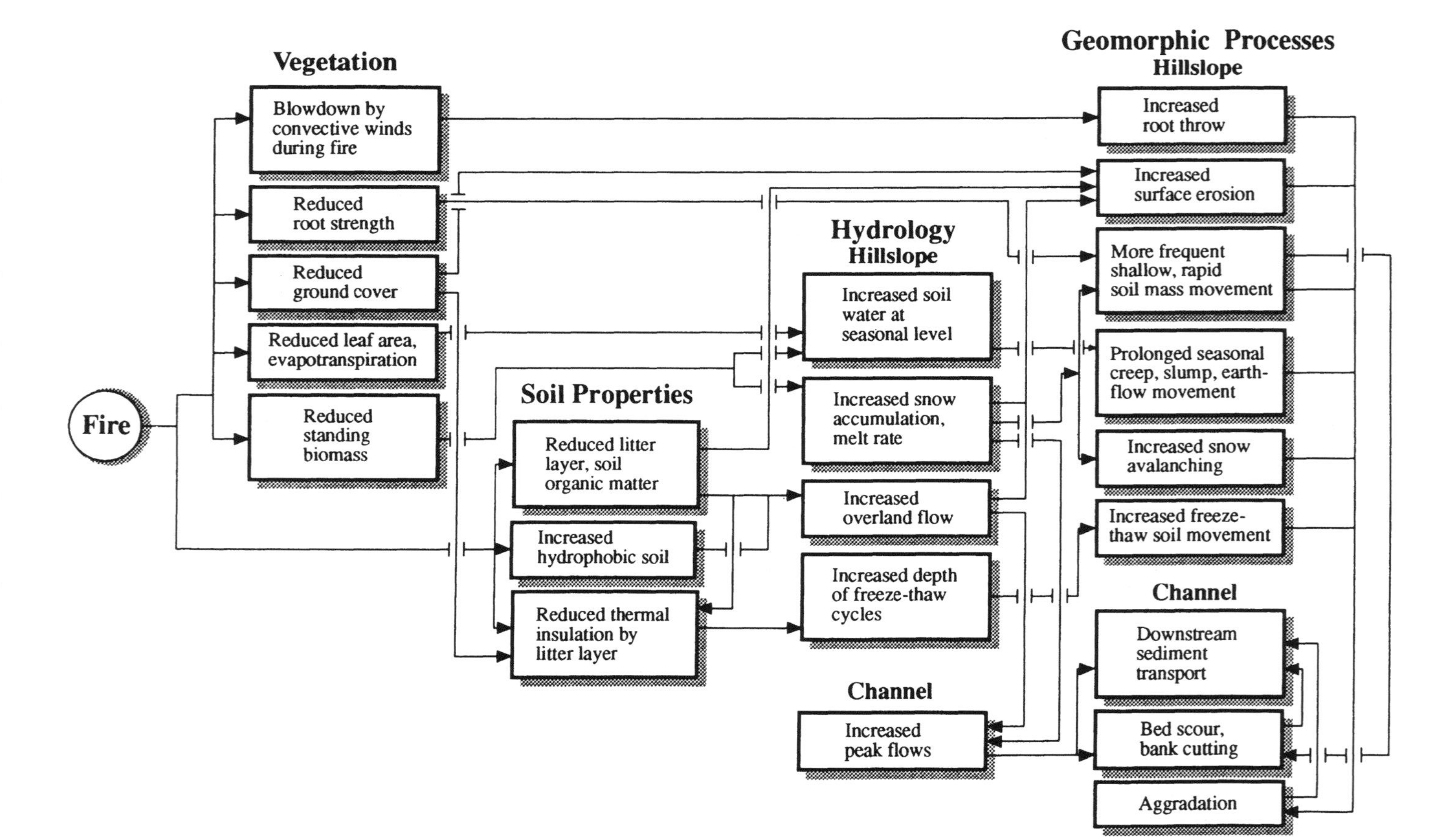

FIG. 6.9. *Effects of fire on vegetation, soil properties, hydrology, and geomorphic processes.* (From Swanson 1981)

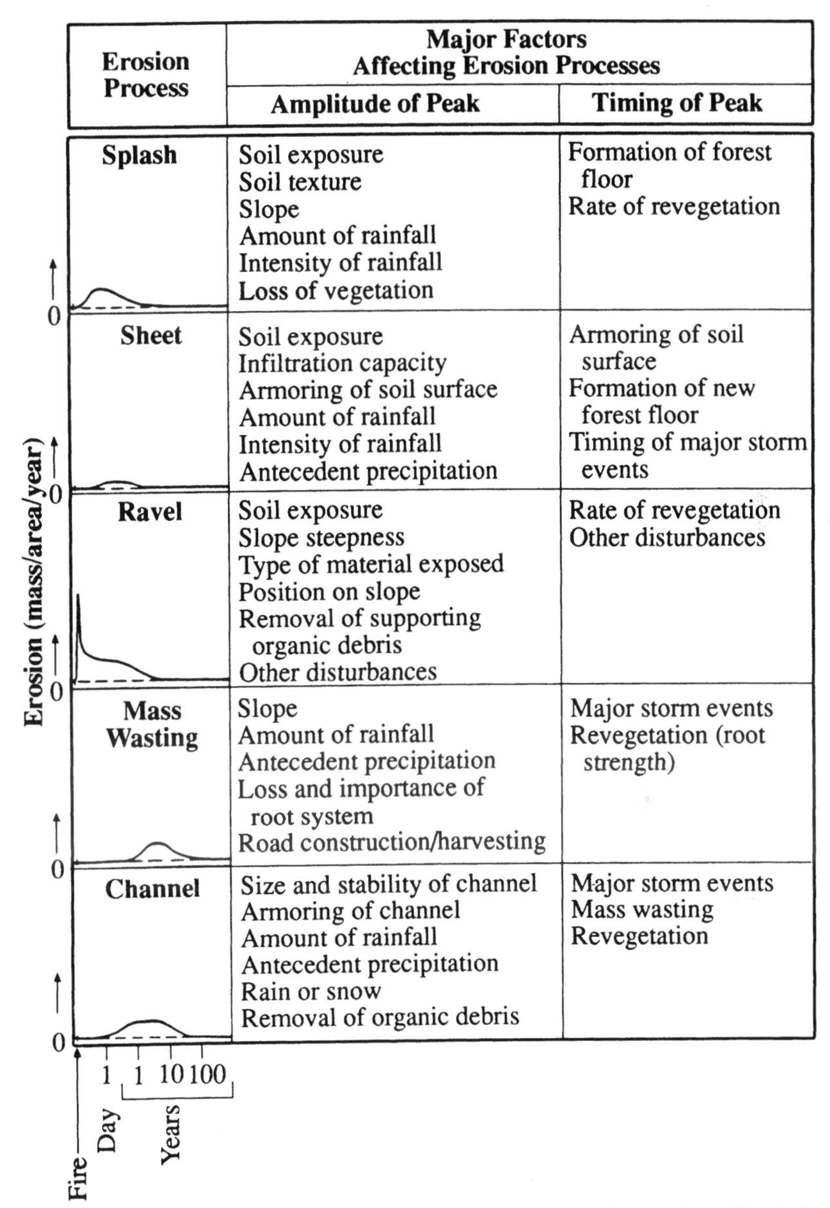

Erosion Process	Major Factors Affecting Erosion Processes	
	Amplitude of Peak	Timing of Peak
Splash	Soil exposure Soil texture Slope Amount of rainfall Intensity of rainfall Loss of vegetation	Formation of forest floor Rate of revegetation
Sheet	Soil exposure Infiltration capacity Armoring of soil surface Amount of rainfall Intensity of rainfall Antecedent precipitation	Armoring of soil surface Formation of new forest floor Timing of major storm events
Ravel	Soil exposure Slope steepness Type of material exposed Position on slope Removal of supporting organic debris Other disturbances	Rate of revegetation Other disturbances
Mass Wasting	Slope Amount of rainfall Antecedent precipitation Loss and importance of root system Road construction/harvesting	Major storm events Revegetation (root strength)
Channel	Size and stability of channel Armoring of channel Amount of rainfall Antecedent precipitation Rain or snow Removal of organic debris	Major storm events Mass wasting Revegetation

FIG. 6.10. *Major factors affecting erosion processes. Important factors that affect both the amplitude and timing of specific types of erosion are listed in conjunction with the relative effect on erosion rate.*
(From McNabb and Swanson 1990)

percentage increase in discharge, and in dry areas this may be a significant beneficial effect of burning for wildlife and aquatic systems.

Various erosion processes may be accelerated by moderate to high severity fires (Fig. 6.10; McNabb and Swanson 1990). The shape of the fire-induced accelerated sediment yield (Fig. 6.11) shows a sharp initial increase, followed by an exponential decline to ambient levels. The increase may be due to a variety of erosion processes accelerated or initiated by the removal of ground litter and vegetation cover, increased water repellency, loss of forest floor, and reduced root strength. These include surface erosion from dry ravel (Mersereau and Dyrness 1972), sheetwash and rill erosion (Sartz 1953), or wind (Blaisdell 1953); rapid mass movement processes

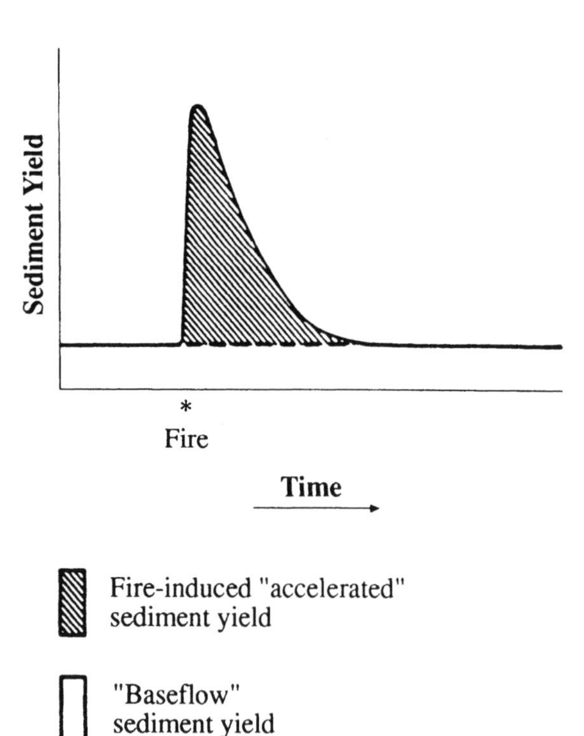

FIG. 6.11. *Hypothetical variation in sediment yield during the period of watershed response to fire. Sediment yield is separated into fire-induced accelerated yield and "baseflow" yield components.*
(From Swanson 1981)

(debris flows and slides; Klock and Helvey 1976); and a variety of sediment storage impacts in stream channels.

The effect of such processes on sediment yield in several fire regimes is shown in Figure 6.12. For the chaparral system (A), the fire-return interval is set at 25 years, and the increase in sediment may be up to 30 times the ambient level, based on records from sediment basins. With such sharp increases in sediment yield, about 70 percent of total sediment is fire-related (Swanson 1981), and 70 percent of that may occur in the first year (Rice 1973). Almost half the sediment yield over time, then, may come in the first year after burning, with associated severe effects on downstream resources and developments. For the Douglas-fir example (B), the fire-return interval is set at 200 years. Based on field studies of debris slide occurrences in clearcut areas and sediment yield from small watershed experiments, sediment output increases by about five times, falling to ambient levels over the next 25 years. Shallow mass movement, caused by decay of tree roots that held marginally stable slopes in place, may increase two to five years after disturbance (Swanston 1974). Roughly 25 percent of the long-term sediment yield from such a nonmanaged watershed is fire-induced.

For the ponderosa pine system (C), the fire-return interval is set at 10 years, and the effects of fire are so benign that fire-induced sediment yields are difficult to separate from ambient levels. More variation in sediment yields is likely to be associated with unusual storms than with repeated minor disturbances, such as fire. If fire is removed for several decades from the ponderosa pine ecosystem, a resulting fire may be of higher severity, with a sediment profile more like the Douglas-fir ecosystem. Every watershed has a unique fire history and geomorphic sensitivity to erosion, so these generalized sediment yield profiles are intended to represent a spectrum of possible impacts rather than specific predictions for a given ecosystem type.

Wildfire Rehabilitation

Fires of high severity remove both canopy and ground cover, and over time they reduce the resistance to surface erosion and soil mass movement. Wildfire rehabilitation efforts have been aimed at re-

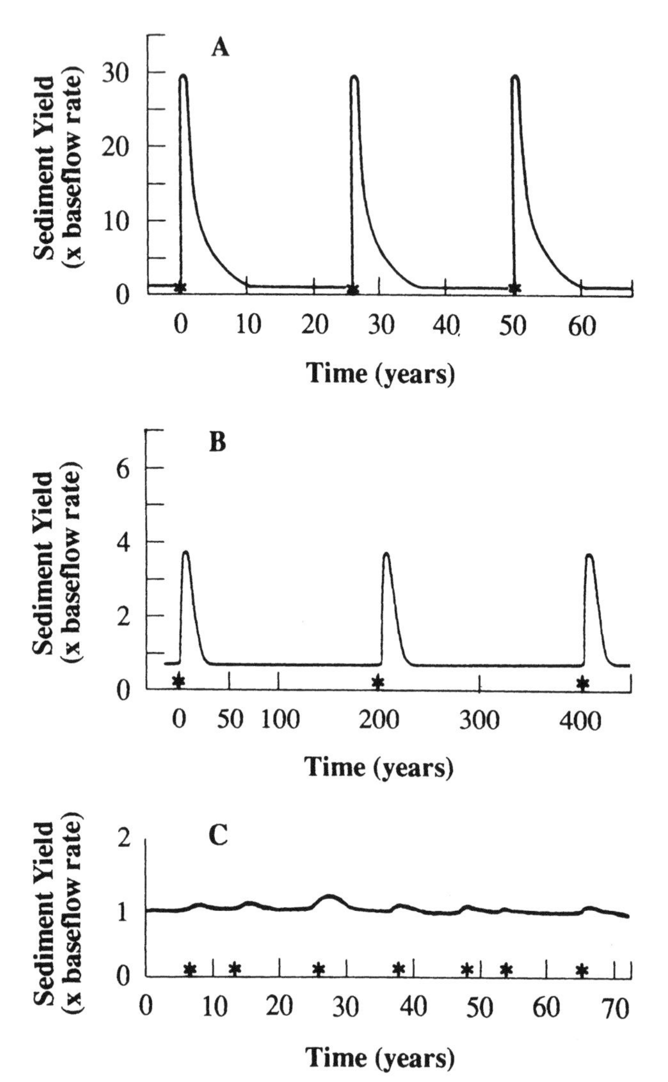

FIG. 6.12. *Hypothetical variation in sediment yield for several fire rotations.* A: *a steep-land chaparral system.* B: *a Douglas-fir western Cascade Mountain system.* C: *a ponderosa pine system with frequent fire. Fires are indicated by asterisks. Note the different scales on time and yield axes.*
(A and B from Swanson 1981)

ducing the effects of wildfires on soil and hydrology. Such efforts have included grass seeding and fertilizing, mulching, contour felling of logs, straw-bale check dams, and rock and log check dams. The major objective of all these treatments is to hold soil in place or nearby until the site is once again relatively stable. The first two or three years after the fire are most important for controlling surface erosion. Shallow mass movement may become more important after two to five years (Swanston 1974) as fine roots in surface soils decay, but it is less capable of being successfully managed.

The mix of treatments on a particular landscape depends on the erosion potential over time, the prefire vegetation, and the values to be protected on and downstream from the burned landscape. Few treatments are cost-effective in wildland situations against mass movement processes, but several are effective in slowing surface erosion or movement of sediment in stream channels. Most treatments are applied only to portions of the burned area. In grassland or brush ecosystems, the selection of treatments will be different than for forested ecosystems. In areas managed for natural values, seeding of alien grasses and forbs may be contrary to management objectives, and on commercial forestland may interfere with natural or artificial regeneration of conifers. These costs must be weighed against the potential benefits of erosion control. The following discussion covers the most common treatments, ignoring very intensive measures, such as channel armoring, construction of debris basins, and emergency road treatment.

GRASS AND FORB SEEDING AND FERTILIZING

The objective of grass seeding is to provide a rapid ground cover over the burned landscape. In an area already dominated by grasses, seeding is usually not needed, as the annual and/or perennial grasses already present usually provide adequate ground cover. In shrublands (a mix of shrubs and grasses), grasses can be seeded in white ash areas, which indicate more intensely burned areas. In brushfields, grasses are often seeded across the entire burn because of the typically more uniform nature of the burn. Forested areas are seeded most commonly in areas that were dominated by trees and shrubs before the fire, because grassy areas usually recover quickly.

The bases of slopes may be seeded most heavily to slow sediment delivery to channels.

The components of the mix depend on the site, but most are nonnative species because they must be obtained in large volume and sowed in a short time after summer wildfires. In California brushfields, annual ryegrass (*Lolium multiflorum*) is most commonly used. It provides rapid cover and can have roots 1 m deep by the following May. Native shrubs cannot compete well, and where ryegrass cover reaches more than 40 percent few shrubs will be found (Schultz et al. 1955). This grass cover can be obtained from seeding rates of 17.5 kg ha^{-1} (Taskey et al. 1989), although much higher seeding rates are often used. In forested ecosystems, a wider mix of species is commonly used, including nitrogen-fixing legumes in attempts to replace site nitrogen lost in the fire. Monitoring often shows that some species are more successful than others. After the Entiat fires of 1970, a mix of perennials was seeded; orchardgrass (*Dactylis glomerata*) and hard fescue (*Festuca ovina*) had higher cover than perennial rye (*Lolium perenne*) and yellow sweet clover (*Melilotus officinalis*) (Tiedemann and Klock 1973). After the 1987 fires on the Trinity National Forest, orchardgrass was the only perennial used in a mix of six species (Miles et al. 1989).

Fertilizing can accompany seeding, typically at rates of 200–300 kg ha^{-1}, but its efficacy is not clear. Often fertilizer is applied as a part of a single treatment, so its effect cannot be studied, or it is applied to only a portion of a burned area without monitoring (Gross et al. 1989). In one of the few experimental seeding and fertilizing treatments, Klock (1971) found little difference in plant cover between different fertilizer treatments in three seeded watersheds within the 1970 Entiat fires. One watershed was not fertilized, one was fertilized with 54 kg ha^{-1} nitrogen as urea, and the third fertilized with 51 kg ha^{-1} nitrogen as ammonium sulfate. Nitrogen applied as ammonium to alkaline ash surfaces after fire may in part ($>$12 percent) be volatilized within days (Crane 1972).

Seeding and fertilizing after wildfires is done under emergency recovery operations. Little useful experimentation has ever been tried, or subsequently monitored, to evaluate the success of various treatments in reducing erosion or the effects on native vegetation.

MULCHING

Direct application of mulch provides immediate ground cover. Mulch can be applied by blower or by hand. Blower application is limited to gentle slopes where the machinery, usually mounted on a truck, can be maneuvered. Hand application is possible in critical areas (firelines, road fills, along streams), but it is slow. Mulch can be blown by the wind but becomes more stable after a rain or two. Mulching appears to control erosion where it is applied at more than 2 t ha^{-1} (Miles et al. 1989).

CONTOUR FELLING

Contour felling (Fig. 6.13A) is designed to produce cross-slope barriers to overland flow and the sediment it may carry. Small trees killed by the fire are felled in a cross-slope direction and secured by stakes or the position of other trees or stumps. Open areas beneath the log must be filled with slash and soil. In southwest Oregon and northern California, contour felling has been judged ineffective (Gross et al. 1989, Miles et al. 1989) because of the cost ($1,200 ha^{-1}), the difficulty of making a good connection with the ground, and the possibility of concentrating runoff if a hole does develop beneath the stem. In local microsites, such barriers have worked to collect and hold sediment four years after the 1988 Dinkelman fire near Wenatchee, so careful prescription of log placement may be critical in using this technique.

CHECK DAMS

Check dams are intended to capture sediment once it has reached a stream channel or to prevent lateral bank erosion. They can be temporary or permanent in design and construction. Each dam should be designed to be integrated with the ones up- and downstream, so that no single dam is overloaded with too much sediment. They should also be tied in, where possible, with natural dams provided by stable coarse woody debris and rocks. Maintenance by emptying sediment during the life of a dam may be important to prolonging its effectiveness.

Straw check dams consist of a series of straw bales secured in small channels with reinforcing bars through the bales. Some sort of energy dissipator, such as a log or rocks, is needed below the dam. Straw dams have a limited life span (two to three years), although they can effectively retain sediment delivered to small channels (Miles et al. 1989). Woven-wire fence can be placed on the downstream side and staked with metal fenceposts (Fig. 6.13B). Sandbags can be used in place of the straw bales for a longer lasting dam (Fig. 6.13C). The most stable check dams are made of 30–40 cm logs keyed into the bank, with a spillway notch in the upper log. As with other types of check dams, an energy-dissipating device is necessary under the spillway. These dams can last 15–30 years (Miles et al. 1989).

THE WILDLIFE COMPONENT

Fire effects on wildlife are complex because they are often indirect, affecting habitat more than individuals. The diversity of wildlife, the diversity of habitats, and the diversity of fire within those habitats make generalizations about fire effects difficult. However, a few statements can be made.

1. Fire is not detrimental to many species of wildlife; conversely, it is not always beneficial to wildlife or of equal effect on all species (Clark and Starkey 1990).
2. Death of large animals directly due to fire is rare (Wright and Bailey 1982, Singer et al. 1989). Death is usually due to suffocation and primarily affects species with small home ranges (Chew et al. 1958, McMahon and deCalesta 1990).
3. Many species ignore the presence of fire, while others are attracted to it because of the availability of prey (Komarek 1969, Bendell 1974).
4. The major effect of fire is on animal habitat: food, cover, and water (Bendell 1974, McMahon and deCalesta 1990).
5. Fire may have different effects over time on an individual species, with immediate beneficial or detrimental effects and later offsetting effects. General reviews of fire-wildlife interactions can be found in Bendell (1974), Lyon et al. (1978),

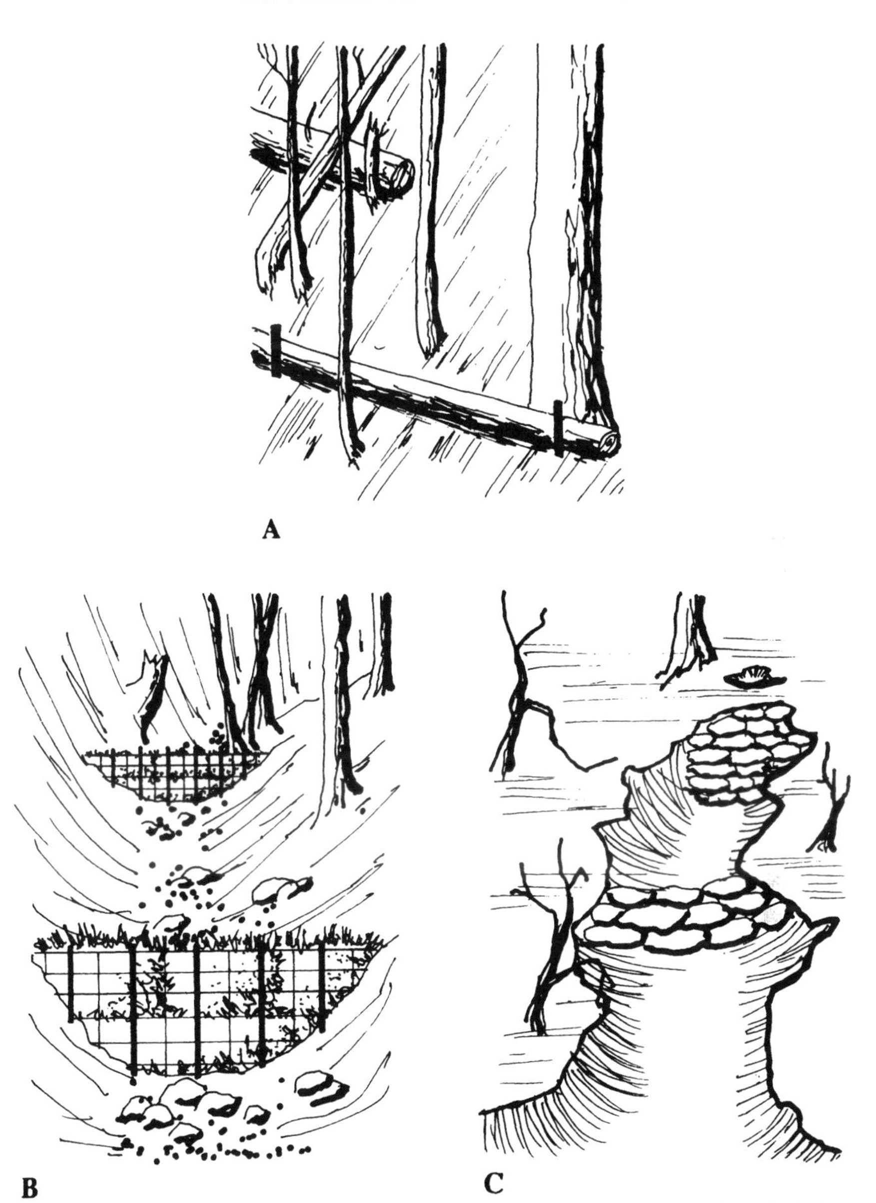

FIG. 6.13. *Watershed rehabilitation measures.* A: *Contour log structure, where the bole of a small tree is placed on the slope, anchored with stakes, and sealed at ground line.* B: *Straw-bale check dam, where bales are wrapped in plastic netting, placed against woven wire fence, and staked.* C: *Sandbag check dam, where rot-proof sandbags are filled onsite and keyed to gully bottom and walls.*
(From Gross et al. 1989)

Wright and Bailey (1982), Lotan and Brown (1985), and McMahon and deCalesta (1990). Small mammals and predators are discussed by Ream (1981), and wetland fire and wildlife are reviewed by Kirby et al. (1988).

Changes in wildlife habitat are the most significant effect of fire on wildlife populations. Some species tend to be "winners" and others "losers" as fire alters the habitat. Many species prized as game animals are favored by habitat changes that reduce forest cover or increase edge: moose (*Alces alces*), deer (*Odocoileus* spp.), elk (*Cervus* spp.), cougar (*Felis concolor*), coyote (*Canis latrans*), black bear (*Ursus americanus*), beaver (*Castor canadensis*), hares (*Lepus* spp.), turkey (*Meleagris* spp.), pheasant (*Phasianus* spp.), bobwhite quail (*Colinus virginianus*), sharptailed grouse (*Pedioecetes phasianellus*), ruffed grouse (*Bonasa umbellus*), red grouse (*Lagopus* spp.), blue grouse (*Dendrogapus* spp.), prairie chicken (*Tympanuchus cupido*), willow ptarmigan (*Lagopus* spp.), and some waterfowl (Lyon et al. 1978). Other animals, such as mountain, woodland, and barren-ground caribou (*Rangifer* spp.), marten (*Martes americana*), red squirrel (*Tamiasciurus* spp.), grizzly bear (*Ursus horribilis*), wolverine (*Gulo luscus*), fisher (*Martes pennanti*), and spruce grouse (*Canachites* spp.), are reported to decrease after burning (Lyon et al. 1978).

Overall species richness may be little affected by fire. A review of 22 studies of birds and mammals by Bendell (1974) showed little change in numbers of species (Table 6.1). For birds, the largest increases were in ground-foraging species, while species foraging in trees showed the largest decrease. This trend is consistent with more recent studies (Huff et al. 1985). For small mammals, overall species richness changed very little. These results suggest that many species may be generalists able to adapt to postfire environments, that unburned patches within a burn may provide sufficient habitat for them, or that fire, particularly if of low intensity, may have little effect on critical habitat parameters.

About half the bird species reviewed by Bendell (1974) showed no change in density, with the remainder showing more increases than decreases (Table 6.2). The mammals had similar proportions of species showing no change, with equivalent proportions showing increases and decreases. The breeding birds and small mammals reviewed are generally those with relatively small home ranges,

TABLE 6.1. CHANGES IN NUMBERS OF SPECIES OF BREEDING BIRDS AND MAMMALS AFTER BURNING

Foraging Zone	*Before Burn*	*After Burn*	*Gained (percent [number])*	*Lost (percent [number])*
Species of birds[a]				
Grassland and shrub	48	62	38 (18)	8 (4)
Tree trunk	25	26	10 (6)	16 (4)
Tree	63	58	10 (6)	17 (11)
Totals	136	146	21 (30)	14 (19)
Species of small mammals[a]				
Grassland and shrub	42	45	17 (7)	10 (4)
Forest	16	14	13 (2)	25 (4)
Totals	58	59	16 (9)	14 (8)

SOURCE: Bendell 1974.

NOTE: Neither predators nor mammals larger than a hare are included.

[a] Sources are cited in Bendell (1974) and repeated in Lyon et al. (1978).

TABLE 6.2. PERCENT CHANGES IN DENSITY OF POPULATIONS OF BREEDING BIRDS AND MAMMALS AFTER BURNING

Foraging Zone	*Increase*	*Decrease*	*No Change*
Birds[a]			
Grass/shrub	50 (33)	9 (6)	41 (27)
Tree trunk	28 (9)	16 (5)	56 (18)
Tree	24 (17)	19 (14)	57 (41)
Totals	35 (59)	15 (25)	50 (86)
Small mammals[a]			
Grass/shrub	24 (9)	13 (5)	63 (24)
Forest	23 (6)	42 (11)	35 (9)
Totals	23 (15)	25 (16)	52 (33)

SOURCE: Bendell 1974.

NOTE: Predators are not included and mammals larger than a hare are excluded. Numbers of species are in parentheses.

[a] Sources are cited in Bendell (1974) and repeated in Lyon et al. (1978).

which can be easily studied. Species with larger home ranges, including most predators, require a landscape approach to evaluate the effects of fire.

The grizzly bear provides a good example of a generalist species with a large home range. Grizzlies may have mean home ranges exceeding 750 km^2 (Picton et al. 1986), so most fires affect only a small portion of the home range. Immediate effects may be detrimental to preferred food items, such as huckleberries (*Vaccinium* spp.), which are topkilled, or whitebark pine seeds, whose trees are often killed by fire (Serveen 1983, Eggers 1986). Longer term effects, however, may be positive, rejuvenating decadent huckleberry stands and preventing subalpine fir (*Abies lasiocarpa*) from replacing whitebark pine on many sites (Eggers 1986, Morgan and Bunting 1990). A landscape where a variety of seral stages is present, rather than one dominated by a single stage, such as recently burned forest or old-growth forest, may be preferred for grizzly habitat, and fire may be an important regulator of that mosaic.

THE AIR COMPONENT

Air quality effects of fire have become a critical limitation to the use of fire in wildland environments. These impacts are by no means new; fire and smoke were common in past centuries and millennia (Table 1.2). Such natural sources of air pollution include volcanoes, emissions from live vegetation (the "smell" of the forest), and forest fires. But while totally "clean" air has not existed as long as physical and biological processes have operated on the earth (Hall 1972), this does not mean that emissions are healthy, can be considered only background pollutants, or can be ignored in fire management planning.

Smoke is found in every fire environment. When a fire occurs, about 90 percent of fire emissions consist of carbon dioxide and water vapor. The portion of the carbon in smoke not converted to carbon dioxide is particulate matter, carbon monoxide, and volatile organic matter. Carbon dioxide is a major contributor to global warming scenarios, although it is not technically considered a pollutant (Sandberg and Dost 1990).

The other gaseous components of smoke are considered pollutants (Ferry et al. 1985). Carbon monoxide is the most abundant air pollutant. Its effect on human health depends on exposure, concentration, and level of physical activity during exposure. Firefighters at the edge of the combustion zone are most at risk, since dilution of carbon monoxide to acceptable levels usually occurs away from the edge of fires. Hydrocarbons—a vast family of chemicals containing hydrogen, carbon, and sometimes oxygen—may remain in the atmosphere as gases, condense into droplets, or be adsorbed onto particles (Sandberg and Dost 1990). Two of the most important classes of hydrocarbons are polynuclear aromatic hydrocarbons (PAHs) and aldehydes. PAHs are associated with cancer, but are emitted in such low quantities from wildland fires that human health risks are minimal. Aldehydes may be produced in sufficient quantity to result in eye irritation for sensitive individuals up to two miles away from wildland fires (Sandberg and Dost 1990). In general, the production and effects of these compounds are fairly trivial compared to the production and effects of particulates.

Particulates are a major cause of reduced visibility and serve as adsorption nuclei for harmful gases. In the 1970s particulate regulation was based on total particulate mass per unit volume of air. Research indicated that the smaller size fractions were most harmful to human health, so standards were developed for particles less than 10 micrometers (called PM10) and more recently for particles less than 2.5 micrometers (called PM2.5). Forest fires produce a majority of particles in the 0.1–0.3 micrometer range, and the proportion of PM2.5 in smoke may be 50–90 percent of the total particulate matter.

Particulate production depends on the type of fuel and how it is burning; flaming combustion emits less particulate matter than smoldering combustion (Table 6.3). Total tonnage consumed per unit area can be estimated from fuel moisture, so that particulate yield per hectare burned can be predicted.

There are three primary control strategies to reduce smoke effects: avoidance, dilution, and emission reduction (Ferry et al. 1985). Avoidance strategies depend on avoiding smoke intrusions into sensitive areas: burning when the wind is blowing away from sensitive areas or burning at high mountainous locations above a layer of stable air below which the smoke will not penetrate. Dilution involves mixing the smoke into larger volumes of air to keep

TABLE 6.3. EMISSION FACTORS FOR PARTICULATE MATTER IN SMOKE

General Fuel Type	*Fire Behavior*	*Particulate Emission* ($g\ kg^{-1}$)
Grass	Flaming dominates	7.5
Understory Vegetation/litter	Flaming, light smoldering[a]	12.5
	Flaming, moderate smoldering[b]	25
	Flaming, heavy smoldering[c]	37.5
Broadcast slash	Flaming dominates	10
	Flaming, smoldering component	20
Piled and windrowed slash	Flaming dominates[d]	12.5
	Flaming, moderate smoldering	25
	Flaming, heavy smoldering[e]	37.5
Brush fuels	Flaming dominates	12.5
	Flaming, moderate smoldering	25
All fuels	Burning where smoldering predominates	75

SOURCE: Ferry et al. 1985.

[a] Backing or heading fires in light fuels with little duff.
[b] Heading fires in litter with heavy loading.
[c] Fires in litter with heavy duff.
[d] Soil free and very dry.
[e] High fuel moisture or heavy mineral soil component.

concentrations low: burning more slowly, burning into thicker stable layers or moderately unstable air, and burning in morning to midday so afternoon winds will blow away the smoke.

The most effective strategy employed recently is emission reduction. The strategy of emission reduction is simply to burn less fuel. Sometimes this is done by carefully weighing the costs and benefits of a burn and deciding that the burn is not necessary. Usually burns can be completed while achieving both site and offsite (air) objectives. Burning under wetter conditions is one way to reduce emissions (Fig. 6.14). For naturally ignited management fires, control over moisture is not always possible, so that the decision criteria include either a *go* (allow the fire to burn) or *no-go* (suppress it) condition on a daily basis. For prescribed fires, air quality concerns can be and usually are a major prescription variable. Computer software is now available to help predict biomass consumption and emissions from individual burns as well as landscape-level emission inventories or annual increments (Peterson and Ottmar 1991).

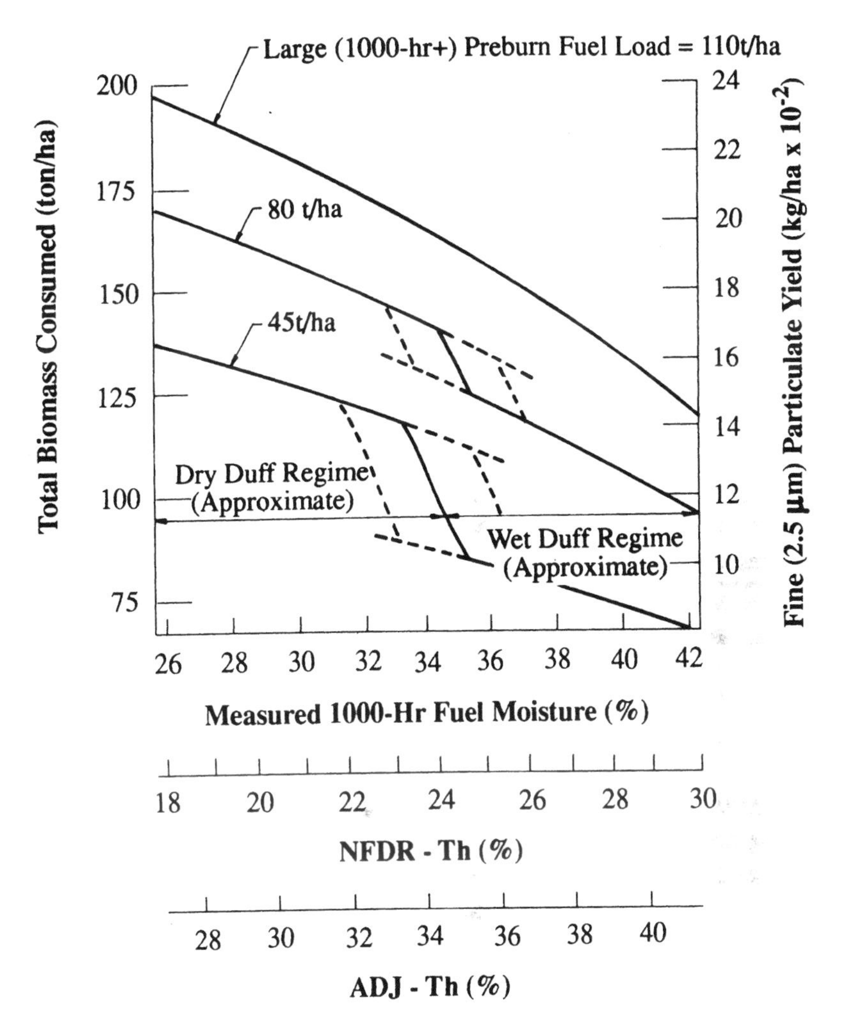

FIG. 6.14. *Nomographs for biomass consumed in broadcast burning. This figure can be used to estimate the total consumed biomass in westside Douglas-fir forests based on the loading of large fuels (>1000 hr + timelag fuels), large-fuel moisture, and duff moisture. NFDR-Th is the 1000-hr fuel moisture estimated from the National Fire-Danger Rating System, and ADJ. Th is a regionally adjusted value for 1000-hr fuels west of the Cascades.*
(From Sandberg and Dost 1990)

THE CULTURAL COMPONENT

Cultural resources of the past are the physical remains of past human occupation. They may include sites, artifacts, objects, structures, art, or altered landscapes. They have scientific significance for researchers attempting to understand past cultures and hold special social or religious value for living people (Lissong and Propper 1990). Cultural resources are often classified by time period. From the prehistoric period (>10,000 ybp) in North America, stone tools are all that survive. The archaic period (10,000–2,500 ybp) is when the modern flora became dominant. Hunting-gathering economies were common, and grinding stones for nuts, plus hunting and butchering tools, survive. Notched spearpoints, as well as both wood and fiber artifacts, can be found from this period. The agricultural period (2,500–400 ybp) is associated with development of crop technology, particularly in the Southwest, with pottery a common artifact. The bow and arrow emerged during this period. The historic period (400 ybp to the present) includes the coming of Europeans to North America and the development of written records. In addition to pottery and stone artifacts, baskets and hides from this era are abundant. This period also includes more recent historical resources, such as mining cabins, corrals, Civilian Conservation Corps developments, and other relatively recent cultural resources.

Fire has a simple impact on cultural resources: the ability to deface or destroy them. In southwestern U.S. wildfires, 67–90 percent of the known cultural resources within the fire boundaries were burned to some extent (Lissoway and Propper 1990). The potential for damage depends on the location of the artifact or resource, as well as the type of fire that occurs and the actions taken to manage or suppress the fire. Indirect impacts include postfire surface or mass erosion and increased trampling from deer and elk on thinner forest floors (Davis et al. 1992).

Fires also may expose previously hidden and undiscovered sites. In the Southwest, surveys after wildfires have shown about one new site per 12 ha. After the 1987 northern California Klamath fires, about 300 new sites were discovered, or about one per 400 ha.

The discovery of new sites has both benefits and costs. Among the costs are direct site damage because of fire management activities or the fire, and vandalism (Traylor et al. 1979). Vandalism sometimes occurs during the fire if the site is discovered by fire crews, or it may occur later as more professional raiders scour the area. Any ground disturbance by fire crews can damage sites by displacing artifacts from proper vertical context in the soil, and artifacts may be destroyed by bulldozer operation or the use of hand tools (Switzer 1974). Some fire retardants can stain and possibly corrode historic structures.

The direct effects of fire depend on the duration and magnitude of temperature increase and can range from no damage to total destruction. Heating can cause ignition of architectural wooden resources or spalling of stone walls and fences. In the Southwest, sandstone building stones are more susceptible to damage than basalt stones (Lissoway and Propper 1990). A significant loss of a wooden structure was the burning of the historic willow corral on the Malheur National Wildlife Refuge, Oregon, by an escaped prescribed fire in early 1992.

Stone artifacts, or lithics, may be damaged depending on the type of stone. Flints and cherts may show bound water loss at >500°C, while quartz shows fewer changes and none below 500°C. Heating of obsidian between 170°C and 350°C causes tinting of the hydration rind (outer shell) from gray to violet, with obliteration of the rind at temperatures over 425°C (Trembour 1979). Prehistoric ceramics are normally structurally safe at temperatures up to 600°C. Above that temperature, older pottery may begin to melt. Organic coating, such as paint, may be oxidized (Switzer 1974) or turn black at temperatures >300°C. Organic deposits from smoke may coat pottery (Seabloom et al. 1991), but these can usually be washed off. Glass artifacts are usually safe up to 500°C. Organic artifacts, such as bones, fibers, and hides, are the most sensitive. Fibers and hides can completely combust, and bones char at 400°C. Pollens and grains can be thermally degraded above 300°C.

Given these temperature thresholds, and the information noted earlier in this chapter about soil heating from fires (Figs. 6.1–6.3), it is clear that surface artifacts are more at risk than buried artifacts from typical wildland fire temperatures (Connor and Cannon 1989). Grass fires are usually less damaging than forest fires where logs and duff can burn or smolder for hours or days. High temperature pulses normally occur for only a short time, however, unless the site is being pile burned. Organic artifacts are most at risk, as much due to age and exposure as to current danger from fire. Buried ones have probably long decayed, so surface artifacts will be all that remain. In wet environments with long fire-return intervals, even surface organic materials have likely decayed. In drier environments, there is a high probability that prehistoric fires in woodland or ponderosa pine forests already crossed these sites many times, and the probability of intact organic artifacts is correspondingly low.

Mitigation of fire effects on cultural resources should include more resource inventory and ground checking of potential sites, especially after fires have occurred. Known sites should be worked into presuppression planning, although specific map locations of sites should not be circulated. An archeologist should be consulted for site clearances before prescribed burning. Before or during fires, some material may be salvageable, while other resources, such as rock art, may be covered with metal shields. Archeologists can even be invited to lead the flagging of fireline to be constructed by hand or bulldozer. Although many cultural resources have survived to the present because of their durability, there is still a need to protect the vestiges of past cultures as far as possible in our current management activities.

CHAPTER 7

SITKA SPRUCE, COAST REDWOOD, AND WESTERN HEMLOCK FORESTS

ALONG THE NORTHWESTERN COAST of North America stand the world's most productive forests. These dense, massive, evergreen forests grow to large size and great age (Waring and Franklin 1979). Three of these low elevation, coastal forest types are considered in this chapter, based on geography (Fig. 7.1) and the presence of western hemlock (*Tsuga heterophylla*) in most plant associations in the three forest types: the *Picea sitchensis, Sequoia sempervirens,* and *Tsuga heterophylla* types. An increase in fire activity in each type is associated with codominance by Douglas-fir (*Pseudotsuga menziesii*).

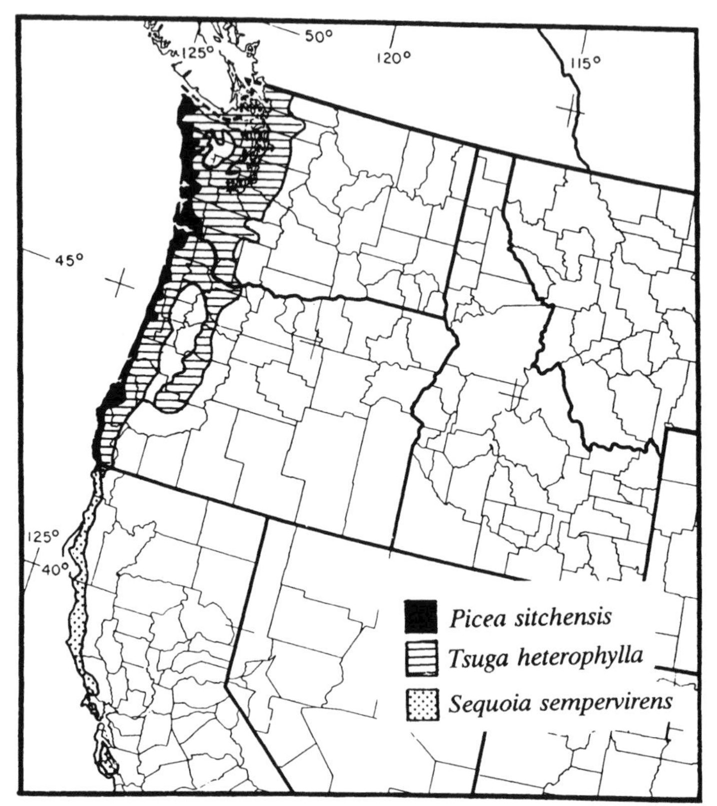

FIG. 7.1. *General locations of the* Picea sitchensis, Sequoia sempervirens, *and* Tsuga heterophylla *zones in the Pacific Northwest. Some scattered stands of* Sequoia sempervirens *along the central California coast and of* Tsuga heterophylla *east of the Cascades are omitted.*
(Adapted from Little 1971)

PICEA SITCHENSIS FORESTS

The *Picea sitchensis* zone is a thin band of coastal forest that extends from southeast Alaska to northern California. It normally exists along the coast adjacent to the more inland *Tsuga heterophylla* type, and is characterized by the presence of Sitka spruce (*Picea sitchensis*).

This forest zone is normally only a few kilometers in width except where it extends up river valleys (Franklin and Dyrness 1973, Fonda 1974).

Environmental Characteristics

Picea sitchensis forests are considered to have the mildest climate of any northwestern forest type, with 200–300 cm annual precipitation, frequent summer fogs and low clouds, and mild year-round temperatures (Franklin and Dyrness 1973). Forests within this type are considered to be the most productive in the world (Fujimora 1971).

Mature forests in the *Picea sitchensis* zone are often codominated by Sitka spruce, western hemlock, and western redcedar (*Thuja plicata*). Douglas-fir and true firs (*Abies* spp.) are found as secondary species, and in southern Oregon Port Orford–cedar (*Chamaecyparis lawsoniana*) will be a common associate. Red alder (*Alnus rubra*) may dominate on disturbed sites, and lodgepole pine (*Pinus contorta*) is found in boggy locations. Black cottonwood (*Populus trichocarpa*) is a common riparian associate.

The Fire Regime

Sitka spruce thrives in the near coastal zone where climatic conditions limit the frequency and intensity of naturally occurring fires. *Picea sitchensis* forests are capable of burning, but fire-return intervals are not well defined for the type. A fire-return interval of 1,146 years was estimated for the "Sitka spruce" type in western Washington (Fahnestock and Agee 1983), based on age-class data applied in a negative exponential model. This is not meant to imply a precise return interval but does indicate that *Picea sitchensis* forests burn rarely. In fact, because the age-class data used as input to the model are more likely associated with wind disturbance, the fire-return interval may be even longer. Similar types of age-class data were used in the same model from the northern and southern Oregon coast, resulting in disturbance cycles of 400 years and 200 years, respectively (Andrews and Cowlin 1934;

J. K. Agee, unpublished data). These estimates include some forest in the *Tsuga heterophylla* zone and clearly include events other than fire.

The major species in the *Picea sitchensis* zone (Table 7.1) are all fire *avoiders*, so the typical severity level of fire is high, even though the fires may be of moderate intensity. Western hemlock, spruce, and western redcedar are shallow-rooted, thin-barked, and shade-tolerant. Douglas-fir might be more common in this type if fires opened larger gaps or patches across the landscape.

TABLE 7.1 RELATIVE ADAPTATION TO DISTURBANCE OF MAJOR TREE SPECIES IN COASTAL FORESTS

SPECIES		ADAPTATION TO:		*Shade*	
	Zone[a]	*Fire*	*Flood*[b]	*Tolerance*	*Longevity*
Coast redwood	R	resister/endurer	endurer	high	extreme
Douglas-fir	SRH	resister	avoider	low	high
Western hemlock	SRH	avoider	avoider	high	medium
Sitka spruce	SR	avoider	avoider	medium	medium
Western redcedar	SRH	avoider	avoider	high	high
Tanoak	R	endurer	avoider	high	low
Pacific madrone	RH	endurer	avoider	low	low
Grand fir	R	avoider	avoider	medium	medium
Red alder	SRH	invader	invader	low	low
Black cottonwood	SH	invader	endurer	low	low
Bigleaf maple	SH	endurer	avoider	low	low

[a] Zones are: S = *Picea sitchensis*; H = *Tsuga heterophylla*; R = *Sequoia sempervirens*.
[b] Flooding impact is silt deposition around tree base.

The extent of past fires is largely unknown in this type. There is one legend of a great fire coming to the Olympic coast about 500 ybp (Anonymous 1983), and scattered 500-year-old stands are common in the upland *Tsuga heterophylla* zone of the western Olympic Mountains. Fires in the 1840s are recorded to have driven Native Americans of the Oregon coast into the sea (Morris 1934b). The 1850 forest stand age map of the Oregon coast (Teensma et al. 1991) shows some of the large fires in the Oregon Coast Range apparently burning into the *Picea sitchensis* zone and sometimes to the edge of the sea. The limited data we now possess indicate that fires were very infrequent and burned wide areas under rare conditions when they were driven into the coastal zone by dry winds from the interior.

Fingers of *Picea sitchensis* forests extend up many of the coastal valleys, interfingering with adjacent *Tsuga heterophylla* forests. Several fires within the upland *Tsuga* forest have extinguished themselves upon entering the *Picea* forests. In these river valleys, the spruce forests are related to terrace development from past glacial action (Fonda 1974) and are moist flats. The 1961 Queets fire and 1978 Hoh fire did not burn into the *Picea sitchensis* terrace forests. However, it is the microclimate rather than the forest that acts as a fire barrier, and *Picea sitchensis* forests will burn under unusual weather conditions.

Wind is a more persistent disturbance factor than fire in these coastal forests (Ruth and Harris 1979). Harcombe (1986) found that small-scale wind events have a larger impact on the forest than large-scale blowdowns. Harcombe compared the canopy turnover time of 119 years for small-scale wind events to 384 years for large-scale events (essentially a mean blowdown return interval, or the time for an area equal to the forest area to blow down from small- or large-scale wind events). In the Olympic Mountains, Henderson et al. (1989) suggest that wind is most important in the coastal zone, where Sitka spruce predominates. Substantial blowdowns have occurred in 1979 (the Hood Canal storm), 1962 (the Columbus Day storm), and 1921 (Boyce 1929). Stand-level information from ridgeline stands at the eastern edge of the zone suggests other wind-associated events in the late 1880s and early 1850s (J. K. Agee, unpublished information). Although no single event affected the entire area, a return interval of about 30 years for major wind disturbances appears to be characteristic of the Olympic peninsula coastal zone.

Stand Development Patterns

Fire in a mature *Picea sitchensis* forest will usually be a stand replacement event because of the fire-sensitive nature of the major tree species. However, protected locations may be missed by the fire. In the mid-1800s fires in Oregon, riparian corridors appear to have been missed in many cases, and scattered older trees are found across the landscape (Greene 1982, Quaye 1982). These residual trees can act as a seed source for newly burned areas. If the fire comes late in the fire season and scorches but does not consume crowns, the frequent and abundant cone crops in Sitka spruce and hemlock (Ruth and Berntsen 1955) may reseed the site immediately.

At least two quite different patterns of postfire tree colonization have been documented: little regeneration and immediate regeneration. The former pattern is due to shrub dominance, which can occur either where seed source is lacking because of a crown fire or where sprouting shrubs are abundant in the understory. Dense shrub communities dominated by salmonberry (*Rubus spectabilis*), salal (*Gaultheria shallon*), red huckleberry (*Vaccinium parvifolium*), or vine maple (*Acer circinatum*) may result. Each of these species can sprout after fire; although they may be damaged by a hot fire, they still may have a relative advantage over trees if nearby seed sources are lacking. In a coastal logged site Tappeiner et al. (1991) found that salmonberry had over 100 km ha^{-1} of rhizomes on treeless sites, so if that species is present and soil heating does not kill the rhizomes, early dominance by salmonberry is not surprising. Successional pathways after shrub dominance have not been studied comprehensively. On logged sites dominated by red alder, the more luxuriant understory associated with alder stands may lead to a semi-permanent brushfield (Newton et al. 1968). Alternatively, shade-tolerant conifers may become established. In the Washington and northern Oregon parts of the *Picea sitchensis* zone, Sitka spruce, western hemlock, and western redcedar are possible replacement species, while in southern Oregon Port Orford–cedar may also be a possible replacement species. Spruce may be favored in coastal settings due to its tolerance of salt-laden ocean winds. Shrubfields may also result from reburns (e.g., Munger 1944), but there is little evidence that frequent reburning occurred in the *Picea sitchensis* zone.

The second and apparently more common postfire successional pattern, rapid tree colonization, is a function of seed source presence and less competition from shrubs. After the Nestucca fire at Cascade Head, Oregon, in 1848, 83 percent of the spruce and 68 percent of the larger hemlocks present in the 1980s germinated within a 20 year period (1851–70) (Harcombe 1986). If Douglas-fir is to be a component of the postfire forest, it will become established in this period, since its lack of shade tolerance will prevent its establishment at other times.

Stand development proceeds relatively rapidly (e.g., Oliver 1981) through the stand initiation stage (new tree stems are established) and stem exclusion stage (establishment stops, and existing stems are thinned through competition) because of excellent growing conditions and subsequent disturbance by wind. After the Nestucca fire around 1848, the major period of stand initiation was 1851–70, followed by slower filling of the stand (1871–1900), stem exclusion (1900–1930), then understory tree reinitiation (1931–50, caused by windthrow), and another stem exclusion phase from 1951 to 1981 (Harcombe 1986). After a wind disturbance, existing understory trees may be released in lieu of new stems being recruited into the stand.

A tree or group of trees that blows down may provide both a gap suitable for regeneration and a substrate for tree establishment. Between 88 percent and 97 percent of tree seedlings in mature *Picea sitchensis* forests grow on logs (McKee et al. 1982). Moss is an important competitive influence for tree establishment in these forests (Harmon and Franklin 1989). Logs on the forest floor represent sites where moss competition may be low enough for establishment of western hemlock and Sitka spruce. Occasionally young western hemlock outgrow the Sitka spruce, which may be due to the spruce being attacked by the Sitka spruce weevil (*Pissodes strobi*) (Wright 1971). The leader of young trees is killed, resulting in forked tops rather than tree death. Associated species that are not affected may be favored by the time necessary for spruce to recover its height growth (Greene and Emmingham 1986).

Wind may be the dominant natural disturbance factor in these stands, creating gaps that allow Sitka spruce to codominate these forests. The pattern of repeated recruitment of tree understory is due to gaps opened by wind disturbance, resulting in a multi-age-cohort stand. The gaps range from very small (essentially single-tree replacement) to quite large (almost stand replacement events). The characteristic "pit and mound" topography—hummocky ground caused by uprooted trees (the pit) and later sloughing of soil from the root wad (the mound)—often indicates repeated windthrow of living trees. In a *Picea sitchensis* forest near Astoria, Oregon, buried charcoal was recovered from cut banks and underneath centuries-old trees (Agee 1989b). Of the 21 samples, 8 were western hemlock, 6 were Sitka spruce, 6 were western redcedar, and 1 was a hardwood (possibly red alder). These samples provide clear evidence that the site did burn at one time, but they also suggest that Douglas-fir was uncommon at the time of the fire, since it did not show up in any samples. This was interpreted to mean that the forest at the time of the fire had probably developed with single-tree or small-gap openings favoring the more shade-tolerant species.

FUELS

Little is known about fuel dynamics in *Picea sitchensis* forests. Because of the moist, mild environment, fine fuels decompose rapidly. Rapid decay of dead fuels was a major reason for early resistance to mandatory slash burning in coastal forests (Lamb 1925). Larger fuels are conifer species that tend to decay relatively fast: Sitka spruce and western hemlock logs (exponential decay constant $k = 0.011$) have a higher decay constant than Douglas-fir ($k = 0.006$) (Graham and Cromack 1982). Half-lives of decay for these k values are 70 years for spruce and hemlock and 115 years for Douglas-fir (Harmon et al. 1986).

Log biomass of 122–165 t ha^{-1} was measured in lower and upper terrace forests along the Hoh River, Washington (Graham and Cromack 1982). However, log biomass of 211 t ha^{-1} was found 120 years after a stand replacement fire in a coastal Oregon *Picea sitchensis* forest (Grier 1978), considerably more than the 150 t ha^{-1} found in a Douglas-fir/hemlock forest 110 years after a stand replacement fire in the Olympic Mountains (Agee and Huff 1987). These levels might only represent site differences but probably also reflect frequent input of coarse woody debris from windthrown trees in *Picea sitchensis* forests; Grier (1978) estimated about 20 t ha^{-1} $decade^{-1}$. The total biomass in both forests was about the same, so less live biomass was present in the *Picea sitchensis* forest due to this thinning effect of wind.

Management Implications

Fire is an infrequent visitor to *Picea sitchensis* forests. There is little use for prescribed fire in management of *Picea sitchensis* natural areas. Lightning is uncommon, and under most conditions a wildfire could not be ignited. There is no documented "fire exclusion" effect, or changes in fire behavior expected because of decades of fire suppression activities. If *Picea sitchensis* forests do burn, they are likely to do so only under severe fire weather in late summer and fall, driven by strong east wind (Randall 1937). Smoke and fire will generally spread in a westerly direction, and fires will likely enter these forests from adjacent *Tsuga heterophylla* zone forests. Fire plays a minor role in the development of *Picea sitchensis* zone forests. Although natural fires have burned and will burn in such forests, their absence for hundreds of years will not create significant "unnatural" impacts in *Picea sitchensis* forests. Wind is a more important disturbance factor in these forests, and like fire it operates at varying frequencies and intensities, significantly affecting forest structure and species composition.

SEQUOIA SEMPERVIRENS FORESTS

Coast redwood (*Sequoia sempervirens*), as its name implies, occupies a coastal strip along northern California, with scattered groves south of San Francisco and a limited distribution into Oregon (Fig. 7.1). This species grows with a wide variety of other conifers and broad-leaved trees, and reaches ages of over 2,000 years and heights exceeding 110 m (Lindquist 1974). These are the tallest trees in the world.

Environmental Characteristics

Sequoia sempervirens forests exist on the moist western end of a steep moisture gradient, and the redwood belt is seldom more than 40 km wide along the transition to the inland mixed-evergreen forest of northern California. Annual precipitation varies from 65 cm to 310 cm (Fowells 1965), and summer fog is common. The fog belt in summer is a fairly consistent indicator of the interior range of the species, since redwood has poor control over transpiration (Daniel 1942). Coast redwood is found from sea level to about 1,000 m elevation, with the best stand development below 700 m. The most superlative *Sequoia sempervirens* forests are on alluvial flats, where periodic flooding in the past has removed its competitors and where the redwood is capable of generating new root systems up into the deposited silt (Stone and Vasey 1968). Cool, moist microclimates have favored large accumulations of biomass, with >3,000 t ha^{-1} measured in one alluvial flat stand (Fujimori 1977).

The Fire Regime

Sequoia sempervirens forests have a moderate severity fire regime, as the dominant vegetation is able to resist the effects of all but the most intense wildfires. The origins of the fires are not well understood. Lightning tends to be uncommon, as with most coastal locations (Sibley, unpublished data, cited in Veirs 1982). Native Americans were known to burn the upland "balds," or prairies (Drucker 1937, Lewis 1973), and they may have contributed to many of the documented fires of at least the last millennium. *Sequoia sempervirens* forests near the edge of prairies may have burned quite frequently (possibly annually) and consequently at low intensity. Critical fire weather is concentrated in July, August, and September; about 70 percent of recent large fires in the two most northwestern California counties have burned in August and September (Gripp 1976).

Fire-return intervals for *Sequoia sempervirens* forests are quite variable. On moist coastal sites at the northern end of the range, fire-return intervals may be as long as 500–600 years (Veirs 1980). A variety of fire-return intervals are possible on intermediate sites, down to intervals of 33–50 years on interior sites (Veirs 1980). Other investigators working to the south where precipitation is lower have found shorter presettlement fire-return intervals. Fire-return intervals based on point frequencies were 31 years in a fire-scarred interior *Sequoia* forest at Humboldt Redwoods State Park

(Stuart 1987), 20–29 years at Salt Point State Park farther south (Finney and Martin 1989), and 22–27 years near Muir Woods National Monument, close to San Francisco (Jacobs et al. 1985). Since point frequencies conservatively estimate fire-return intervals, these data imply a pattern of fairly frequent presettlement fire throughout many of the *Sequoia sempervirens* forests before European settlement.

The events appear to have included fires of low and moderate severity, with rare high severity fires evidenced by scattered "ghost forests" of charred, undecayed redwood snags. Low intensity fires of 1880 and 1974 were documented by Veirs (1980). Four "severe" fires per century over the last millennium were apparent to Fritz (1932) in a 12 ha coast redwood stand near the Eel River in Humboldt County, but a reanalysis of the data (Veirs 1982) suggests an all-aged stand without major regeneration pulses, and therefore less evidence for very severe fires. Fires burning in ericaceous shrub understories, such as evergreen huckleberry (*Vaccinium ovatum*) or western rhododendron (*Rhododendron macrophyllum*), may locally be very flashy because of the waxy nature of the leaves, in contrast to alluvial flats with a sword fern (*Polystichum munitum*) or oxalis (*Oxalis oregana*) understory.

The extent of past fires in *Sequoia sempervirens* forests is not well documented. Most studies of fire history have concentrated on defining point frequencies rather than areal extent. Stuart (1987) attempted to define fire size in a 3,500 ha inland coast redwood study area and found mean fire sizes of 918–1,748 ha, 1,097–2,018 ha, and 786–1,629 ha for conservative and liberal estimates of the postsettlement (>1898), settlement (1875–97), and presettlement (<1875) eras. Given the large mean fire size relative to the size of the study area, these mean sizes probably underestimate true mean fire sizes, which more likely overlapped the study area. An index to the local variability in fire microclimate is provided by Pillers and Stuart (in review), who measured relative humidity on slope and valley forests for coastal (Prairie Creek) and inland (Humboldt Redwoods) sites. These data can be converted to equilibrium fuel moisture content for 1 hr timelag fuels (Table 7.2). The higher number of days with lower fine-fuel moisture on upland sites and interior forests in contrast to valley and coastal forests is apparent and is consistent with shorter fire-return interval data for the former locations.

TABLE 7.2. ESTIMATED DURATION (NUMBER OF DAYS) OF LOW FINE-FUEL MOISTURE CONTENT IN COAST REDWOOD FORESTS

Fuel Moisture of 1 Hr Timelag Fuels	PRAIRIE CREEK		HUMBOLDT REDWOODS	
	Flat	*Slope*	*Flat*	*Slope*
At or below 9%	0	27	46	172
At or below 7%	0	0	2	37

SOURCE: Adapted from relative humidity information in Pillers and Stuart (in review).

NOTE: Prairie Creek is a coastal location, while Humboldt Redwoods is an inland river valley location.

Wind is the other major disturbance factor in *Sequoia sempervirens* forests. In coastal Humboldt County, recent average timber losses from windthrow exceeded the combined losses from fire, insects, and disease (Oswald 1968); the 1962 Columbus Day storm caused extensive windthrow of individual trees and tree groups in *Sequoia sempervirens* forests. Wind-flagging of trees is common along the coastal rivers of the redwood belt (Zinke 1988).

STAND DEVELOPMENT PATTERNS

The major species in *Sequoia sempervirens* forests include coast redwood, Douglas-fir, western hemlock, and tanoak (*Lithocarpus densiflorus*). Other associated species include grand fir (*Abies grandis*), Pacific dogwood (*Cornus nuttallii*) and California bay (*Umbellularia californica*), Sitka spruce on moist sites, red alder on moist and disturbed sites, and Pacific madrone (*Arbutus menziesii*) and chinquapin (*Castanopsis chrysophylla*) on drier sites. The autecological characteristics of the species (Table 7.1) help in understanding stand development patterns in the presence of periodic disturbance. Three disturbance scenarios are presented to span the range of fire interactions with stand development.

Light Fire through Mature Upland Stand

The first scenario is a light fire (<300 kW m^{-1}, flame length <1 m) through a mature upland stand. The stand basal area may be up to 150 m^2 ha^{-1}, with about 50 percent in coast redwood, 35 percent in

Douglas-fir, and 15 percent in other species. The low intensity fire kills few overstory redwood or Douglas-fir, but most hemlocks and tanoaks and grand fir are killed, along with thin-barked young trees of all species. Top-killed redwood (Finney 1991), tanoak, and madrone and many understory shrubs will resprout from basal burls. Few, if any, overstory gaps are created, so that only shade-tolerant species will establish from seed. Little Douglas-fir will invade the site, while tanoak and western hemlock, and occasionally grand fir, will seed in again. This pattern is shown by the reconstruction of an 1880 low intensity fire (Veirs 1980), where little mortality in redwood or Douglas-fir occurred and at age 100, 1,500 stems ha^{-1} of western hemlock that came in after the 1880 fire were still present. In a 1974 fire, Veirs (1980) found similar overstory effects, with topkill of tanoak. The regeneration of roughly 2,900 trees ha^{-1} after six years was about 40 percent redwood, 40 percent hemlock, and 20 percent Douglas-fir (much of which may be suppressed and soon die). Coast redwood seedlings are susceptible to damping-off and root rot fungi (Davidson 1971), and successful establishment is associated with mineral soil after disturbance from fire and flood. In closed stands, low levels of regeneration will be found on logs and root wads.

Moderate Severity Fire through Mature Upland Stand

The second scenario is a higher average intensity fire (600 kW m^{-1}, flame length 1.5 m) in the same stand conditions. Some gaps are created by torching of the crowns over limited areas (perhaps 1 ha). Hemlock and true fir are almost all killed by the fire. Douglas-firs that are crown scorched die, because thick bark does not effectively resist crown damage from scorch. Coast redwood that is scorched will resprout a new crown, known as a fire column, along all portions of stems that were scorched but where heat was not sufficient to kill buds beneath the bark. A mosaic of regeneration patterns is likely: a pattern similar to that described for the low intensity fire in areas that lightly burn, and in the larger gaps where fire was more intense at least partial dominance by a cohort of Douglas-fir. Douglas-fir grows rapidly in these higher light levels, and its pulses of establishment serve as a marker for the patch of more intense fire that opened its growing space.

An example of this second scenario is Veirs's (1982) Plot 12 (Fig. 7.2). Coast redwood of all ages is found, while Douglas-fir is represented by about five cohorts, and western hemlock and grand fir are represented only in the most recent postfire cohort (note that tanoak is not represented on the graph although present on this site).

Moderate Severity Fire in Moist Alluvial Flat

A third scenario is provided by assuming a fire similar to the second scenario but in a moist alluvial flat. These flats burn occasionally, but fire scars may be covered over time as the elevation of the flat increases with overbank depositions of silt. One tree on the Tall Trees Flat on Redwood Creek, Humboldt County, has a fire scar that extends 2 m below the current ground surface. In this scenario, the hemlock and grand fir are killed either by fire or as root platforms are buried by sediment. Because of low light levels, Douglas-fir rarely establishes itself or grows to large size, leaving the coast redwood as a sole conifer dominant with a hardwood understory of tanoak and bay, all of which sprout. Redwood seedlings may carpet

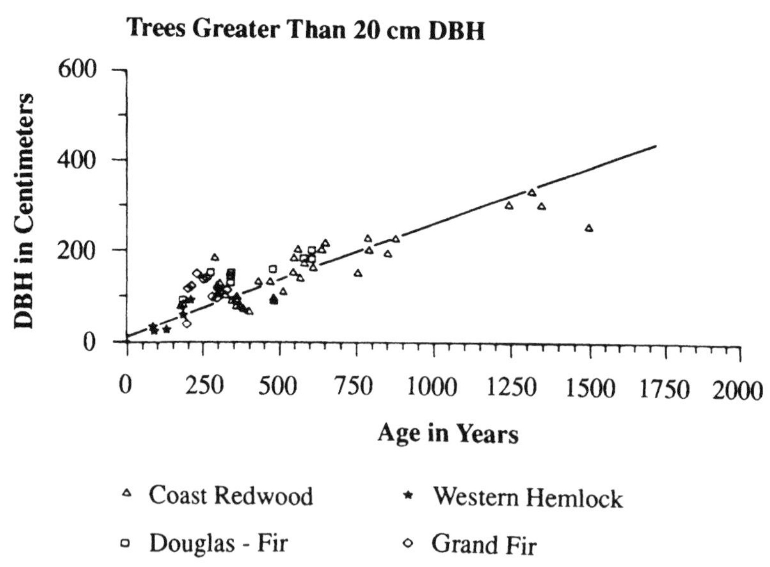

FIG. 7.2. *Diameter and age of trees greater than 20 cm dbh in a low elevation, interior redwood stand.*
(From Veirs 1982)

the deposited silt by the next year (Zinke 1981), but most do not survive; Douglas-fir has insufficient light, and the other conifers appear to be outcompeted by other understory plants. Moist conditions and disturbances favor good growth in pure stands of redwood (Fig. 7.3) and create spectacularly large specimens, including most of the tallest trees in the world. Periodic fires on the flats have effects mostly on understory trees, although fire scars on the larger trees may occasionally enlarge (Finney 1991) and sometimes cause immediate or delayed windthrow.

The three scenarios described here may occur in the same fire event, since a ridgetop fire might move downslope and upslope during day and night, at windy and calm periods, and in wet and dry periods during late summer and fall.

Because coast redwood appears to be favored by disturbance, it has sometimes been called a fire-dependent species (Cooper 1965, Stone and Vasey 1968). Yet age-class data (Fig. 7.4; Fritz 1929) suggest an all-aged condition. Without fire, redwood can become established on downed logs or mineral soil, and once established it can withstand damage to the stem and shading. Even as an understory tree it may sprout from a basal burl. Once a canopy opening occurs understory redwoods, which may be 100–150 years old, will emerge as canopy dominants and may live for another 1,000 years. Veirs (1982) concludes that upland coast redwood forests are a climax type codominated by coast redwood, western hemlock, and tanoak, and that coast redwood is not a fire-dependent species.

Fuels

Coast redwood has long been favored in building construction because of its resistance to decay. It is relatively free of volatile oils and resins, making it somewhat fire-resistant (Lindquist 1974). These qualities suggest that in some sites, coarse woody debris accumulations may be substantial. However, generally frequent fire-return intervals in these forests allow multiple fires to consume large redwood logs incrementally. If the first fires do not result in total log consumption, later fires complete the process, so upland sites should not have unusually high loads of coarse woody debris. Riparian areas, being more protected from fire, are expected to main-

FIG. 7.3. *Interior of a coastal redwood stand at the edge of an alluvial flat. Char at the base of the stem of the large redwood tree is partially obscured by understory vegetation.*

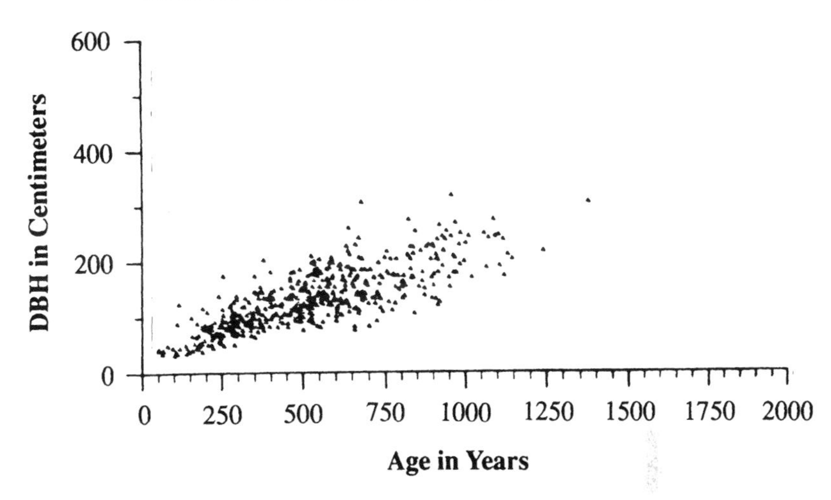

FIG. 7.4. *Diameter and age of coast redwood trees in a low elevation, interior redwood forest. Some smaller redwood and all other species are omitted from the graph. An all-aged distribution is evident.*
(From Veirs 1982, after Fritz 1929)

tain larger inventories of coarse woody debris. Harmon et al. (1986) reported riparian accumulations of coarse woody debris in *Sequoia* forests of up to 1,800 t ha^{-1}, about two or three times the maximum reported in other temperate forests. Coarse woody debris biomass outside riparian areas seems relatively similar to that of other northwestern moist coniferous forests. Finney (1991) found only 10 t ha^{-1} in a young-growth sprout forest at Annadel State Park, while 280 t ha^{-1} was found in another young-growth forest at Humboldt Redwoods State Park, and 200 t ha^{-1} of coarse woody debris was found in an upland old-growth forest (Bingham and Sawyer 1988).

Litter and duff loading is a complex balance of input and decomposition. Values in old-growth forest of 14.3–22.3 t ha^{-1} (Pillers and Stuart in review) were a result of average annual inputs of 3.1–4.7 t ha^{-1} and instantaneous decomposition constants of 0.14–0.26 based on mass balance analysis. The relatively short time required to reach 95 percent of the eventual equilibrium mass of forest floor varied among old-growth sites from 11 to 21 years. Others have shown more mass in the forest floor: the 16 sites examined by

Finney (1991) measured 30–60 t ha^{-1}, and Greenlee (1983) reported 186 t ha^{-1} in the Santa Cruz Mountains. The reasons for such wide ranges are not clear, but they likely represent differential decomposition and disturbance history.

Management Implications

Sequoia sempervirens nature preserves are small, with substantial edge effect. This is true even for Redwood National Park and is the result of the coastal nature of the species. Flexibility in fire management is therefore reduced. Natural fires, which appear to have burned for weeks to months in past years (Stuart 1987), are not capable of being contained under all fire weather conditions. The ignition source for many "natural" fires of the past is not yet clear. Although prescribed fire may be easily used in oak woodlands on balds within the coast redwood type, its application is not as simple within closed redwood forest. The more intense end of "natural" burning conditions is difficult to mimic in a controlled fire. Douglas-fir seems much more fire-dependent than coast redwood in this forest type. Redwood will regenerate and persist without fire. Yet without stand openings caused by more intense fires, which were clearly a part of the stand development patterns of the past, Douglas-fir will slowly disappear as a canopy dominant in the old-growth forests.

In young-growth forests that have been logged, the opposite problem occurs in the short run: excessive Douglas-fir in the seeded or planted young stands, and very high total tree densities. In nature preserves established after logging of old-growth redwood, this cohort of the Douglas-fir will persist for centuries. It may be necessary to lower the proportion of Douglas-fir to redwood by thinning, planting, and adjusting structure (using such techniques as variable-height pruning) in order to provide a mimic of the composition and architecture of young natural stands. Prescribed fire may be a possible tool but may be less desirable in young stands than manual/mechanical methods.

Another constraint is the political turmoil likely to be caused by any plans to disturb forests for whose preservation a substantial public investment was made. If preservation of redwood is the only management criterion, the inclusion of fire in future management may be less important than if the other tree species in the ecosystem are considered. Clearly, innovative management strategies for fire will be necessary in noncommercial *Sequoia sempervirens* forests.

TSUGA HETEROPHYLLA FORESTS

Tsuga heterophylla forests are quite varied in character but have western hemlock in common as the late successional dominant. Western hemlock may be an early seral invader on moist sites but may be absent or suppressed on other sites, occasionally for centuries. Much of the area covered by the *Tsuga heterophylla* series is called the Douglas-fir region because of the dominance of Douglas-fir, but Douglas-fir is a late successional or climax species over only a small portion of the Douglas-fir region and over none of the *Tsuga heterophylla* series, where western hemlock is the climax dominant. The dominance of Douglas-fir in the *Tsuga heterophylla* zone at the time of European settlement was largely due to disturbance, primarily by fire, for many centuries before such settlement. Pollen records establish that Douglas-fir dominance in past millennia often coincided with charcoal peaks in the pollen profile (Brubaker 1991).

Environmental Characteristics

Tsuga heterophylla forests are widely distributed in the lowlands of western Oregon and Washington. The area has a mild, maritime climate with greater extremes of moisture and temperature than the environment of adjacent *Picea sitchensis* forests (Franklin and Dyrness 1973). Summers are relatively dry, and the resulting moisture stress is reflected in the spectrum of plant associations across this zone (Zobel et al. 1976). *Tsuga heterophylla* forests abut a variety of other Pacific Northwest forest types. On the moist coastal edge *Picea sitchensis* forests are found, while *Abies amabilis* forests occur at higher elevation. To the south, drier valleys have *Pseudotsuga menziesii* as a climax dominant. Each of these adjacent types has a quite different fire regime, and *Tsuga heterophylla* forests reflect parts of those adjacent fire regimes in ecotonal areas.

In *Tsuga heterophylla* forests, Douglas-fir is the most fire-adapted of the common trees (Flint 1925, Starker 1934). Recognition of its superior ability, relative to other species, to adapt to the presence of fire is critical to understanding the ecological impacts of disturbance (e.g., Rowe 1981, Wright and Bailey 1982; Table 7.1). The success of Douglas-fir in comparison to its competitors depends on which others are part of the stand. Where western hemlock and western redcedar are the only other species in a mature stand, and a fire kills all the trees but does not consume all the organic mat (which would encourage the invader red alder along with Douglas-fir), Douglas-fir is likely to be a dominant in the new stand.

Listing a species as an *avoider, endurer,* or *resister* treats both species and fire as constants. Yet young Douglas-fir is an *avoider,* having thin bark and a low crown; it becomes a *resister* only when mature. In addition, the frequency or intensity of fire can create different ecological effects. Over the *Tsuga heterophylla* zone, there is considerable variability in the age of stands that burn, as well as in fire frequency, intensity, and extent, which creates a variety of postfire effects.

The Fire Regime

Anthropogenic ignitions are known to have been important in certain forest and grassland types of the Pacific Northwest. However, west of the Cascades in the *Tsuga heterophylla* zone, the role of aboriginal burning is much less clear. Native Americans apparently burned the valley bottoms of at least some of the major tributaries of the Willamette River (Teensma 1987). Some of the earliest evidence for Native American burning in *Tsuga heterophylla* forests was collected by Morris (1934), who first documented accounts of large historical fires in the Pacific Northwest. Coastal Oregon tribes were the victims of some of these fires, having been driven to the Pacific Ocean to survive (Morris 1934b). Whether lightning or humans were the source of the fires is not known. Oregon tribes of the northern coast were reported to burn Neahkahnie Mountain and the hills near Bay City every spring to stimulate browse and attract deer and elk (Sauter and Johnson 1974). It is reasonable to assume that some human-caused prairie fires must have burned adjacent *Tsuga heterophylla* forest, and some must have burned intensely and extensively during east wind events in late summer and autumn. At present, the case for widespread aboriginal fires throughout *Tsuga heterophylla* forests is not convincing.

The variability in regional patterns of lightning ignition is illustrated by the application of a fire cycle model based on climate (Agee and Flewelling 1983), which generates fire ignitions and sizes based on climatic parameters of the site. Only the ignition portion of the model is regionally applicable, since the size (and fire cycle) portion of the model is based on empirical relationships between local fire size and climate in the western Olympic Mountains.

The model requires four inputs, each used in 1 of 12 10-day periods during the fire season. These are the probability of long-term drought, of a rain exceeding 0.25 cm, of a thunderstorm occurrence, and of an east wind event. The model was applied to three sites in the *Tsuga heterophylla* zone: the western Olympic Mountains, the Wind River area in southern Washington, and the McKenzie River vicinity in the central Oregon Cascades (Fig. 7.5A). The probability of long-term drought, defined as below-average annual precipitation, was set at 0.5, with additional cumulative years of drought at lower probabilities (Agee and Flewelling 1983). Short-term drought was defined as (1 - the probability of significant precipitation) during 10-day summer periods and was gathered for each area based on records from nearby stations (Munger 1925). Thunderstorm probability was determined from Pickford et al. (1980) for the Olympics and from Morris (1932) for the other sites.

As for any stochastic model, long-term expected averages are best calculated using many model repetitions; in this case, each simulation is based on averages from a 10,000 year run of the model. This represents not the Holocene but only a long-term average (or expected value) of the effects of portions of twentieth-century weather on ignition patterns. Results are relative rather than absolute, although the model did predict proportional seasonal distribution of Olympic fires well (Fig. 7.5B).

There is a pattern of increasing probability of ignitions exceeding 1 ha in size as location shifts east and south from the western Olympics (Fig. 7.5C). The Wind River area has twice as many igni-

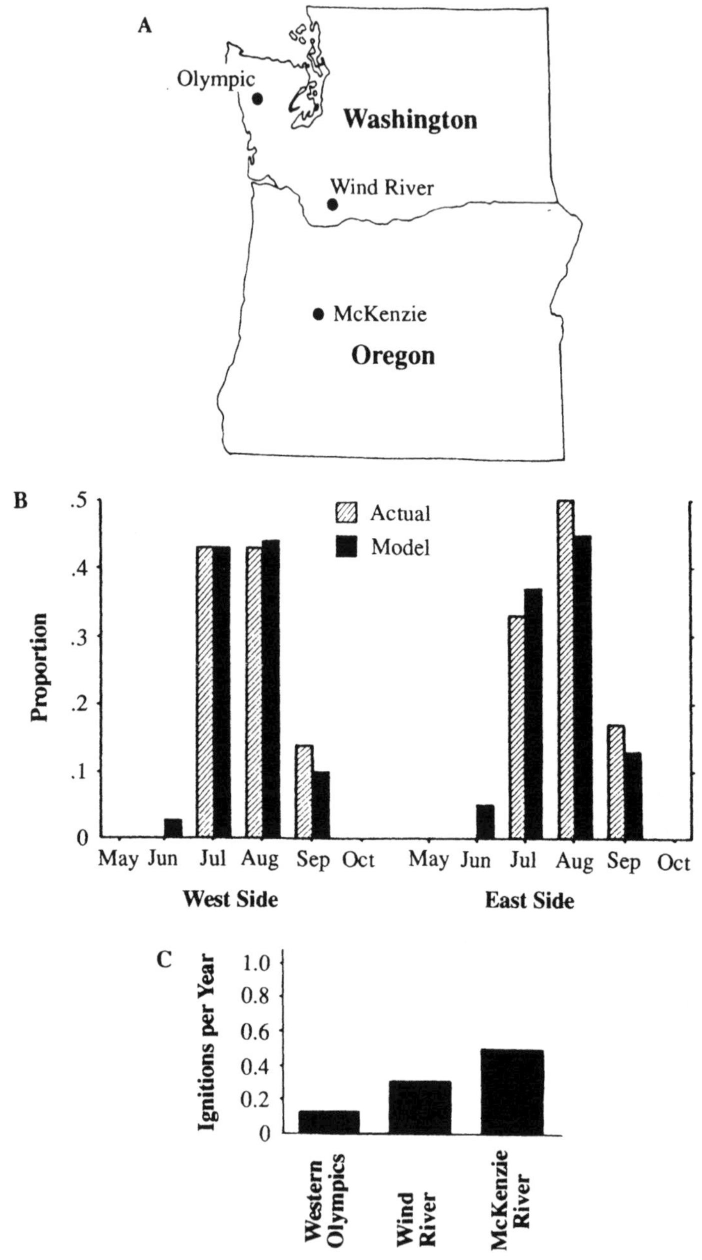

FIG. 7.5. A: *Location of the sites used to simulate lightning ignitions in the fire cycle model.* B: *Seasonal distribution of simulated and actual* (*twentieth-century*) *lightning ignitions for the western and eastern Olympics.* C: *Simulated lightning ignitions for the regional locations shown using the fire cycle model.*
(From Agee and Flewelling 1983, Agee 1991a)

tions as the western Olympics, and the McKenzie River experiences about 60 percent more than Wind River. The differences are largely due to increased lightning frequency and decreasing summer precipitation patterns from northern Washington to central Oregon. In each area, July and August are the months of greatest ignition activity, but September has a higher proportion of total ignitions in the McKenzie area than in the two northerly areas. The model, which was initially designed for the Olympic Mountains, does not include October ignitions, which would likely increase the total ignitions to the south even more. The patterns of ignition clearly support the hypothesis of higher fire activity toward the southern end of the *Tsuga heterophylla* zone.

A regional average fire-return interval for the Douglas-fir zone has been estimated at 230 years, based on an analysis of 1930s forest survey records (Fahnestock and Agee 1983). There is considerable spatial variability in this estimate, as suggested by the fire cycle model. Significant temporal variability is also characteristic of these fire-return intervals, so that the 230 year average is of limited utility as a parameter of fire frequency across the region. The notion of a fire cycle, or a return interval of regular frequency, is not as meaningful as in drier forest types, where with some measure of variability a roughly cyclic occurrence of fire can be assumed (for example, mixed-conifer forest [McNeil and Zobel 1980]). Rarely in a *Tsuga heterophylla* forest is the fire record long enough or regular enough to infer a cyclic pattern of fire, particularly in the presence of climatic shifts that would alter any "cycle" in operation. *Episodic* is a better descriptor of fire-return intervals in these forests.

In the moist *Tsuga heterophylla* forests of the Coast Range of Oregon, the Washington Cascades, and the Olympics, most forests are first-generation postfire forests less than 750 years old. This would suggest a fire-return interval less than 750 years. The fire cycle model of Agee and Flewelling (1983) could not reproduce a natural fire rotation (essentially a fire cycle) of less than 3,500 years using twentieth-century climate patterns, and even with significant alteration in climate input to the model, fire-return intervals could only be brought down to about 900 years. They suggested that perhaps events much larger than average may have occurred in the past (also suggested by Henderson and Peter 1981, for the southeastern Olympics) as a result of short-term but extreme changes in two or more of the climate parameters that drive the model. Additional ignitions by Native Americans possibly explain the shorter fire-return intervals seen on the landscape.

Our knowledge of dates of establishment of old-growth forest is so weak as to preclude firm hypotheses about disturbance pulses of the past. A series of stands with an origin of ca. A.D. 1230 has been identified in the southern and western Olympics (Agee, personal observation). A series of stands of ca. 1330 and/or ca. 1480–1530 origin is apparent from the data of Henderson and Peter (1981) in the southern Olympics, Franklin and Hemstrom (1981) and Yamaguchi (1986) in the southern Washington Cascades, and Huff (1984) in the western Olympics. Many stands in the eastern Olympics date to 1670–1700 (Fonda and Bliss 1969, Henderson et al. 1989). Although the forest age-class data are sparse, these are also times of sunspot minima identified by Stuiver and Quay (1980), using tree-ring analysis of carbon-14 activity. If large fire events are associated with these periods of general global cooling, they may represent periods when altered synoptic weather patterns, particularly during the growing season, contained higher lightning frequency and foehn (east) wind patterns, as well as altered precipitation regimes.

Recent work suggests a higher fire frequency in the drier *Tsuga heterophylla* forests of the Oregon Cascades, reinforcing the output of the fire cycle model (Fig. 7.5C). A site transitional to the *Abies amabilis* zone in the western Oregon Cascades (Stewart 1986) near the H. J. Andrews Experimental Forest has experienced four fires since ca. 1530. Some of these were in the settlement period and probably reflect settler-caused fires of that period, which were used to create and maintain clearings. Over a broader area several kilometers to the southeast in the *Tsuga heterophylla* zone, Morrison and Swanson (1990) suggest a natural fire rotation (see chapter 4) of 95–145 years over the last five centuries, well below that of the moist Douglas-fir forests of Washington. Another fire frequency analysis was completed by Teensma (1987) near the area studied by Morrison and Swanson. Using conservative methods that recognized only fires that resulted in substantial regeneration or fire scarring, Teensma estimated a natural fire rotation of 100 years over the last five centuries. Other stands 500 years old or older exist without much evidence of fire. These studies strongly indicate that a variable fire regime with much higher frequency than the typical moist Washington *Tsuga heterophylla* forest occurs in the central Oregon Cascades and in other mesic to dry *Tsuga heterophylla* forests.

Fire intensity is often high under conditions that allow fire spread in moist *Tsuga heterophylla* forests. Extensive stand-destroying crown fires in old-growth forest have occurred, as in the Tillamook fire (Pyne 1982) or other historic accounts (Morris 1934b), but extensive mortality can occur from fires that scorch the crown but do not consume it (Agee and Huff 1987). Evidence of fires in 100-year-old stands during the late 1400s (Henderson and Peter 1981) suggests either that independent crown fire behavior occurs in this thick-canopied stage or that an extended shrub stage prolonged the period of accentuated flammability in these forests. Toward the southern part of the *Tsuga heterophylla* zone, fire severity becomes patchier. Stewart's (1986) site near the H. J. Andrews Experimental Forest experienced three partial mortality fires after a stand replacement event ca. 1530. The patchiness of these fires is illustrated by a fire severity map from Morrison and Swanson (1990; Fig. 7.6). A similar fire pattern was noted by Means (1982) on dry sites in the western Oregon Cascades. Working in the H. J. Andrews area, Teensma (1987) calculated a mean fire-return interval for stand

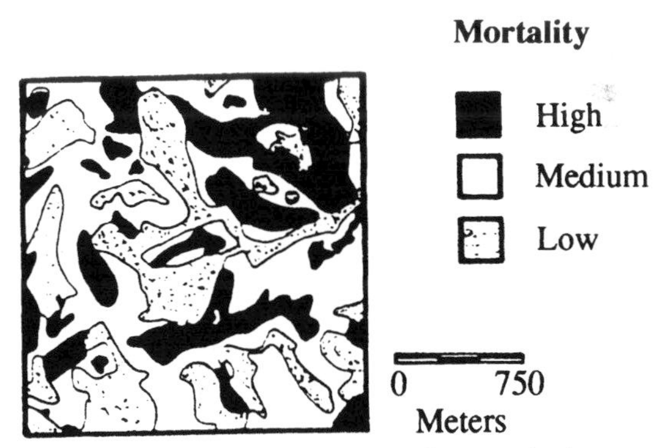

FIG. 7.6. *Reconstructed fire severity for a site in the central western Cascades. More area in this particular fire burned with moderate to low severity than high severity, but high severity fire remains an important disturbance process in the mesic to dry portion of Douglas-fir forests.*
(From Morrison and Swanson 1990)

replacement fires of 130–150 years, about 50 percent longer than the fire-return interval that included moderate severity fire (both estimates excluded low intensity fires). A moderate severity fire regime is characteristic of the southern portion of the *Tsuga heterophylla* zone.

There are few regional data to permit comparison of patterns of fire extent in *Tsuga heterophylla* forests. Modern fire suppression and timber-harvesting activities have largely prevented fires from burning free in the forest as they once did. The pattern of fire spread in the Olympic Mountains typically features relatively rapid spread during periods of east wind events until the onset of significant precipitation. Large fires are clearly part of the historic record (e.g., 1902 Yacolt fire, 1933 Tillamook fire), spreading over several hundred thousand hectares, and Henderson et al. (1989) have reconstructed a large fire event covering much of the eastern Olympics about A.D. 1700. Although longer droughts and longer burning fires have likely occurred in the past, the typical fire of this century has burned for a few days before being naturally suppressed by onshore marine airflow and precipitation. Although there are no long-term records in any area of the region, the lower probability of precipitation toward the south probably allows fires to "die down" without being extinguished during periods of low winds or moderate weather, so that they remain capable of renewed spread under patterns of windy or warmer weather.

Stand Development Patterns

In moist *Tsuga heterophylla* forests, long early seral tree recruitment (e.g., 75–100 years for Douglas-fir) has been documented after disturbance by fire (Franklin and Hemstrom 1981). This pattern is not characteristic of all prehistoric fires. For example, Huff (1984) shows a 60 year recruitment interval after a fire of A.D. 1465 in the western Olympics, while Yamaguchi (1986) shows that about 95 percent of Douglas-fir was recruited within 40 years after a fire in about 1300 near Mount St. Helens. Even on these sites, however, the reestablishment period is decades long and probably represents some regeneration from trees that initially colonized the burn and grew large enough to produce viable seed to help completely restock the stand.

Lack of seed source, brush competition, and/or reburns have been identified as factors delaying regeneration after stand-replacing fires (Franklin and Hemstrom 1981). Early season crown fires, or crown scorch fires in poor seed years, may be associated with a lack of early regeneration after a fire. Late season fires, even if they are crown fires, occur when seed is mature, and Douglas-fir seed in cones can survive crown fire events (Hofmann 1924). Patterns of reburns on the Tillamook fire of 1933 at six year intervals (1939, 1945, 1951; Pyne 1982), at Mount Rainier in the late nineteenth century, and at the southern Washington Yacolt burn of 1902 (Gray 1990) are evidence that these sites will reburn. Herbs and shrubs, once established, may extend the stand initiation stage for decades. High surface fire potential during early succession in Douglas-fir forest was identified by Isaac (1940) as a "vicious cycle" of positive feedback, encouraging rhizomatous bracken fern (*Pteridium aquilinum*). Given sufficient sources for reignition (the original Yacolt and Tillamook burns and all reburns are thought to have been human-ignited), the reburn hypothesis is likely to be true in certain areas. It is not clear, however, whether reburns were a common event before European settlement in moist *Tsuga heterophylla* forests.

The process of tree regeneration after fire was debated extensively in the 1920s. At the time, cutover areas were burned again and again by escaped slash fires (Lamb 1925), destroying natural regeneration. One of the arguments against slash burning was that seed of Douglas-fir remained viable in the forest floor for more than two years (Hofmann 1917a, 1924) and that burning depleted this supply. By the 1930s, further research determined that seed did not remain viable but was blowing in from residual trees or neighboring stands; the forest floor was not a source for Douglas-fir seed (Isaac 1935). Isaac (1943) found that Douglas-fir regenerated well when partially shaded but without significant root competition and that severely burned areas were not ideal spots for Douglas-fir regeneration.

Tree regeneration may be immediate if the fire intensity is such that it remains on the surface or in the understory, leaving current-year canopy seed intact (Agee and Huff 1980). In the first year after the 1978 Hoh fire, seedling density was 73,000 ha^{-1}, with about 80 percent western hemlock. By year 2, about 1,500 of the first-year seedlings were left, and 75 percent of these were Douglas-fir (Huff 1984). Another 20,000 seedlings ha^{-1} established themselves in year 2, with about 75 percent western hemlock. By year 3, herbaceous competition was significant, and total seedling density was about 8,400 ha^{-1}, with roughly 60 percent Douglas-fir. As crown closure occurred on older fires, almost all the small trees were western hemlock. Fire is an important process opening growing space for shade-intolerant Douglas-fir. Where Douglas-fir is a seral species, it is a common practice to date disturbance events by Douglas-fir age classes (Agee 1990).

In stand replacement fires, a relay floristics model of succession was once assumed typical (Munger 1940). The stand would be initially dominated by Douglas-fir, and after 100 years or so hemlock would begin to enter the understory. This may be a pattern on drier sites in the *Tsuga heterophylla* zone, but on mesic to wet sites an initial floristics model is more common, in which hemlock and Douglas-fir colonize together (Huff 1984). Because hemlock grows more slowly in open conditions, it may appear to be younger than the Douglas-fir, but small hemlocks or western redcedars may in fact be the same age as larger Douglas-fir (Stubblefield and Oliver 1978).

If the fire is less than a stand replacement event, sufficient overstory may remain to encourage regeneration of western hemlock and western redcedar but not Douglas-fir (Hofmann 1924). Larger Douglas-fir are resistant to fire because of thick bark, so a ground fire may only eliminate thin-barked or young trees and understory vegetation and consume litter and duff. Hofmann (1924) documented the effect of heat on Douglas-fir in laboratory experiments, and his empirical results closely resemble the model results of Peterson and Ryan (1986). As long as crown scorch is not severe, older trees may survive even two fires in a 10–20 year period (Hofmann 1924). A pattern of low intensity fire having little overstory effect has been locally noted in moist *Tsuga heterophylla* forest in the Skokomish and Hoh River valleys. It more commonly occurs in the drier southern portion of the *Tsuga heterophylla* zone.

Understory successional patterns after fire are a function of disturbance intensity and the life history traits of the species involved, a combination of deterministic and stochastic processes. Many of the species in these forests are persistent after disturbance (Huff 1984, Halpern and Franklin 1990). Halpern (1989) studied 20 years of succession after logging and burning, and although his sites were seeded or planted with Douglas-fir, the trends are still applicable to natural forest succession. He divided the early successional species into two broad species groups, invading and residual (Table 7.3; Fig. 7.7). Fugitive annual species (I1) peaked first in abundance and were very prominent on burned microsites. A single perennial colonizer (I2, *Epilobium angustifolium*) had rapid establishment and expansion; it persisted as a major herb species although percent cover began to decline after year 6. Wind-dispersed composites (I3) were a minor component with maximum cover in years 3–5. Other minor herbs and shrubs (I4) reached maximum cover in years 3–9 and had extended bell-shaped curves for cover over time. Species reproducing from buried seed (I5) were represented by the tall shrubs, redstem ceanothus (*Ceanothus sanguineus*) and snowbrush (*C. velutinus*). Both species had variable patterns over space and time, representing different burn intensities (favored by "heavy" burns) and possibly different stand histories. Invading perennials (I6) with numerous sources of origin, had slowly increasing constancy and cover over time. Some of these species spread through deeply buried rhizomes or were dispersed by animals.

Residual species, those present before disturbance, were separated into five groups. Trailing blackberry (R1, *Rubus ursinus*) had increasing constancy over the study period and stable cover after year 5. It would likely persist until crown closure. Some species (R2) were temporarily released by burning: Starflower (*Trientalis latifolia*) persisted by means of short, thickened tubers; trail plant (*Hieracium albiflorum*) recruited through wind-borne seed; and whipple vine (*Whipplea modesta*), a trailing subshrub that roots at its nodes, was least common on heavily burned sites. Predisturbance dominants (R3) showed population reductions after fire but were generally recovering through vegetative regeneration. A diverse group of subordinate forest species showed either less impact of disturbance with minor long-term changes in cover (R4) or else little recovery after disturbance (R5). Similar patterns were documented for herb and shrub frequency by Huff (1984) in a *Tsuga heterophylla*/*Polystichum munitum* chronosequence.

TABLE 7.3. CHARACTERISTICS OF ELEVEN SERAL SPECIES GROUPS IN THE *TSUGA HETEROPHYLLA* ZONE OF THE CENTRAL OREGON CASCADES

Species Group	*Growth Forms*	*Common Species*
I1	H	*Senecio sylvaticus, Epilobium paniculatum, Conzya canadensis*
I2	H	*Epilobium angustifolium*
I3	H	*Agoseris* spp., *Cirsium* spp., *Gnaphalium microcephalum, Lactuca serriola*
I4	H,S	*Anaphalis margaritacea, Rubus leucodermis, Collomia heterophylla, Vicia americana, Bromus* spp.
I5	S	*Ceanothus velutinus, Ceanothus sanguineus*
I6	H,S,T	*Pteridium aquilinum, Rubus parviflorus, Salix scouleriana, Prunus emarginata*
R1	H	*Rubus ursinus*
R2	H	*Trientalis latifolia, Whipplea modesta, Hieracium albiflorum*
R3	H,S,T	*Acer circinatum, Polystichum munitum, Gaultheria shallon, Rhododendron macrophyllum, Berberis nervosa, Corylus cornuta, Tsuga heterophylla*
R4	H,S,T	*Coptis laciniata, Viola sempervirens, Vaccinium parvifolium, Castanopsis chrysophylla, Oxalis oregana, Rubus nivalis, Acer macrophyllum, Cornus nuttallii*
R5	H,T,	*Chimaphila umbellata, Thuja plicata, Goodyera oblongifolia, Synthyris reniformis, Taxus brevifolia*

SOURCE: Halpern 1989.

NOTE: Species groups: I = invaders, R = residuals. Growth forms: H = herbs and low shrubs, S = tall shrubs, T = understory trees.

Understory response after fire may be partly deterministic, as these patterns suggest. Patterns for some species, however, varied by site, disturbance intensity, predisturbance population conditions, postdisturbance propagule dispersal, and local weather. At the local scale, then, stochastic processes may also be operating in postfire understory dynamics. The vegetative recovery of R3 species, those characteristic of the preburn forest, suggests that over longer time periods the process of recovery may be relatively deterministic for these *Tsuga heterophylla* forests.

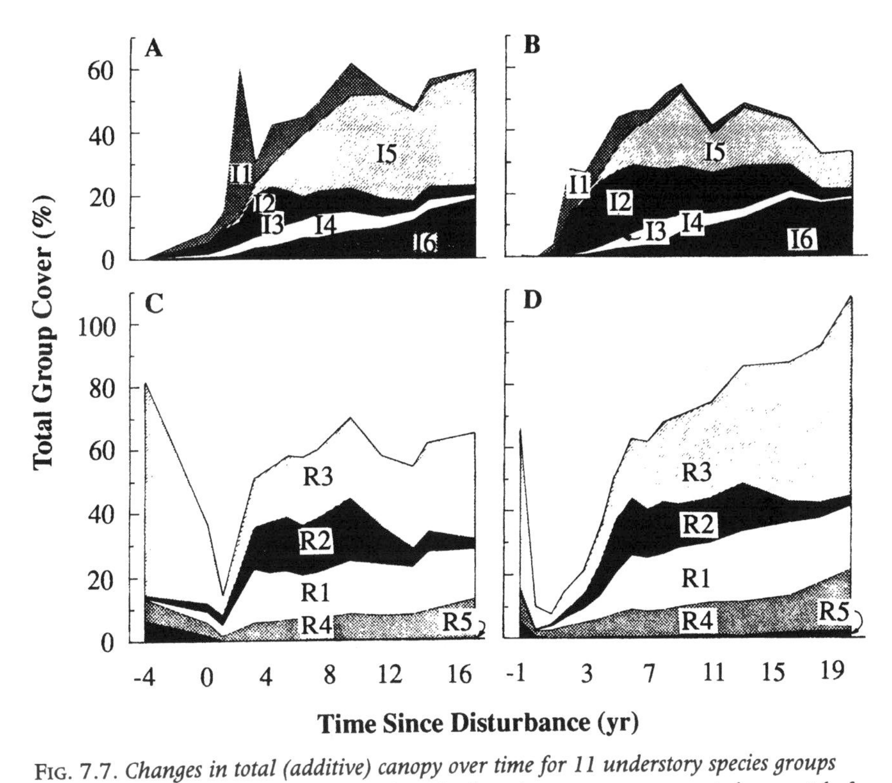

FIG. 7.7. *Changes in total (additive) canopy over time for 11 understory species groups after disturbance by fire. Invading and residual species groups are presented separately for clarity.* A,B: *Groups of invading species on study sites WS-1, WS-3.* C,D: *Groups of residual species on study sites WS-1, WS-3.*
(From C. B. Halpern, "Early Successional Patterns of Forest Species: Interactions of Life History Traits and Disturbance," Ecology 70 [1989]: 704–20. Copyright 1989 by Ecological Society of America. Reprinted by permission.)

As stands enter the stem exclusion stage, overstory tree density begins to decline. In a *Tsuga heterophylla/Polystichum munitum* chronosequence, Huff (1984) measured declines in tree density from 700 ha^{-1} in a 110-year-old stand to 450 ha^{-1} in a 515-year-old stand. The proportion of western hemlock increased from about 66 percent to about 95 percent. However, the proportion of the 85 m^2 ha^{-1} basal area dominated by Douglas-fir decreased much less (from 60 to 40 percent), indicating fewer but much larger Douglas-fir in the old-growth stage. Distributions of tree diameters show a bell-shaped curve for Douglas-fir that becomes narrower with age, while a negative exponential curve for western hemlock becomes extended over time as regeneration takes place in the understory reinitiation and old-growth stages. Projecting successional patterns past the typical lifespan of Douglas-fir results in a stand dominated almost exclusively by western hemlock (Dale et al. 1986). This transition to climax western hemlock forest has rarely occurred, due to the probability of fire returning to these forests before the transition is complete.

Huff (1984) has summarized the species response to disturbance regimes for a range of wet to dry *Tsuga heterophylla* zone forests (Fig. 7.8). If fire is absent for 700–1,000 years on wet sites, Douglas-fir will drop out of the stand, and western hemlock will be the primary seed source for postfire regeneration. On sites with fire-return intervals in the 300–600 year range, well within the longevity of individual Douglas-fir, mixed dominance of Douglas-fir and western hemlock or Pacific silver fir will result from a typically severe stand replacement fire—the *fire-resilient* community concept proposed by Fonda (1979). A stand development sequence will occur as illustrated in Figures 7.9 and 7.10.

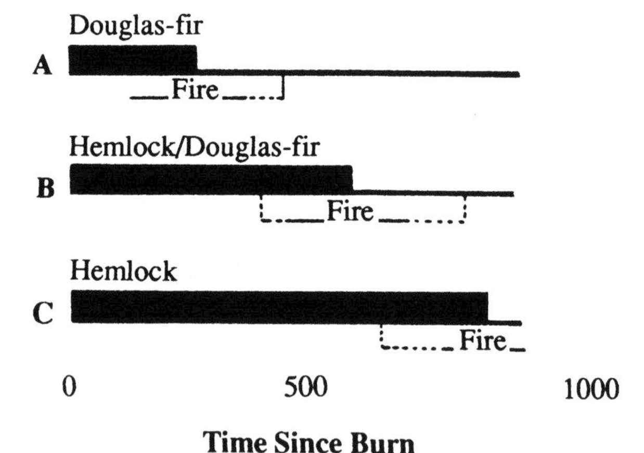

FIG. 7.8. *Species responses to changes in fire frequency. These fire regimes represent fire-return intervals of about 100+ years* (A) *to about 1,000+ years* (C). *The solid/dashed line underneath each thick line represents the average range of fire-return intervals. The species name is the one likely to dominate after disturbance.*
(From Huff 1984)

FIG. 7.9. *Photographic sequence of stands typical of a chronosequence after fire in the western Washington* Tsuga heterophylla *zone. Numbers represent postfire age of stand.*

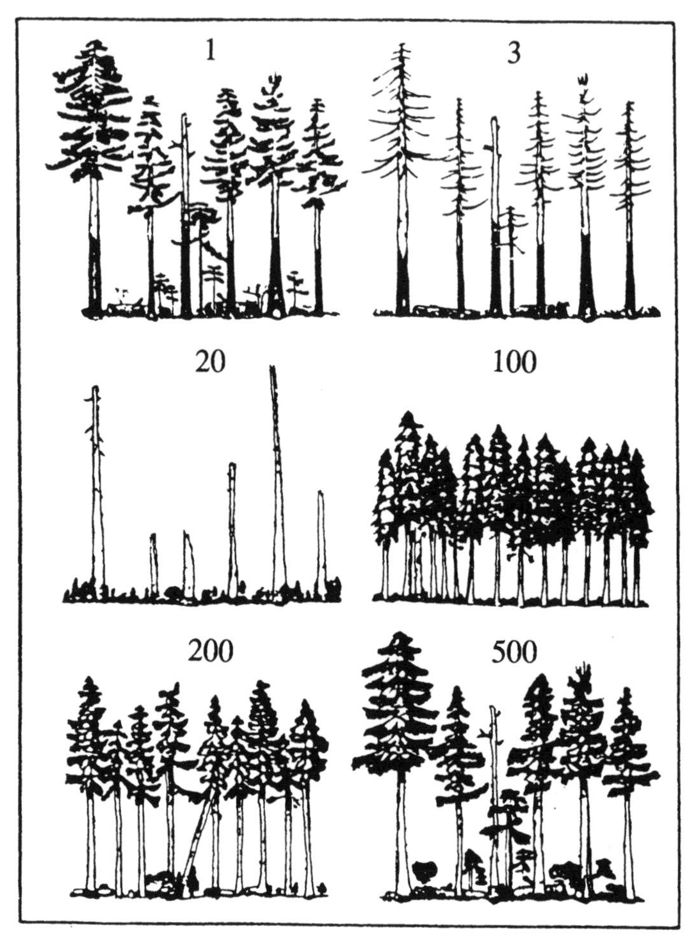

FIG. 7.10. *Stand development sequence for wet Douglas-fir forests, illustrating the creation of old-growth character. Fire-return interval is 500 years or more, and the old-growth character is found after about 200 years. Even after 500 years, the old-growth character remains a function of the previous disturbance, which initiated the Douglas-fir.* (Adapted from Huff 1984)

FUELS

Changes in biomass of fine and coarse woody debris over time are a function of predisturbance biomass, the frequency and severity of the disturbance, and successional patterns after disturbance. Fires in *Tsuga heterophylla* zone forests may consume 0–10 percent of stems and branches, 20–100 percent of foliage, 20–75 percent of understory vegetation, 20–30 percent of snags and down logs, and 60–80 percent of the forest floor (Fahnestock and Agee 1983), with considerable variation around these figures.

Postfire fuel dynamics can be characterized by considering the various size classes of fuel. The fine fuels (1, 10, and 100 hr timelag size classes) generally peak in the early stages of succession, particularly after stand replacement events that scorch but do not consume crown fuels (Agee and Huff 1987). Herbaceous fuels, such as bracken fern, may add considerable fine fuel before tree crown closure occurs (Isaac 1940). The 1,000 hr timelag size class (7–20 cm diameter) peaks in mid-succession during the self-thinning stem exclusion phase, and the large log biomass peaks in the later successional period (Fig. 7.11; Agee and Huff 1987). Fire behavior mod-

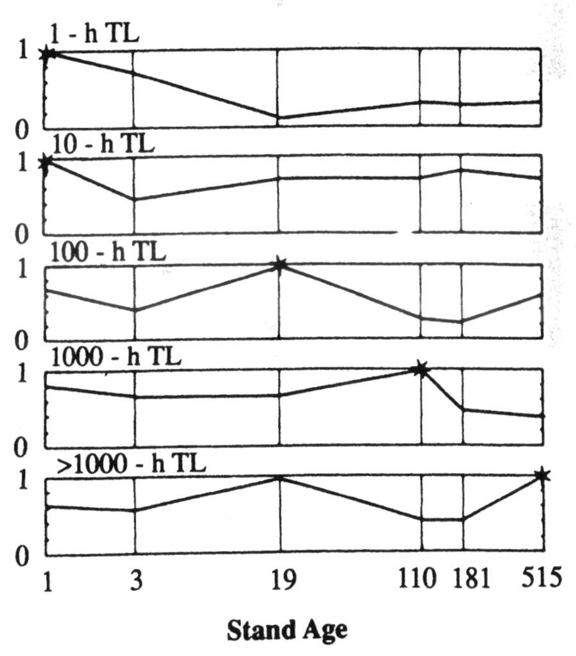

FIG. 7.11. *Normalized fuel loadings by size class for dead and down fuels by stand age. Maximum fuel loading within a size class is scaled to 1.0. Loadings of that size class at other stand ages are shown as proportions of the maximum level.*
(From Agee and Huff 1987)

eled from these changes indicates a peak in surface fire behavior potential during early succession (Fig. 7.12; Agee and Huff 1987). After disturbance, and over time, potential surface fire behavior declines, particularly after crown closure, and then gradually increases in the old-growth seral stage.

Large log biomass trends depend on the intensity and frequency of fire. Three patterns of biomass trends over time were identified by Spies et al. (1988; Fig. 7.13). Pattern A is a stand replacement fire, where the entire prefire stem biomass is added at once to the coarse woody debris (logs and snags) category, as occurred in the Hoh fire (Agee and Huff 1987). Pattern B, where some residuals are left by the fire and slowly die over the next several hundred years, may be locally applicable at the edge of stand replacement burns in moist *Tsuga heterophylla* forests. A similar pattern exists in southern *Tsuga heterophylla* forests, except that fire-return intervals

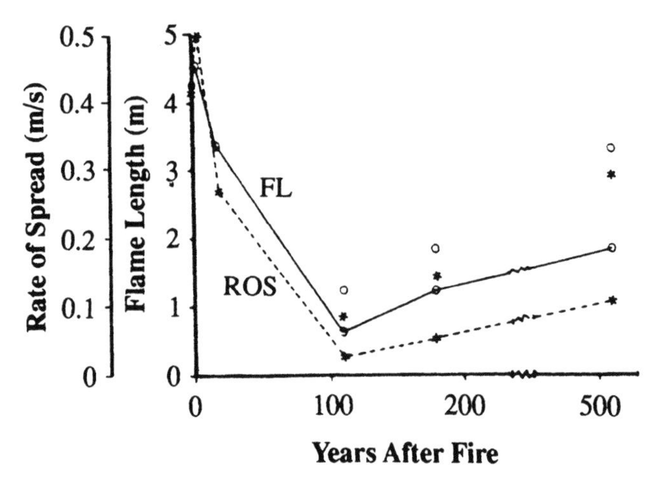

FIG. 7.12. *Rate of spread and flame length by stand age for a sere in the western Olympic Mountains. Wind speed is 268 m/min, and dead fuel moisture contents are 6, 7, and 8 percent for 1 hr, 10 hr, and 100 hr timelag fuels at stand ages 1, 3, and 19. Connected points represent microclimate buffering at stand ages 110, 180, and 515: Wind speed is reduced to 134 m/min, and fuel moisture increases to 10, 11, and 12 percent. Unconnected points at these ages represent results if constant microclimate is assumed across all sites.*
(From Agee and Huff 1987)

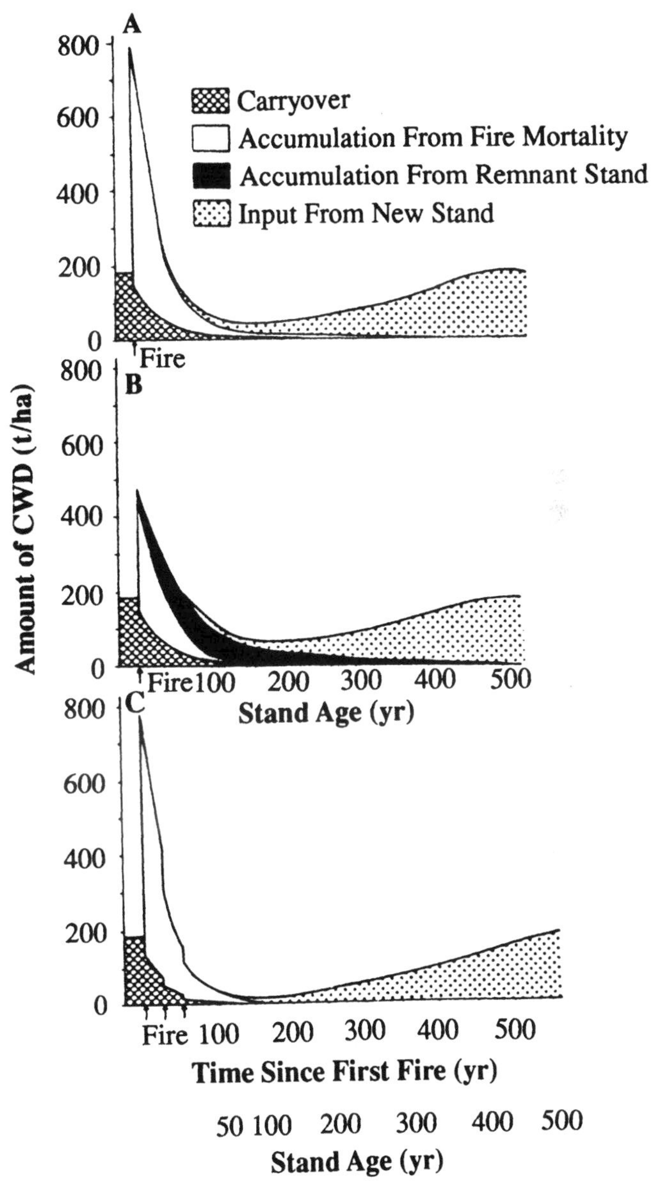

FIG. 7.13. *Predicted changes in mass of coarse woody debris following fire in a 450-year-old* Pseudotsuga-Tsuga *forest:* A: *Catastrophic fire.* B: *Partial burn.* C: *Catastrophic burn and two fires 25 and 50 years later.*
(From Spies et al. 1988)

may be a century or less (Morrison and Swanson 1990), repeating the pattern shown in the graph for the first 100 years. Pattern C, where reburns occur, may be applicable to the Yacolt area of southwestern Washington or the Tillamook fire of northwestern Oregon, where fires appear to be infrequent, but on areas once burned the increased surface fire potential may beget reburns until crown closure is finally able to occur.

Management Implications

The historic role of fire in *Tsuga heterophylla* forests is critical to our understanding of how these forests developed. *Tsuga heterophylla* forests are most typically sandwiched between two zones, the more moist *Picea sitchensis* and cooler *Abies amabilis* forests, that burn less frequently. Experience suggests that the microclimates of these adjacent forests are responsible for the lack of fire activity. Where major topographic barriers exist to the east of such forests, as in the western Olympics and western Cascades, fires burning under east wind conditions will often move upslope-downslope as much as upvalley-downvalley, because local upvalley winds offset the gradient winds. Where topographic barriers are absent or lower—as in the Lake Crescent corridor in the Olympics (see Fig. 2.6), the Columbia Gorge, or portions of the Oregon Coast Range—or where the topographic barrier is to the west, as in the eastern Olympics, a much stronger trend for fire spread to the west is likely. Such fires may stop at forest zone boundaries or under exceptional conditions may burn until fuel is exhausted at water or rock barriers.

The extent to which such patterns and processes provide a blueprint for future management in natural areas is less clear. No natural areas are large enough to allow totally free-ranging fire, and few have management objectives that will allow prescribed natural fire (lightning ignitions allowed to burn under certain conditions in certain zones). Additionally, much of the protected *Tsuga heterophylla* forest is within conservation areas for the northern spotted owl (*Strix occidentalis* spp. *caurina*) and is being managed for old-growth structure to preserve owl habitat.

Historic fire patterns hold several implications for remaining natural stands of *Tsuga heterophylla* forests. Today's forests tend to be more fragmented as a result of timber harvest and the legal boundaries and small size of many preserves. Based on historic patterns of fire, increased fragmentation compared to natural conditions should be more significant to the north, which apparently had larger blocks of single-cohort stands. The increasingly smaller size of old-growth stands will create drier, windier microclimate along stand edges (Chen 1991), accelerating windthrow along edges as well as potential fire behavior.

To suggest that fire is an appropriate process in the preservation of old-growth forests seems paradoxical. Yet it is apparent that almost all the natural forests of the *Tsuga heterophylla* zone are first- or multigeneration stands born of fire. Fire is responsible for their destruction, yet it is also responsible for their creation and maintenance. Without fire, the proportion of Douglas-fir in natural stands will decline, particularly on more mesic sites, and western hemlock will assume a more important role.

CHAPTER 8

PACIFIC SILVER FIR AND RED FIR FORESTS

THE UPPER MONTANE FORESTS of the Pacific Northwest dominated by Pacific silver fir (*Abies amabilis*) or red fir (*Abies magnifica*) are transitional between drier, warmer forest types at lower elevation and moister, colder, subalpine forests at higher elevation. *Abies amabilis* and *Abies magnifica* forests are discontinuous (Fig. 8.1); *Abies amabilis* forests extend from British Columbia south to about 44°N latitude, and *Abies magnifica* forests begin at about 43°N latitude and extend south into the southern Sierra Nevada (Franklin and Dyrness 1973). Red fir is taxonomically more closely related to noble fir (*Abies procera*) than to Pacific silver fir (Franklin et al. 1978). Their inclusion here in the same chapter is related less to taxonomic similarities than to mid-montane locations. The average elevation of these montane forest zones increases from north to south: 450–1,100 m in British Columbia (Krajina 1965), 900–1,500 m in central Oregon, 1,600–2,000 m in the southern Oregon Cascades, and up to about 2,200 m in the Siskiyou Mountains (Whittaker 1960). In California *Abies magnifica* forests are found at 2,000–2,500 m in

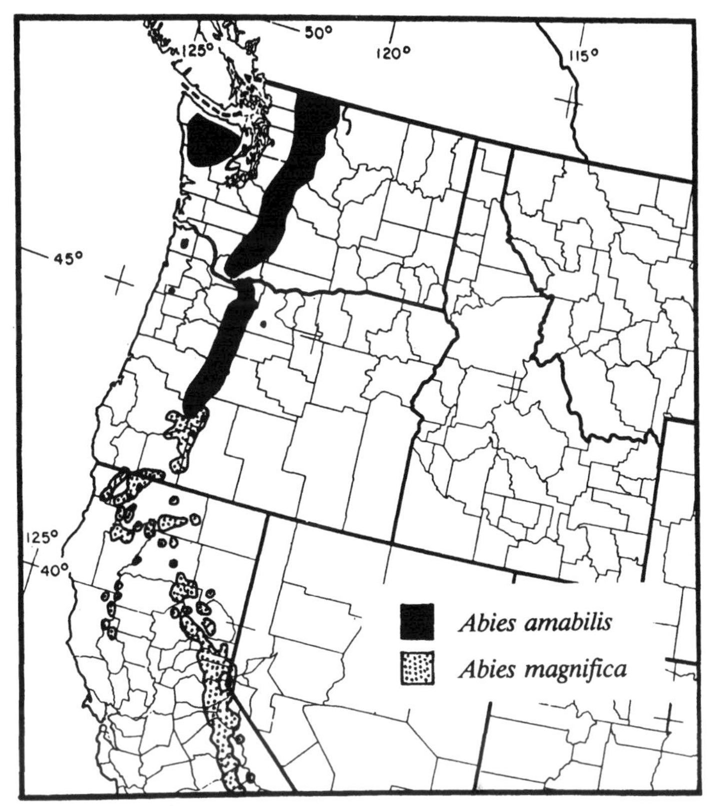

FIG. 8.1. *The range of Pacific silver fir and red fir forests.* (Adapted from Little 1971)

the central Sierra Nevada (Oosting and Billings 1943) and at 2,100–3,050 m in the southern Sierra Nevada (Pitcher 1987).

ABIES AMABILIS FORESTS

Abies amabilis forests, because of their transitional nature, may be dominated or codominated by many of the species found at lower and higher elevation (Table 8.1). At lower elevations Douglas-fir (*Pseudotsuga menziesii*) and western hemlock (*Tsuga heterophylla*) are the most common associates, while at upper elevations mountain

TABLE 8.1. COMMON TREE SPECIES AND SUCCESSIONAL STATUS IN PACIFIC SILVER FIR FORESTS

Tree Species	*Successional Status*
Pacific silver fir	major climax
Subalpine fir	seral
Noble fir	seral
Alaska yellow-cedar	seral
Lodgepole pine	minor seral
Western white pine	seral
Douglas-fir	seral
Western hemlock	minor climax
Mountain hemlock	minor climax

hemlock (*Tsuga mertensiana*) and Alaska yellow-cedar (*Chamaecyparis nootkatensis*) are more often codominant. Where climate is more continental, as in east Cascades forests transitional to the subalpine zone, subalpine fir (*Abies lasiocarpa*) and lodgepole pine (*Pinus contorta*) may codominate (Franklin and Dyrness 1973).

ENVIRONMENTAL CHARACTERISTICS

The cool, temperate climate of *Abies amabilis* forests is characterized by moderate winter temperatures, substantial winter snowpack, and relatively limited summer drought (Packee et al. 1981). Average January temperature in the Olympic Mountain *Abies amabilis* series is about 0°C (Henderson et al. 1989), with a 1–3 m snowpack (Franklin and Dyrness 1973). Pacific silver fir is known to have low drought resistance (Minore 1979), and summer precipitation where this species is found usually exceeds 15 cm.

FIRE REGIMES

The fire regime for *Abies amabilis* forests is characterized by infrequent fires of high severity. Mature Pacific silver fir seldom survives fire because of its thin bark and shallow roots. Because few fire-scarred trees survive, fire-return intervals must be calculated from stand-age analysis. At the moist end of *Abies amabilis* forest environments, fire-return intervals range from 300 to 600 years (Table 8.2), particularly for valley bottom types such as the *Abies amabilis/ Oplopanax horridum* plant association. Lower elevation and drier types may have fire-return intervals of about 100–300 years. Fires are most frequent in eastern North Cascades forests where Pacific silver fir exists only on north aspects and is surrounded by forests with drier and warmer environments. Where natural fire rotations have been determined, the variability of fire-return intervals may be best described by a Poisson distribution (see Fig. 4.16, Mount Rainier).

Fires in *Abies amabilis* forests usually occur under unusual conditions of summer drought and east wind (Agee and Flewelling 1983). They tend to be of high intensity and kill most or all of the trees on the site. Underburned stands at the edge of wildfires have also shown complete mortality. Low intensity prescribed fires in stands of Pacific silver fir and noble fir left 58 percent of the residual

TABLE 8.2. FIRE-RETURN INTERVALS FOR PACIFIC SILVER FIR FORESTS

Community Type	*Location*	*Fire-Return Interval (years)*	*Source*
Silver fir/devil's club	Mount Rainier	535	Hemstrom (1982)
Silver fir/Alaska huckleberry	Mount Rainier	474	Hemstrom (1982)
Silver fir/beargrass	Mount Rainier	323	Hemstrom (1982)
Silver fir/Oregon grape	Mount Rainier	295	Hemstrom (1982)
Western hemlock/ silver fir transition zone	central Oregon Cascades	149	Morrison & Swanson (1990)
Silver fir a primary dominant	east-north Cascades	192	Agee et al. (1990)
Silver fir a secondary dominant	east-north Cascades	108	Agee et al. (1990)
Silver fir/mountain hemlock type	east-north Cascades	137	Agee et al. (1990)

stems dead or severely damaged after one year (Pickford 1981). On drier *Abies amabilis* forest sites, the more fire-tolerant Douglas-fir may be an early seral codominant, and fires may occur often enough (every 100–200 years) that Douglas-fir may survive with either charred bark or a fire scar. In the central western Cascades, Morrison and Swanson (1990) found pre-1800 fires in this forest type to be predominantly stand-replacing, but between 1800 and 1900 only about 25 percent of the area burned was of high severity, with 32 percent in the moderate class and 43 percent in the low class. Although no data showing the overstory tree structure were presented, the surviving species from which fire severities were inferred was predominantly Douglas-fir (P. Morrison, personal communication, 1986). In a nearby forest, mature Pacific silver firs showed growth releases after fire, suggesting they might be unusual fire survivors or were directly adjacent to stems that were killed by the moderate severity fire (Stewart 1986). Early spring crown fires of limited extent may occasionally occur in *Abies amabilis* forests (see chapter 9).

The areal extent of natural wildfires in *Abies amabilis* forests varies considerably (Fig. 8.2). At Mount Rainier National Park, very large fires were reconstructed for the periods ca. A.D. 1230 and ca. 1400, ca. 1500, and ca. 1630, with little area burning at other times. The earliest event was estimated to have burned about 47 percent of the forested area of the park (25,000 ha), while the later events burned about 25 percent each (Hemstrom and Franklin 1982). At the southern end of the range of Pacific silver fir, Morrison and Swanson (1990) found all historic fires to have overlapped their study boundaries, making total area burned difficult to reconstruct and suggesting widespread fire activity. Recent fires since 1800 appeared to be patchier in nature, with areas of low, moderate, and high severity each ranging from <10 to 50–150 ha. This may be because more recent events can be more precisely reconstructed.

The cool, moist conditions in *Abies amabilis* forests compared with forests at lower elevations have acted to dampen and extinguish many historic fires. In the glaciated valleys of Vancouver Island, Schmidt (1960) noted that fires burning in western hemlock forests on steep valley walls did not carry into Pacific silver fir forests on gentler slopes at higher elevation. A similar pattern is evident on the 1961 Queets fire in Olympic National Park. Henderson and Peter

FIG. 8.2. *Two stands of Pacific silver fir (Cayuse Pass Rd., Mount Rainier), showing the infrequent occurrence of fire in these forests. The stand at the left is 750 years old, while the one at the right is 150 years old. They are separated by an avalanche track that may have expanded after the fire of 150 years ago.*

(1981), in reconstructing fires over the last 700 years on the Shelton District of Olympic National Forest, found that three of four large fires in the *Tsuga heterophylla* series stopped roughly at the ecotone with the *Abies amabilis* series. There are likely a few scattered areas where there is no evidence of fire either in the current stands or in the soil (Hines 1971).

STAND DEVELOPMENT PATTERNS

Most *Abies amabilis* forests experience a long fire-free interval between burns. Forest understories are dominated by ericaceous genera, such as *Chimaphila, Pyrola, Gaultheria, Menziesia, Rhododendron,* and the typical dominant *Vaccinium* (Franklin and Dyrness 1973), and all but the first two genera sprout after fire. Many, including thinleaf huckleberry (*Vaccinium membranaceum*), Alaska huckleberry (*Vaccinium alaskaense*), salal (*Gaultheria shallon*), and beargrass

(*Xerophyllum tenax*), can create substantial competition for tree species (Henderson et al. 1989).

The rather gradual change in species composition from the *Tsuga heterophylla* forests to *Abies amabilis* forests, together with the long lives of most coniferous trees in the area, makes classification of plant associations and successional implications quite difficult. Climatic changes, such as the Little Ice Age, have caused the boundary of this zone to shift upslope and downslope over centuries, so that a stand established 300 years ago may have a different successional endpoint now than when it was first established (Henderson et al. 1989). Persistent winter snowpack is associated with the lower boundary of the zone. Persistent snow shortens the summer drought period and also favors Pacific silver fir seedlings, which shed the accumulated litter in the snowpack when it melts much better than western hemlock seedlings (Thornburgh 1969). After a fire or series of fires about 3,600 ybp at Mount Rainier, a shift to cooler and/or wetter conditions (Dunwiddie 1986) was inferred from more subalpine fir and Pacific silver fir, and less noble fir and Douglas-fir, in seral postfire communities.

Although *Abies amabilis* forests are generally cool and moist, postfire environments may be too harsh for immediate colonization by some tree species. The removal of canopy increases any frost-prone tendencies and may also increase daytime maximum temperatures. Microenvironmental studies on clearcuts have shown lethal temperatures exceeding 54°C on both south and north aspects in July and August (Emmingham and Halverson 1981). Sites burned without being logged have the advantage of shade from snags; protected microsites appear necessary for substantial regeneration to occur. In wetter cold-air drainages, Engelmann spruce may be most successful, while noble fir and western white pine will be found on frost-prone mesic to dry areas (Emmingham and Halverson 1981). Noble fir may do well in exposed, partial shade and complete shade, while Pacific silver fir may depend on at least partial shade for establishment on recently disturbed areas (Thornburgh 1969). Douglas-fir may be dominant on warmer, drier sites within the Pacific silver fir zone. Since noble fir, western white pine, and Douglas-fir grow faster than Pacific silver fir and are longer lived, their age is commonly used as a marker of the minimum age of the last major disturbance to the site. However, because of competition from shrubs, the early seral stand initiation stage may last for decades, as shown by the extended establishment of Douglas-fir in the *Abies amabilis* series at Mount Rainier (Hemstrom and Franklin 1982; Fig. 8.3). Thus, the age of early seral dominants, unless sampled intensively, will often underestimate the date of the fire event because of this delayed regeneration pattern.

Mature forest stands in the Pacific silver fir zone are increasingly dominated by Pacific silver fir over time. In western Washington, Henderson (1981) showed early dominance of subalpine fir on an *Abies amabilis/Vaccinium membranaceum* plant association, with Pacific silver fir continually increasing in cover over time. Where *Valeriana sitchensis* is the understory dominant, Henderson (1981) showed no such change in dominance over time, although Pacific silver fir remained present, indicating that the plant series was *Abies lasiocarpa*. At Mount Rainier, Franklin et al. (1988) argued that the plant series for this plant community type should be *Abies amabilis* based on the presence of Pacific silver fir in the understory. This discrepancy may reflect long-term changes in climate that temporarily shift the ecotone between the *Abies lasiocarpa* and *Abies amabilis* zone upslope or downslope. Disturbance may result in a species dominance that reflects the environment directly after the disturbance, while later tree understory dynamics will reflect the shade-tolerant species best suited to the climate during the time an

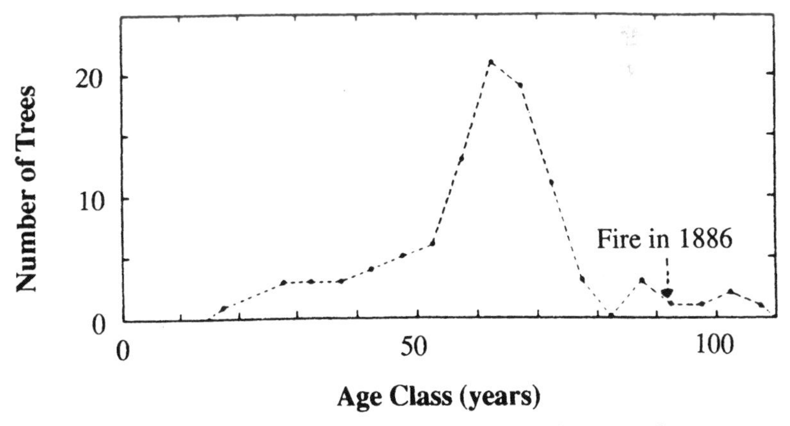

FIG. 8.3. *Recruitment of Douglas-fir after disturbance by fire in an* Abies amabilis *series forest at Mount Rainier National Park, extending over many decades.*
(From Hemstrom and Franklin 1982)

understory develops (e.g., the understory reinitiation stage of Oliver 1981).

After fire in *Abies amabilis* forests, stand basal area steadily increases over time. In stands of *Abies amabilis/Tiarella unifoliata* and *Abies amabilis/Menziesia ferruginea,* the basal area of shorter lived seral species, such as noble fir, declined significantly over several hundred years, while that of longer lived seral species, such as Douglas-fir, slightly increased (Franklin et al. 1988), representing continued growth by the few large Douglas-firs in the stand. Meanwhile, the basal area of Pacific silver fir doubled from about 20 to 40 m^2 ha^{-1}, a significant portion of the total basal area of 90–100 m^2 ha^{-1} in these mature mixed-species stands.

Fire-return intervals in *Abies amabilis* forests are long enough that an understory reinitiation stage (Oliver 1981) may occur before another fire event, at least in the northern part of this forest type. Individual overstory trees may die, opening up growing space on the forest floor. Shrub and herb richness increased from 9 to 16 species as Pacific silver fir stands increased in age from 150 to 350 years in the Washington Cascades (Thornburgh 1969). In the Olympic Mountains an *Abies amabilis* forest initiated after a fire ca. 1480 (Huff 1984) was a natural fire barrier for the 1978 Hoh fire, because of its moist microclimate on a north aspect (Agee and Huff 1980). Much of the current overstory in this forest became established after an event of ca. 1850 (Fig. 8.4), which was probably a windstorm. In the 1880s, another windstorm affected the stand. What has since been identified as bear damage resulted in basal scars on three of the sample trees and many others of that cohort in the 1920–30 period, which may have implications for basal decay and wind damage to those individuals.

The combined effect of wind and fire on Pacific silver fir forests on the northwestern flank of Mount Rainier was studied by Sugita (1990). He used bracken fern (*Pteridium aquilinum*) spores and charcoal in small bogs as an indicator of disturbance. The presence of both fern spores and charcoal was used to infer a fire; bracken spores alone were interpreted as disturbance by wind. By fine separation of core sediments, Sugita dated many fire and wind disturbances over the last 800 years: 1188 (fire), 1266 (wind), 1300 (fire), 1428 (wind), 1630 (fire), 1857 (fire), and 1804–1920 (wind). Not every disturbance affected all of his three closely spaced bog sam-

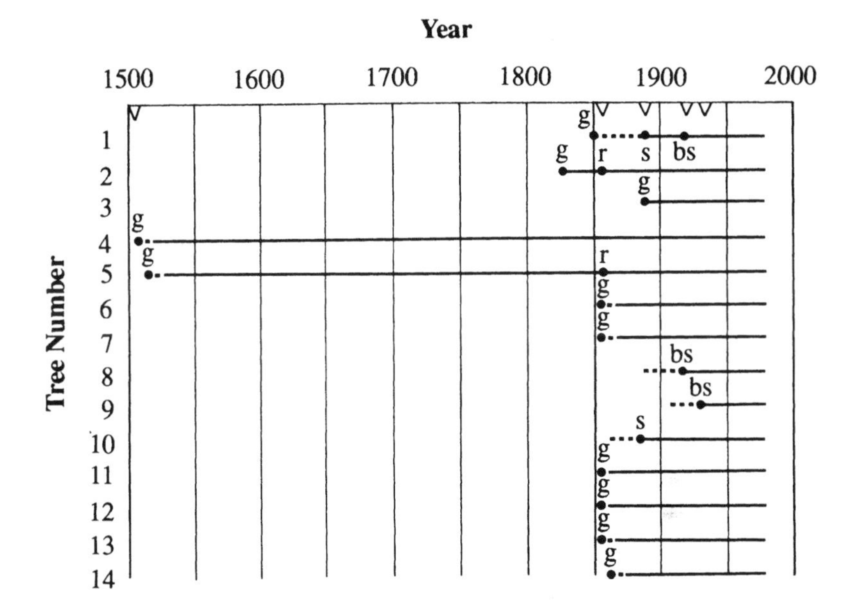

FIG. 8.4. *Patterns of germination and growth release of Pacific silver fir after wind disturbance in the Olympic Mountains.*

ples. Fires drastically reduced coniferous tree pollen, while windstorms showed less significant changes in tree pollen deposition rates, indicating that fire is generally a more severe disturbance to Pacific silver fir forests than is wind.

The effects of fire on forest structure have been documented in two relatively dry *Abies amabilis* forests: Desolation Peak in the North Cascades, where the *Abies amabilis* forest is surrounded by drier forests, and in the central Oregon Cascades, near the southern limit of continuous *Abies amabilis* forests. At Desolation Peak in the North Cascades, near the eastern edge of its local range, Pacific silver fir is found only on protected north aspects. Stands average 20–45 m^2 ha^{-1} basal area, primarily because they are younger stands recovering from past fires (Agee et al. 1990), in contrast to the basal area of 60–70 m^2 ha^{-1} in closed canopy *Abies amabilis* stands to the west (Agee and Kertis 1987).

The lower elevation stands at Desolation Peak burned in 1851 and 1926, and are composed primarily of Douglas-fir and Pacific silver fir in all forest layers. The faster height growth of Douglas-fir has allowed it to emerge as the dominant canopy tree. The stands are still relatively open, and regeneration of both species continues. Over time, Douglas-fir will disappear from the understory, either growing into the upper layer or being suppressed and dying in the lower layers. As a result, Pacific silver fir will be the only conifer in the understory. If past fire-return intervals (Table 8.2) continue, the stand will burn long before Douglas-fir disappears from the overstory. At higher elevation, *Abies amabilis* forests with mountain hemlock have a heterogeneous species composition. Pacific silver fir is the most constant and dominant conifer on these sites, which have burned least frequently of all forest types on Desolation Peak. Understory tree layers are a mix of silver fir, mountain hemlock, and subalpine fir.

The size of an opening created by disturbance may have significant effects on later stand development patterns in drier Pacific silver fir forests. In small gaps created by less severe disturbance, such as a small disease pocket or windthrow, shade-tolerant species will be favored. The architecture of two stands transitional from the *Tsuga heterophylla* series was analyzed on the Willamette National Forest in the central Oregon Cascades (Stewart 1986). In the first, Douglas-fir was dominant, with two cohorts: 450–360 years old and 70–90 years old (1890), both initiated by previous forest fires. Noble fir (*Abies procera*) was the most successful invader after the last fire. Although Pacific silver fir and western hemlock were represented in all size classes, fir had established itself more recently than hemlock. Both of the latter species grew in small gaps opened by the 1890 fire and more recently in gaps opened by windthrow of large old Douglas-fir. In the second stand, which burned in 1860 and 1890, large gaps were present, and regeneration after the 1890 fire was dominated by Douglas-fir, similar to the pattern at Desolation

Peak. The groupings of trees in a stand that die from fire may have a significant impact on the species capable of successfully regenerating after the fire.

Fuels

The stand-replacing nature of most forest fires in the *Abies amabilis* forests results in coarse woody debris patterns resembling a U-shaped curve over time (see Fig. 7.13). This is similar to the pattern of stand replacement fires at lower elevation (Harmon et al. 1986, Agee and Huff 1987). To the preexisting coarse woody debris load, much of which remains after the fire, is added the newly created snags. As they slowly decay, and the stand ages to the point where the relatively short-lived fir die, coarse woody debris loads slowly increase again. One of the few studies showing coarse woody debris loads in *Abies amabilis* forests documented 157 t ha^{-1} in a 180-year-old stand (Grier et al. 1981). Old-growth stands probably have a total down woody fuel load somewhere between that found in lower elevation *Tsuga heterophylla* (450 t ha^{-1}; Agee and Huff 1987) and higher elevation *Abies lasiocarpa* (135 t ha^{-1}; Huff et al. 1989) or *Tsuga mertensiana* forests.

Prescribed fire has been used to reduce the loading of woody fuels in these forests after partial canopy removal. Over 80 percent of the 1 and 10 hr timelag fuels, 50 percent of the 100 hr timelag fuels, and 30 percent of the larger fuels were removed even with fuel moisture of the finer fuels above 20 percent (Pickford 1981). Later mortality of residual trees resulted in additional woody fuels added after the burn, creating a pattern similar to that just described.

Implications for Fire Management

Montane forests dominated by Pacific silver fir do not burn frequently. They may act as a microclimatic barrier to fire moving across drier, lower elevation areas, as long as weather conditions are not severe. Topographic factors that influence the distribution of forest communities, such as aspect or slope, act as natural fire barriers under some conditions. However, management plans should not rely on such "barriers," as severe fires have moved through most *Abies amabilis* forests every few hundred years.

In natural areas, prescribed natural fires in *Abies amabilis* forests may be an allowable disturbance if old-growth conditions are not mandated and the stands are surrounded by natural barriers, such as cliffs or snowfields. Historically, fires tend to be brief and intense under the restricted conditions under which they will burn, at least in the northern part of the range. However, where *Abies amabilis* forests are surrounded by drier forest types, longer burning fires in those types result in repeated ignitions and fire spread into the *Abies amabilis* forests. This pattern may be typical in the southern and eastern portions of the range of Pacific silver fir.

That fires are infrequent in *Abies amabilis* forests does not diminish the difficulty of managing natural fires in this type. At Mount Rainier National Park, where these forests are the dominant coniferous vegetation, they are continuous with drier, lower elevation forests that extend outside the park. Finding natural fire barriers that will be effective under all conditions along this continuum may prove difficult. Fires moving from outside the park upslope may be stopped as they reach the Pacific silver fir zone, but fires in the zone may not be impeded from burning through lower elevation forests unless the weather conditions abate.

Prescribed fire is not a useful management option. Under controllable conditions, prescribed fires will not spread. As a fuel reduction tool, prescribed fire will only increase dead fuel loadings, because almost all the dominants in these forests are fire *avoiders*. If fuels are to be manipulated, mechanical or manual techniques are preferable so that dead fuels can be treated without creating even more dead fuel.

ABIES MAGNIFICA FORESTS

Red fir is part of a taxonomic complex that grades into noble fir at the north (Franklin et al. 1978). Red fir has two distinct taxonomic units: California red fir (*Abies magnifica*) and Shasta red fir (*Abies magnifica* var. *shastensis*). In this chapter, both are referred to as red fir. Forests in the *Abies magnifica* zone often form nearly pure stands, but because they normally occur between forests of higher and lower elevation, they commonly have a wide variety of associates in the ecotonal areas (Table 8.3). In the Klamath Mountains, up to 16 conifer associates are found with red fir (Sawyer and Thornburgh 1977).

Environmental Characteristics

These forests occupy cool sites with substantial winter snow: winter snowpack of 2 m or more (Waring 1969) in the Siskiyous, and 1.8–3.5 m in the Crater Lake area (Chappell 1991). Summer maxima for July range around 20°C, and summer environments for *Abies magnifica* forest are fairly dry (June–August precipitation about 10 cm; NOAA 1982), with precipitation from occasional thunderstorms.

The Fire Regime

Abies magnifica forests have a moderate severity fire regime, with fire frequencies and intensities intermediate to those of other Pacific Northwest forests. Fire-return intervals in the *Abies magnifica* zone have not been extensively studied because of the difficulty in determining fire history: fire scars can be grown over, true firs tend to rot after being scarred, and evidence of fire is often obscured by the benign nature of some fires. One of the first studies of fire history in

Table 8.3. Major Tree Species and Successional Status in Red Fir Forests

Tree Species	*Shade Tolerance*	FIRE STRATEGY *Young*	FIRE STRATEGY *Old*	*Successional Status*
White fir	high	avoider	resister	major climax
Red fir	high	avoider	resister	major climax
Lodgepole pine	low	avoider	invader/evader	major seral
Western white pine	medium	avoider	resister	minor seral
Mountain hemlock	high	avoider	avoider	minor climax

NOTE: A variety of minor species, including Douglas-fir, whitebark pine (*Pinus albicaulis*), Jeffrey pine (*Pinus jeffreyi*), Alaska yellow-cedar, sugar pine (*Pinus lambertiana*), and several others also occur in red fir forests.

these forests was conducted at Sequoia National Park (Pitcher 1987). A fire-return interval of 65 years was calculated for the pre-1886 period on plots of 0.25–0.65 ha. In northeastern California, at Swain Mountain Experimental Forest, Taylor and Halpern (1991) found a mean fire-return interval of 40–42 years on 0.5–1.0 ha plots. Using similar techniques at Crater Lake National Park, McNeil and Zobel (1980) and Chappell (1991) calculated mean fire-return intervals of 42 years and 39 years. All sites had wide variability in fire-free intervals: 5–126 years at Sequoia, 5–65 years at Swain Mountain, and 15–157 years at Crater Lake. If low intensity fires were present, their traces may have been undetected using the fire-scar/age-class analysis techniques of these studies. Nevertheless, these studies indicate that fires are a common form of disturbance in red fir forests and that short-interval reburns occur.

A modern example of reburning in *Abies magnifica* forests occurred at Crater Lake National Park. A fire was allowed to burn in 1980 at Crater Peak under a prescribed natural fire management plan. This site had experienced small fires in 1951 and between 1890 and 1913 (Chappell 1991). Ignited by lightning, a 1980 fire covered roughly 115 ha, and two other lightning-ignited fires within the 1980 perimeter were suppressed in 1986 and 1989. Had the two most recent fires been allowed to burn, substantial reburning might have occurred because the 2 ha 1986 fire spread largely through fuels created by the 1980 fire.

Lightning has been the primary cause of fires in the *Abies magnifica* zone in recent decades. In Yosemite National Park, the red fir type covers 8.4 percent of the park but contains 16.2 percent of the park's fires (van Wagtendonk 1986). On a per-area basis at Yosemite, the *Abies magnifica* forest type has the highest fire incidence of all vegetation types. Native American ignitions are not documented for this forest type, although their burning at lower elevations may have spread upslope into *Abies magnifica* forests.

Available evidence suggests that fires burning in red fir forests span a wide range of intensities, resulting in variable fire severity in space and time. An example of the spatial variability is the Crater Peak fire (Fig. 8.5), where severity levels were equitably distributed in total area and interspersion (29 percent low severity, 32 percent moderate severity, 34 percent high severity, 5 percent nonforest; Fig. 8.6; Chappell 1991). Temporal variation in fire severity is illus-

FIG. 8.5. *The 1978 Crater Peak fire at Crater Lake.* A: *The fire burning.* B: *Six years later. Three categories of fire severity include low* (C), *where even understory trees survived and fire spread was patchy; moderate* (D), *with scorch line shown by white line and understory trees largely killed; and high* (E), *shown about eight years after the fire, with total mortality at the time of the fire.*

trated by the Swain Mountain stands, which burned with different severity with time (Taylor and Halpern 1991).

Fires in the *Abies magnifica* zone have ranged from small to large. One of the first prescribed natural fires in Sequoia National Park burned from July through September and covered about 4 ha (Kilgore 1971). At Crater Lake, numerous natural fires in the red fir zone have reached an estimated 200 ha, although Chappell (1991) noted that these sizes were, in retrospect, overestimates. It appears that most fires beginning in June or July have the capability of burning until the first autumn snows (Kilgore and Briggs 1972).

Wind is another important disturbance factor in red fir forests (Taylor and Halpern 1991). Because of the generally small scale of wind disturbance, there appears to be little interaction of wind with increased fuel loads leading to intense fires, although a fire burning in such an area would likely be more intense. Kilgore (1971) noted

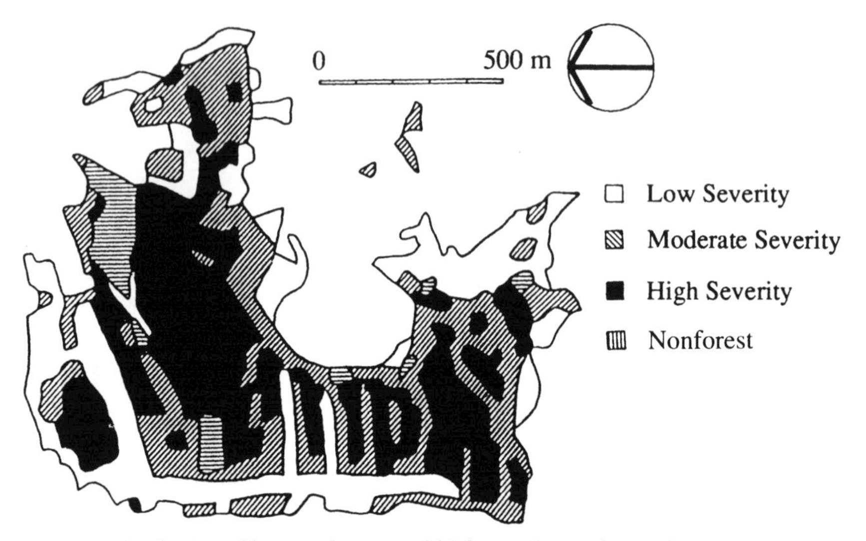

FIG. 8.6. *Distribution of low, moderate, and high severity patches in the 1978 Crater Peak fire shown in Figure 8.5.*
(From Chappell 1991)

increased fire intensity in downed trees mixed with dense thickets of small trees.

Stand Development Patterns

Stand development in red fir forests is characterized by both episodic and continuous recruitment, a function of periodic disturbance and the ability of red fir to colonize small gaps in the forest. Early workers assumed red fir forests to be even-aged (Schumacher 1928) or uneven-aged (Oosting and Billings 1943) based primarily on size structure of stands and the presence or absence of red fir regeneration. The small patch structure of these forests was universally recognized. More recent work has shown that classification into either even-aged or uneven-aged forest is too simplistic, that both patterns may exist in the same stand, and that poor size-age correlations are found in these forests (Taylor and Halpern 1991).

The complexity of stand development is due to the ability of red fir to establish itself either in the partial shade of small canopy gaps or in the more severe microclimates of larger gaps. Seedling establishment may be negatively correlated with high irradiance, high surface temperature, and long-duration sunflecks at midday (Ustin et al. 1984, Selter et al. 1986), and first-year seedlings do well with partial shading (Gordon 1970). These data suggest a late successional, shade-tolerant role for red fir, implying that it is unlikely to be a dominant after disturbance. However, it also establishes well on bare mineral soil (Gordon 1970) and on thin litter layers (Barbour and Woodward 1985), which suggests that disturbance encourages establishment of red fir.

On some sites the stand development sequence can be traced to a high severity fire in past centuries that killed most of the overstory (e.g. Pitcher 1987, plot 2; Taylor and Halpern 1991, plot 1). One stand development scenario after high severity fire, which can extend over hundreds of hectares, is brush dominance in the gap or patch (Parsons, personal communication, cited in Taylor and Halpern 1991). Chappell (1991) found that large patches of high severity fire are associated with poor reproduction of red fir; lodgepole pine, which might have colonized the site, was locally absent, so that tree establishment on the site will occur over many decades. On such sites, mortality of red fir seedlings from drought stress and herbivory from cutworms and vertebrates may be high (Barbour et al. 1990). Presence of charcoal on the forest floor increases germinant mortality by raising daytime surface temperatures (Chappell 1991). Shrubs such as snowbrush (*Ceanothus velutinus*), manzanitas (*Arctostaphylos* spp.), and gooseberries (*Ribes* spp.) may be favored (Gordon 1970, Chappell 1991). Over time, these shrubs may ameliorate the microclimate for trees to become established. Where lodgepole pine seed source is absent, true fir (red and/or white fir) will recolonize; stand initiation periods of over 50 years have been reported (Pitcher 1987, Taylor and Halpern 1991).

A second stand development scenario may occur if a lodgepole pine seed source is present. Lodgepole pine is a major associate through much of the range of red fir. It can often establish itself under temperature and moisture conditions too severe for other conifers to tolerate (Cochran and Bernsten 1973). The variant of lodgepole pine associated with red fir is generally nonserotinous, although some cones do hold seed for more than one year. Where a seed source for lodgepole pine is present, the abundant seed may help lodgepole pine to establish profusely on intensely burned microsites, with highest densities where a stand-replacing fire has occurred (Fig. 8.7). It grows rapidly and can overtop the shrubs that may be competing for growing space. This scenario was envisioned by Zeigler (1978) at Crater Lake, where he saw many lodgepole pine stands with emerging red fir understories. Pieces of charcoal in these stands included *Abies* and *Tsuga* spp., implying that they were once fir-hemlock forests, but postfire colonization by lodgepole pine had been very successful. On a smaller scale, this scenario of fire creating small, high severity gaps was assumed by Rundel et al. (1977) to be the cause of groups of lodgepole pine that dominated small areas within *Abies magnifica* forest in the Sierra Nevada. These small patches with mineral soil seedbed seemed to "fortuitously" catch a seed crop of lodgepole pine. Fir regeneration under the mature pines indicated that such sites would eventually revert to red fir dominance.

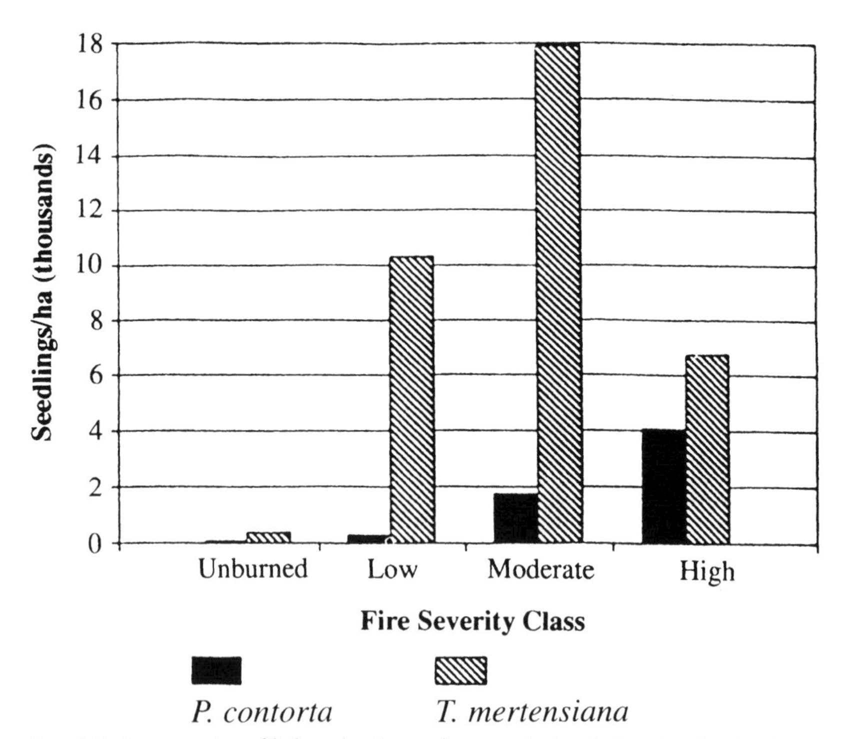

FIG. 8.7. *Regeneration of lodgepole pine and mountain hemlock a decade after fire in patches of varying fire severity, Crater Lake National Park.*
(From Chappell 1991)

Western white pine (*Pinus monticola*) appears to do well on sites burned by moderate to severe fire. Because white pine lives longer than lodgepole pine, its presence will serve for many centuries as a marker of a past event of moderate to high severity.

The stand replacement fire, although present in almost every study location, may not be the dominant disturbance in these stands. Fires of lower severity appear to make up most of the area of recently monitored fires (Kilgore and Briggs 1972, van Wagtendonk 1986, Chappell 1991). Red fir tends to be a dominant residual; with its thick bark it is better adapted to fire than any of its associates (Chappell 1991). Fires under such conditions leave bare mineral soil for regeneration and partial to almost full shade from the residual canopy. Some understory trees may be left due to the mosaic nature of fire spread in low severity patches.

Red fir has a majority of good seed years at less than three year intervals (Gordon 1978). Thousands of seedlings ha^{-1} may colonize patches burned with moderate to low severity fire (Fig. 8.8), where

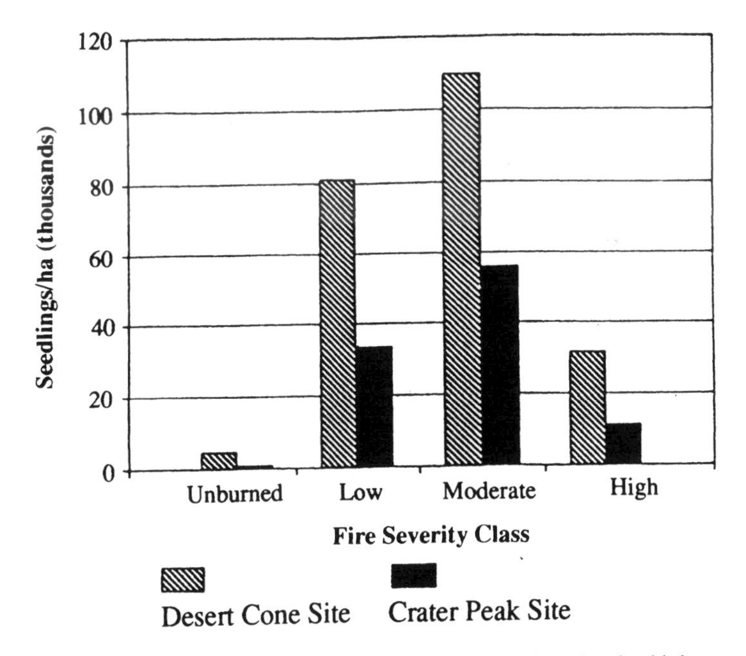

FIG. 8.8. *Regeneration of red fir after two fires roughly a decade old, by patches of varying fire severity, Crater Lake National Park.*
(From Chappell 1991)

bare soil and partial shading occur. Chappell (1991) noted a lag of three or four years after disturbance before considerable seedling establishment on two fires burned in different years, but could not establish the cause: postfire weather, or perhaps a fire effect on cone crops. Moderate severity fires appear to be associated with substantial regeneration, although seedling height is depressed as overstory canopy cover increases (Chappell 1991). Understory red fir averaged 33 years to grow from 30 cm to breast height in the Sierra Nevada (Pitcher 1987).

As a stand matures, the larger trees develop more fire resistance because of thicker bark. A stand that has passed through the stem exclusion stage has reached maximum fire resistance for the type. At the 1986 Castle Point fire near Crater Lake, young, closed canopy *Abies magnifica* stands experienced significant reductions in basal area, while mature and old-growth stands were much less affected (Table 8.4). The young stands were composed of trees primarily 50–80 years old, associated with a previous stand replacement event. The mature stands had red fir up to 300 years old, with almost uniform recruitment of trees 100–300 years old and occasional dense cohorts of trees 80–100 years old. The old-growth stands had red firs of various ages up to 525 years, with a pronounced 30–60-year-old cohort that might represent a group release after opening of a canopy gap. In the same fire, young lodgepole pine stands rarely burned, but half the older pine stands experienced more than 50 percent mortality.

In the southern Sierra Nevada, multiple age classes in a stand are

TABLE 8.4. PERCENT AREA BURNED IN FOUR FIRE SEVERITY CLASSES IN RED FIR FOREST AT CASTLE POINT FIRE, CRATER LAKE

	FIRE SEVERITY LEVEL MEASURED BY BASAL AREA REMOVAL			
Stand Development Stage	*Low (0–20%)*	*Moderate (20–50%)*	*High (50–80%)*	*Extreme (80–100%)*
Stem exclusion	41	6	29	24
Understory reinitiation	69	15	13	3
Old-growth	68	16	10	6

SOURCE: J. K. Agee, unpublished data.

common (Pitcher 1987). The recruitment periods are sometimes as long as the disturbance-free intervals. Fires were associated with most distinctive age cohorts, but also burned with no later cohort initiation. Four fires in the nineteenth and twentieth centuries at Swain Mountain in the southern Cascades (Taylor and Halpern 1991) were associated with light, patchy reproduction of red and white fir (*Abies concolor*). Significant growth releases and growth suppressions were also noted, associated with release from competition and damage to trees from the fires. In the northern Sierra Nevada, Oliver (1986) found that released saplings with high live crown percentages grew much better than those with small crowns. The trees with small percentages of live crown also suffered the highest rates of sunscald, which creates entry courts for fungi. Later, such scars may be mistaken for fire scars.

In the intervening periods between fire or large-scale wind disturbance, fir recruitment can occur under canopy gaps (Ustin et al. 1984). Gaps may be the result of individual windthrows, or they may be created by insects or disease. The most important small-scale disturbance factors that can result in red fir regeneration or the release of understory red fir include Indian paint fungus (*Echinodontium tinctorium*) or fir engraver beetle (*Scolytus ventralis*) on red fir, pine beetles (*Dendroctonus* spp.) on lodgepole and western white pine, and laminated root rot (*Phellinus weirii*) on mountain hemlock. It is no wonder that unraveling the dynamics of *Abies magnifica* forests is an effort that has only just begun.

Fuels

Little work has been done on the fuel changes associated with burning in *Abies magnifica* forests. Given the wide range of fire severity, it is likely that some fires, or some portions of each fire, result in decreases to the dead woody fuel load, while others show increases. The fire consumes some portion of the dead woody fuel on site, but it also creates new fuels by killing live trees and shrubs. In the southern Sierra Nevada, Kilgore (1971) noted in a prescribed fire that about two-thirds of the roughly 25 t ha^{-1} of the <100 hr timelag (flash fuels) biomass and about 50 percent of the roughly 40 t ha^{-1} of duff were consumed by the fire. Large downed logs (about

10–15 t ha^{-1} before burning) were reduced 30–50 percent by mass. Some fires spread primarily by log-to-log smoldering—as described in chapter 11 for *Pinus contorta* climax forests—so that large woody fuels are largely consumed.

Generalization about the impact of fire on fuels is difficult due to the variability of fire in this forest type. A stand replacement fire will clearly create a great deal of woody fuel, but after a crown fire little fine fuel remains for reburns. As evidenced from lightning fires after the Crater Peak fire, reburns in heavy fuels created by an initially high severity fire are possible. At the other extreme, a low severity fire may consume forest floor fuel without affecting the overstory, resulting in a net loss of dead fuel after fire. Intermediate scenarios, resulting in substantial fuel consumption and creation by the fire, are also possible.

Management Implications

Most red fir forests have not experienced major alterations in forest species composition or structure due to fire suppression policies of this century. However, there have probably been minor landscape-level shifts in the forest mosaic due to the absence of fire. The period of absence is approaching twice the average fire-return interval found in several studies, but each of these studies also showed ranges of fire-free intervals longer than what has occurred due to fire exclusion policies. Because of these minor alterations to the natural fire regime, prescribed natural fires have been allowed to burn in several national parks (Crater Lake, Sequoia–Kings Canyon, Yosemite) without pretreatment of fuels and will likely be allowed to burn in the future.

The variable severity of fire regimes in *Abies magnifica* forests suggests that under most weather conditions, fires will remain within controllable intensities. When weather becomes extreme, larger patches of high severity fire are possible. If well contained within a natural fire zone, this behavior should still produce effects within the range of presettlement conditions and remain within prescription for that zone. However, if the fire is near the edge of the zone, difficulties in containment may occur. At the 1988 Prophecy fire at Crater Lake National Park (Jones 1990), the fire burned at low to moderate severity for several weeks under daily monitoring. Late in August, it was blown by a strong wind during a single day over 800 ha, about half of which was adjacent Forest Service land. To avoid such situations, it may be advisable in selected locations to manipulate fuels through prescribed fire or by thinning stands to reduce the hazard near the edges of such zones. This manipulation may seem "unnatural," but it is a realistic alternative to the otherwise unavoidable fire suppression policy.

Abies magnifica forests comprise a classic mosaic of patches associated with historical variation in disturbance intensity, primarily from fire. The species- and landscape-level diversity associated with past fires is substantial, although it is partly masked by recruitment of red fir during disturbance-free intervals. This mosaic can be allowed to perpetuate itself through intelligent management of fire in natural areas, if we recognize that the landscape mosaic is dynamic and that no particular percentage of any stand age, structure, or species composition is likely to be a stable landscape goal.

CHAPTER 9

SUBALPINE ECOSYSTEMS

SUBALPINE FORESTS and associated parkland and alpine zones cap the crest of the major mountain ranges in the Pacific Northwest (Fig. 9.1). The elevation range of the subalpine forest zone increases from north to south and from west to east, and the dominant tree species differ along this gradient. The wetter, more coastal *Tsuga mertensiana* forests are found up to 1,700 m in northern Washington and to 2,000 m in the southern Oregon Cascades. *Abies lasiocarpa* forests, the continental analog of the coastal forest, often begin their lower limit at those elevations (Franklin and Dyrness 1973).

ENVIRONMENTAL CHARACTERISTICS

Annual precipitation varies from 160–300 cm in *Tsuga mertensiana* forests to 100–200 cm in *Abies lasiocarpa* forests. Prolonged winter snowpack is characteristic of these subalpine zones, with 7–8 m of snowpack in wetter locations. These forest zones are the coolest in

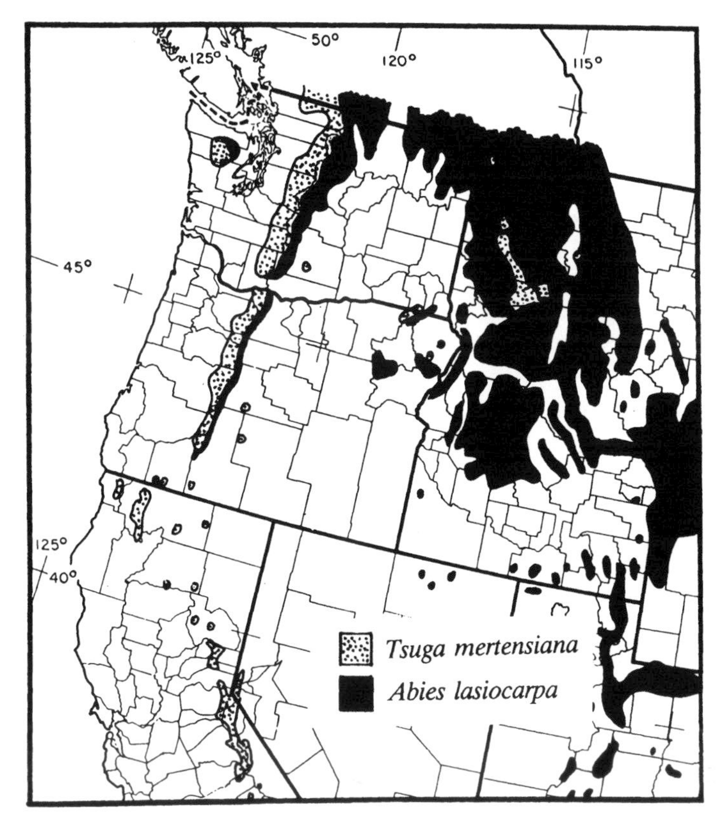

FIG. 9.1. *General location of the subalpine forests of the Pacific Northwest.*
(Adapted from Little 1971)

the Pacific Northwest. The short growing season has significant effects on the fire regime and on postfire successional patterns.

Tsuga mertensiana forests are characterized by the presence of mountain hemlock as a major climax dominant. The *Tsuga mertensiana* zone is the highest elevation forest zone west of the Cascade crest; it is usually found above the *Abies amabilis* forests (north) or *Abies magnifica* forests (south). Either Pacific silver fir or subalpine fir or both can be minor climax dominants in the northern part of the zone (Table 9.1). In the Klamath Mountains (Sawyer and Thornburgh 1977) and southern Cascades, red fir replaces Pacific silver fir as a common ecotonal associate. Lodgepole pine (*Pinus contorta*) is a common early seral species in the southern part of the *Tsuga mertensiana* zone.

TABLE 9.1. COMMON TREE SPECIES AND SUCCESSIONAL STATUS IN THE SUBALPINE ZONE OF THE PACIFIC NORTHWEST

Species	*Life History Strategy to Fire*	*Successional Status*
Tsuga mertensiana forests		
Mountain hemlock	avoider	major climax
Subalpine fir	avoider	minor/major seral
Pacific silver fir	avoider	minor climax
Alaska yellow-cedar	avoider	minor climax
Western white pine	invader	minor seral
Lodgepole pine	evader/invader	major seral (south only)
Abies lasiocarpa forests		
Subalpine fir	avoider	major climax
Engelmann spruce	avoider	major climax
Lodgepole pine	evader/invader	major seral
Whitebark pine	resister/invader	minor seral
Subalpine larch	avoider	minor seral

Abies lasiocarpa forests are found throughout the Rocky Mountains, in the Cascades of Washington and Oregon, and in the northeastern Olympic Mountains. These forests experience very cold winters but can have dry, warm summers. Freezing temperatures during the night can occur throughout the summer, but daytime heating can create conditions for significant fire spread. Lodgepole pine is the most ubiquitous early seral species in this zone (Table 9.1); the codominance of Engelmann spruce (*Picea engelmannii*) increases to the east.

Whitebark pine (*Pinus albicaulis*) and subalpine larch (*Larix lyallii*) are the two most common conifers at timberline east of the Cascade crest. Whitebark pine, at the maritime end of its range in Oregon and Washington, is characteristic of cold, dry environments. Its environment includes mean January temperatures of -5°C with summer maxima of 26–30°C (Arno and Hoff 1989). It is often found on southerly or westerly aspects in shallow, rocky soils. Strong winds and blizzards are common. In contrast, the environment of subalpine larch stands is generally cooler throughout the summer, and the species often is found in late melting snow basins, where snowmelt drainage may augment available soil moisture (Arno and Habeck 1972). Early warm temperatures combined with above-average annual precipitation (represented as snowpack) is associated with decreased growth in subalpine larch. This is hypothesized to be due to lethal transpirational demand on new needles at a time when roots are too cold to transport sufficient water to the crown (Graumlich 1985).

Subalpine tree species are not well adapted to survive fire (Table 9.1). All the species in the northern part of the *Tsuga mertensiana* zone are fire *avoiders;* in the southern part lodgepole pine is an *evader/invader,* and red fir is a *resister.* In the *Abies lasiocarpa* zone, all the species are fire *avoiders* except for lodgepole pine. In eastern Washington and Oregon, lodgepole pine is in a transition from nonserotinous ssp. *murrayana* to serotinous ssp. *latifolia.* This dichotomy is less real than apparent, since *latifolia* has some nonserotinous cones and *murrayana* has some that are serotinous. In timberline environments, whitebark pine is a weak *resister,* having thin bark but subject to fires of patchy intensity and extent, while subalpine larch is a true fire *avoider.* Subalpine forests exist in a marginal environment for tree establishment and growth, so a fire disturbance that kills most or all of a stand can create almost permanent meadows or open parklands that persist for decades to centuries. Fire has been an important factor in the creation of subalpine meadows (Kuramoto and Bliss 1970).

THE FIRE REGIME

Although the subalpine ecosystems of the Pacific Northwest have much in common, they are heterogeneous in their fire regimes and reaction to disturbance. The variation in fire regimes will be described by forest type.

FREQUENCY

Fires are infrequent in the subalpine zone. Because subalpine forests grade into parkland and alpine areas, a clear delineation of the zone is difficult, and areal estimates of fire-return intervals are often not possible. The high mortality associated with most fires leaves few fire-scarred trees, so that point estimates of fire-return intervals are seldom possible. The estimated fire-return intervals shown in Table 9.2 are arranged from longest to shortest. Subalpine larch forests usually occur in rocky talus areas or in moist sheltered cirques and rarely burn even though they occur in the drier, more fire-prone eastern Cascades and northern Rocky Mountains (Arno and Habeck 1972). These forest are rarely contiguous over large areas. Mountain hemlock forests are widely distributed in the western Cascades and Olympics. Some forests appear to have been fire-free for over a millennium (Lertzman and Krebs 1991), while others burn rarely (Fahnestock and Agee 1983). Fire-return intervals for mountain hemlock forests of the Oregon Cascades are not well documented. In the central Oregon Cascades, Dickman and Cook (1989) found that at least half of their 18,000 ha study area had burned over in the last 500 years, again suggesting that fire is infrequent in these forests.

The *Abies lasiocarpa* zone east of the Cascade Crest or in the rainshadows of mountains west of the crest has burned more frequently (Table 9.2). The rainshadow forests of Mount Rainier (Franklin et al. 1988) and Desolation Peak (Agee et al. 1990) have fire-return intervals much shorter than those of their associated westside forest types. The fire-return interval of 250 years for subalpine fir in the Pasayten Wilderness (Fahnestock 1976) is probably

TABLE 9.2. FIRE-RETURN INTERVALS (FRI) IN SUBALPINE FORESTS OF THE PACIFIC NORTHWEST

Forest Type	*FRI (years)*	*Method*[a]	*Source*
Subalpine larch	none	—	Arno and Habeck (1972)
Mountain hemlock	1,500+	—	Lertzman and Krebs (1991)
Washington subalpine	800	NE	Fahnestock and Agee (1983)
Subalpine fir	275	NFR	Franklin et al. (1988)
Subalpine fir	250	—	Fahnestock (1976)
Subalpine fir	109–137	NFR	Agee et al. (1990)
Whitebark pine	50–300	C	Arno (1980)
Whitebark pine	72–94	C (1)	Arno and Peterson (1983)
Whitebark pine	30–41	C (300)	Arno (1986)
Whitebark pine	29	C (100)	Morgan and Bunting (1990)

[a]NE = negative exponential fire cycle; NFR = natural fire rotation; C = composite fire interval over (number of hectares). Dash indicates method not described.

typical of the east Cascades subalpine zone. In the central Washington Cascades, Woodard (1977) documented fires in 1835, 1850, and 1890 over an 11 ha area, suggesting that short-interval reburns can occur in these forests.

Most of the fire history research in whitebark pine forests has been conducted in the northern Rocky Mountains. Fire-return intervals grade from those typical of the lower elevation subalpine fir forests to composite fire intervals of about 30 years within large stands. Not all scars on whitebark pine may necessarily be fire scars, so careful interpretation is required (Morgan and Bunting 1990).

INTENSITY

The estimation of fire behavior in subalpine forests is complicated by the erratic, often weather-driven nature of these fires. The interaction of live crown fuels as a heat source is important in crown fire generation. In *Tsuga mertensiana* forests of the Olympic Mountains, the Hoh fire (1978) and Chimney Peak fire (1981) were high intensity crown fires (Agee and Smith 1984). Fire behavior estimates based on dead fuel loads on the forest floor surface generally underestimate the fire potential of subalpine forests. For example, Despain and Sellers (1977) suggested that the frequency of major fire events was dependent on development of a fuel complex capable of supporting crown fire behavior, and Romme (1982) suggested this will take up to 300–350 years on a typical Yellowstone site. Abundant dead fuels and development of an understory to carry fire into the overstory formed part of this hypothesis. This fuel-limited hypothesis was supported by much of the 20 year history of the natural fire program at Yellowstone (Romme and Despain 1989). Yet the 1988 Yellowstone fires were largely weather-driven and burned across almost all successional stages of forests. In normal to somewhat unusual years, a fuel-limited situation for crown fire development exists, but in very unusual years, perhaps separated by centuries, large-scale crown fires will burn regardless of the fuel situation.

Most fires are stand replacement events because of the lack of fire resistance of the major tree species. Lower intensity fires may occur in whitebark pine forests, as fire scars are often present in these stands. Woodard (1977) suggested that some fires in the central Washington Cascades were of a moderately low surface fire intensity, since residual lodgepole pine were fire-scarred rather than killed. The scarring might be associated with the type of log-to-log burning pattern observed in climax lodgepole pine forests (see chapter 11).

EXTENT

The extent of subalpine fires is partly a function of fire weather, but mainly depends on the distribution of the subalpine forest. Subalpine forest is often patchy and may grade into herbaceous parkland or rock or snow, which may not allow fire to carry across to the next forest patch along continous burning fuels or by spotting. Lower elevation forests may be less affected by strong winds and less capable of sustaining long fire runs, although fuels may be more continuous. In the *Tsuga mertensiana* zone of Washington, fires have tended to be confined to individual slopes (e.g., Agee and Smith 1984) or river valleys due to the discontinuous nature of the forests. In the *Tsuga mertensiana* zone of Oregon, Dickman and Cook (1989) found historic fires >3,200 ha, although a majority of the stands they studied were patches of <500 ha each. Larger patches of burned forest are noticeable as the extent of the *Abies lasiocarpa* zone increases east of the Cascades (Fahnestock 1976), and perhaps the best examples of large subalpine fires are the 1988 Yellowstone fires and possible precursors of earlier centuries in the same area (Romme and Despain 1989). In whitebark pine forests, most fires appear to be small (Tomback 1986, Morgan and Bunting 1990), and many may burn only one clump (Fig. 9.2).

SEASONALITY

Crown fires can occur in subalpine forests when foliar moistures are low and may be aided by lichens draped within the canopy. Late summer crown fires have occurred in Olympic Mountain subalpine forests with foliar moisture between 90 percent and 107 percent (Huff et al. 1989). Prescribed crown fires in early spring have been

FIG. 9.2. *Whitebark pine clump struck by lightning, burning through the evening at Crater Lake National Park, 1980. Within this 875 m^2 clump and environs, a previous fire had burned another portion of the clump.*

ignited by helitorch when foliar moisture was at 92 percent (Woodard et al. 1983) but snow was on the ground. In 1986 and 1987, similar spring crown wildfires occurred in the Cascades of Oregon and Washington with 1–2 m of snow on the ground. Each occurred after five to seven days of exceptionally warm weather, which apparently caused needle moisture loss while the tree roots were too cold to absorb and translocate moisture to the foliage (e.g., Hinckley et al. 1985). Lichen moisture can closely parallel that of fine dead fuels, except that the lichens have a much higher ratio of surface area to volume, making them exceptionally flammable. Lichen moisture contents have ranged from 6 percent to 9 percent during summer crown fire events in the Pacific Northwest (Morris 1934b; Agee, unpublished data). The presence of lichens (Huff 1988) and the concentration of dead fuels around subalpine fir clumps (Taylor and Fonda 1990) can encourage clump-to-clump torching, while the intervening shrubland areas are left unburned.

In whitebark pine forests, where fire-scarred trees are available, the position of the seared cambium within the annual growth ring helps determine the season of historical burns. In a Wyoming study area (Morgan and Bunting 1990), fires occurred all through the growing season, but the largest fires (those scarring multiple trees across the landscape) occurred in late season.

Synergism with Other Disturbances

Because subalpine forests usually burn infrequently, it is not surprising that other disturbances may be important in forest stand dynamics. In long undisturbed mountain hemlock forests, mortality of individual standing trees may be important (Lertzman and Krebs 1991), while in other mountain hemlock forests, laminated root rot (*Phellinus weirii*) may cause doughnut-shaped rings of mortality several hectares in size (McCaughley and Cook 1980). White pine blister rust (*Cronartium ribicola*), an introduced canker disease, may have deleterious effects on western white pine (Fonda and Bliss 1969) and whitebark pine (Arno and Hoff 1989). Mountain pine beetles (*Dendroctonus ponderosae*) may be important in stand development patterns by selectively eliminating pines (Woodard 1977), while spruce budworm (*Choristoneura occidentalis*) may kill true fir and spruce bark beetles (*Dendroctonus rufipennis*) may kill Engelmann spruce (Miller 1970, Furniss and Carolin 1977). Where these disturbances create large patches of dead trees, the flammability of the site may increase.

STAND DEVELOPMENT

Subalpine fires may burn forest and nonforest communities, altering successional patterns in both. In this chapter, patterns of forest stand development are described separately for *Tsuga mertensiana* and *Abies lasiocarpa* forests, but as a group for nonforest communities.

TSUGA MERTENSIANA FORESTS

Fire is the primary large-scale disturbance agent in *Tsuga mertensiana* forests. Most other disturbances operate at the scale of a tree or small stand. None of the tree dominants is resistant to fire or adapted to grow well in open, recently burned environments. Postfire communities may be dominated by shrubs or herbs for a century or more. South-facing slopes are much more likely to burn than north-facing slopes (Agee and Smith 1984).

In subalpine portions of the 1978 Hoh Fire, the early August burning period occurred before the current year's seed was mature (Agee and Huff 1980), so very little of the abundant seed was viable. Although mountain hemlock and subalpine fir were present before the fire, neither was represented in the first-year seedling population of 955 ha^{-1}, all of which was Pacific silver fir. Total seedling density increased to 1,592 ha^{-1} by year 3, but all the seedlings were one or two years old, indicating that the first year's seedlings had died (Agee and Smith 1984). By year 7, subalpine fir and mountain hemlock represented two-thirds of the tree density of 167 ha^{-1}. These were well-established seedlings, unlike the delicate, small seedlings of the earlier surveys. A single *Pseudotsuga menziesii* seedling was the tallest (31 cm) tree, while mountain hemlock and subalpine fir averaged 11 cm and 9 cm height. Some of each species were browsed by elk. Shrub species tend to be persistent. Both *Vaccinium deliciosum* and *V. membranaceum* sprouted after the fire, as well as *Xerophyllum tenax* and other herbs.

Tree establishment is a function of seed sources and climate. Most subalpine fir seed falls within 20 m of the seed tree, while mountain hemlock seed travels about 40 m (Huff and Agee 1991). Early colonizing trees may play a crucial role in restocking more central areas of a burn as they mature and are able to produce seed. Postfire residuals are rare after fire in the *Tsuga mertensiana* zone, but when they do occur, they can have a significant impact on species composition of the new stand (Table 9.3).

The interior of larger burns will remain treeless longer because of the seed source effect. At a 55-year-old, 100 ha burn (Mount Wilder), tree density near the southern edge was 864 trees

TABLE 9.3. CORRESPONDENCE BETWEEN RESIDUAL SPECIES COMPOSITION AFTER FIRE AND PRESENT SPECIES COMPOSITION (PERCENT) IN *TSUGA MERTENSIANA* FOREST

	HIGH DIVIDE RESIDUALS <1891		MOUNT WILDER EDGE PLOTS RESIDUALS <1924			
			Plot A		Plot B	
Species	*Residual*	*1979*	*Residual*	*1979*	*Residual*	*1979*
Subalpine fir	100	96.6	87.9	90.8	28.6	18.1
Mountain hemlock	0	1.6	0	0.6	67.3	53.5
Pacific silver fir	0	0	0	2.2	0	2.4
Alaska yellow-cedar	0	0	12.1	5.8	4.1	3.8
Western white pine	0	0	0	0.6	0	17.3
Douglas-fir	0	0.2	0	0	0	5.6

SOURCE: Agee and Smith 1984.

ha^{-1}, while tree density in the center was 72 trees ha^{-1}; the center was still a shrubland. The rate of tree establishment over time appears to be a site-specific process. Subalpine fir is dominant at one edge plot (Fig. 9.3, Edge A), but not at another (Edge B; Agee and Smith 1984). While Edge B has nearly continuous low-level recruitment over time, Edge A is discontinuous with a large pulse beginning in the late 1960s. Springs and summers that are drier than normal have less recruitment than normal to wet years (Agee and Smith 1984). Near Mount Rainier, Little (1992) found a positive association between warm springs and subalpine fir establishment, likely due to an extended early establishment period when snowpacks melted early.

Stocking of trees of seed-bearing age has a significant effect on later recruitment (Table 9.3; Agee and Smith 1984), suggesting that early colonizers will have a large effect on later successional patterns. Again, during normal to wet periods, recruitment is greater than during dry periods. On a 90-year-old fire in the Olympic Mountains, total tree density was 1,948 stems ha^{-1}, but only 400 were taller than 5 m and 200 exceeded 10 cm dbh (Agee and Smith 1984). This site was still quite open and will not be closed forest for decades.

Once closed, the forest canopy may remain closed for further centuries. If mountain hemlock is the primary species, its long life span (occasional cored specimens in the Olympic Mountains exceed

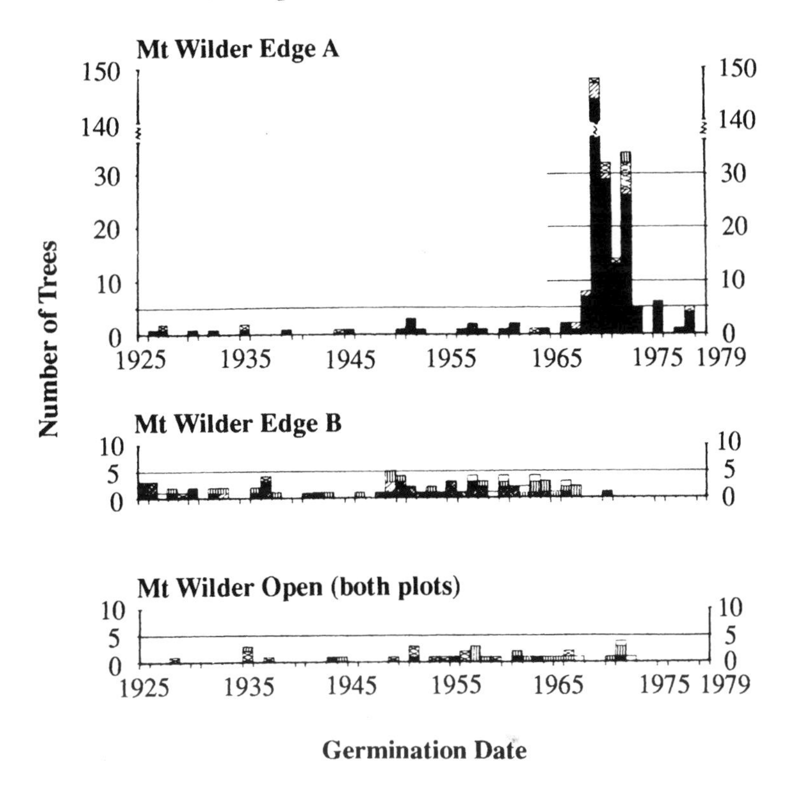

FIG. 9.3. *Tree establishment rates at the 1924 Mount Wilder fire in Olympic National Park. Plots are 0.25 ha.*
(From Agee and Smith 1984)

600 years, and stumps east of Vancouver, B.C., exceed 1,200 years) may allow it to dominate the site until the next major disturbance. In British Columbia forests at the lower edge of the *Tsuga mertensiana* zone, Lertzman and Krebs (1991) commonly found gaps of 20–200 m^2 from standing mortality of single trees. These gaps are

usually filled by shade-tolerant Pacific silver fir. Farther south in the Oregon Cascades, wave-form dieback of pure stands of mountain hemlock is related to infection of roots by *Phellinus weirii* (McCaughley and Cook 1980, Matson and Boone 1984). The wave expands outward at about 0.3 m yr^{-1}; within the "bare zone," nitrogen mineralization increases. Higher nitrogen availability apparently allows mountain hemlock to recolonize the site and resist *Phellinus* infection. The openings are large enough, however, to allow lodgepole pine to invade as well, so that lodgepole pine ages in these stands are not a unique indicator that fire has occurred.

The presence of lodgepole pine in the Oregon *Tsuga mertensiana* zone is usually due to fire. In young (60-year-old) postfire stands, Dickman and Cook (1989) found 4,000 stems ha^{-1} of lodgepole pine, with few mountain hemlock. By age 115, mountain hemlock is the dominant understory tree, and lodgepole pines, aged 100–200, are dying from bark beetle attacks and other agents. By age 250, the stand is primarily mountain hemlock, with a few lodgepole pines found in 360-year-old stands. By age 460, lodgepole pine is no longer present. Fire may play an important role in breaking up *Phellinus* centers, by creating conditions more suitable for the *Phellinus*-resistant lodgepole pine. The pine may then competitively exclude mountain hemlock from the site until the *Phellinus* inoculum is present only in large isolated stumps, remnants of the former stand that have not yet fully decayed (Hansen 1979). Dickman and Cook (1989) suggest three interactions between fire and fungus that depend on fire-return interval: (1) a fire-return interval of 200 years or less, resulting in dominance by lodgepole pine and disfavoring *Phellinus;* (2) a fire-return interval of 600 years, which they believe fosters mountain hemlock stands infested with *Phellinus,* mixed with other stands dominated by lodgepole pine, much like the present situation; and (3) absence of fire as a disturbance agent, which disfavors lodgepole pine and increases the role of *Phellinus,* creating a forest landscape much different than the one seen today.

A recently discovered pattern of crown fires in spring months (Huff 1988) has some unique implications for succession in the transitional (from *Abies amabilis*) *Tsuga mertensiana* zone. Crown desiccation after four to seven days of foehn wind conditions creates a phenomenon similar to "red belt" or "parch blight" observed in winter dieback (Bega 1978), resulting in low foliar moisture. Later lightning storms can result in fires burning over many snow-covered hectares before advancing weather fronts extinguish the fires hours to days later (Gary Mills, Willamette National Forest, personal communication, 1986; Robert Dunnagan, Mount Rainier National Park, personal communication, 1987). Mature mountain hemlock is more affected than Pacific silver fir, because the fir is more efficient at translocating water under conditions of low soil temperature under snow (Hinckley et al. 1985). About two-thirds of trees below 2 m height (75 percent Pacific silver fir) were unaffected by fire (Huff 1988). Consequently, species composition shifted to the more shade-tolerant and late successional Pacific silver fir, essentially advancing succession in terms of species composition on this transitional site. This pattern is an analog to small trees at Mount St. Helens that were protected from the eruptive heat and wind blast because they were covered by snow (Means et al. 1982). It is not yet clear how common this pattern of spring subalpine fires might be in the Pacific Northwest.

Abies lasiocarpa Zone

Subalpine fir is a *fire avoider*, possessing thin bark, a low crown with persistent branches, and shallow rooting habit, making it very sensitive to fire. Rating systems (Starker 1934) and fire effects models (Peterson and Ryan 1986) place it at the bottom of almost any list of fire resistance in coniferous trees. Where associated trees have some advantage in the presence of fire, such as the serotinous (or semi-serotinous) cones of lodgepole pine east of the Cascades, subalpine fir may be replaced on the site. It may persist as a suppressed understory species or recolonize the site so slowly that before subalpine fir can replace the pine another fire will allow another generation of lodgepole pine to dominate the site (Romme and Knight 1981). At higher elevation sites in the Rocky Mountains, trees may be eliminated for a century or more (Stahelin 1943, Clagg 1975), depressing timberline below the physiological limits of tree growth (Arno and Hammerly 1984). Some sites may revert to ribbon forests, where forest regeneration is associated with windward areas of trees blown free of snow (Billings 1969). On wetter Cascade Mountain sites with little or no lodgepole pine, none of the tree species is adapted to the postfire conditions, and meadows are the likely result of a fire on such sites.

Seedling survival for subalpine fir, lodgepole pine, and Engelmann spruce is likely to be higher where seeds contact bare mineral soil (Day 1964, Woodard 1977). Ash seedbeds do not necessarily favor one of these species over the others (Woodard and Cummins 1987). Decayed wood is sometimes a good substrate (Day 1963), but needle litter is not preferred (Gracz 1989). Subalpine fir will do better than Engelmann spruce on needle litter (Knapp and Smith 1982), because spruce is susceptible to fungal attack (Daniel and Schmidt 1962). The literature is full of conflicting reports on the effects of charcoal and ash on seedling establishment for these species (Woodard 1977). Successful establishment may also require a moderated environment, or safe sites (e.g., Harper 1961), consisting of shrubs, existing small trees, or logs. On burned sites, successful seedling establishment may require protection from full sunlight, browsing, or frost damage provided by safe sites. In the western Cascades and eastern Olympic Mountains, safe sites are important for successful establishment of subalpine fir (Gracz 1989, Little 1992). Logs are the most consistent safe site identified, although shrubs, trees, krummholz, rootthrow mounds, and herbaceous clumps are also found. Snags may serve a similar role on recently burned sites. Both snags and resprouting shrubs will be available safe sites within a few years, but safe sites of live trees and downed logs may not be available for decades after a fire.

A series of studies in *Abies lasiocarpa/Vaccinium membranaceum* plant associations in Olympic, North Cascades, and Mount Rainier National Parks documented patterns of postfire tree recruitment in these forests (Huff and Agee 1991). These are modal *Abies lasiocarpa* forests on drier westside Cascade sites that do not contain mountain hemlock or lodgepole pine and are transitional to the *Abies amabilis* zone. Subalpine fir constitutes 40–90 percent of the tree density. The sites cover an age range: early (<25 years), young (40–100 years), mature (125–250 years), and old (>300 years). With the single exception of a young site, all have relatively similar species composition in the overstory and understory. One early site, four young sites, five mature sites, and two old sites were compared.

Little tree recruitment is present on the one early site. The four young sites (Fig. 9.4) show patterns of increasing tree recruitment over time. Apparent decreases in the frequencies below ages 15–20 are a result of not sampling trees that were less than 20 cm tall, so that reliable patterns are those above 20 years of age. The Deadwood Lakes and Eagle Point (Waterhole Canyon) fires had some residual trees (>100 years old), which may have aided subsequent tree recruitment. The mature sites (Fig. 9.5) show early recruitment patterns that have some similarity to the young sites, with some significant troughs and peaks during mid-succession. The old sites (Fig. 9.6) have some old trees together with significant recruitment in recent decades.

The patterns of stand development defined by Oliver (1981) are not clearly shown by these data. Stand development in these *Abies lasiocarpa* forests is affected by two major factors: relatively slow tree recruitment after disturbance and relatively short-lived dominant species. On a young burn site such as Eagle Point, the rate of recruitment increases for almost a century, documenting a very long stand initiation stage. Similar patterns are found at Upper Crystal and Deadwood Lakes. The Beebe Mountain site, although clearly open in appearance, has a recent lack of recruitment suggesting that it is entering the stem exclusion stage. In this case, the stem exclusion stage may be temporary, perhaps due to a recent climatic shift of some kind, or it may indicate presence of a limiting growth factor on the site (in the sense used by Oliver 1981) other than light.

All of the mature sites (125–250 years old) have roughly bell-shaped age-class distributions. This suggests that there was a time in the past when tree recruitment peaked, usually 75–125 years after disturbance. The recruitment was then sustained for a while (Old Perry and Trout Pond), declined slowly (Beebe Mountain), or peaked at several different times (Mount Seymour). These patterns reveal individualistic tree recruitment by site, suggesting that the stem exclusion stage is not well defined in these forests.

The shape of the age-class distributions on the old sites appears to be closer to the young than to the mature sites. Rather than bell-shaped distributions, the Stillwell Creek and Steeple Rock sites show substantial recent recruitment, indicative of the understory reinitiation stage of stand development. The old-growth stage does not really have time to develop, because the dominant old

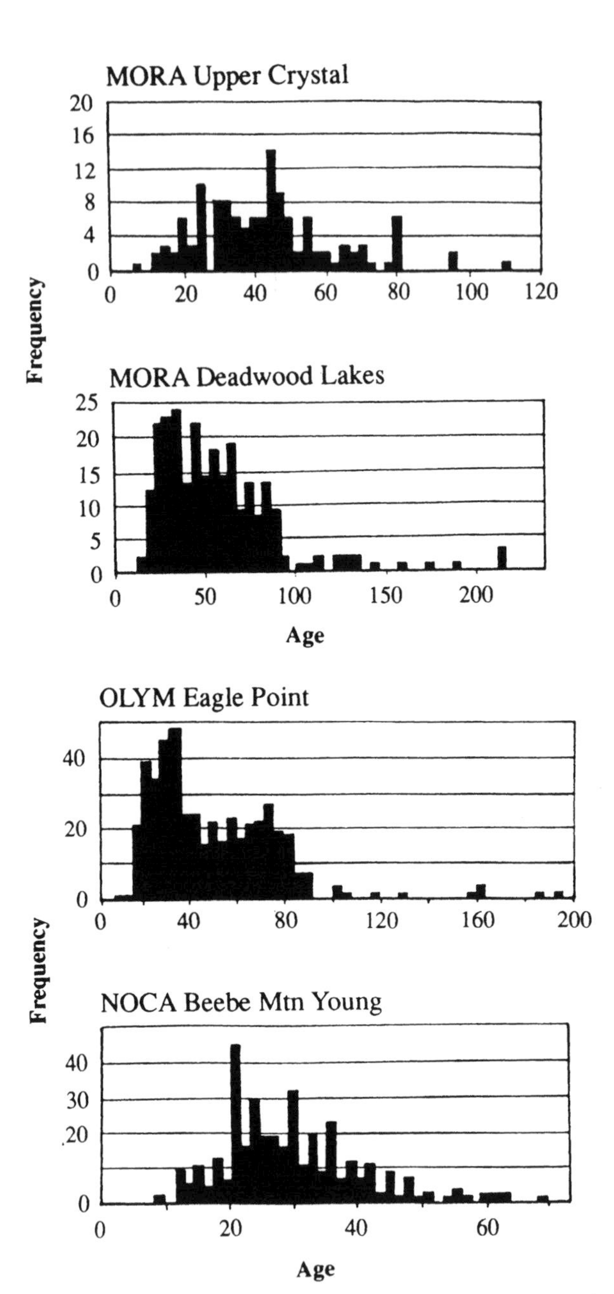

FIG. 9.4. *Age-class distributions after fire for young (40–100-year-old) burns in* Abies lasiocarpa/Vaccinium *forests.* (From Huff and Agee 1991)

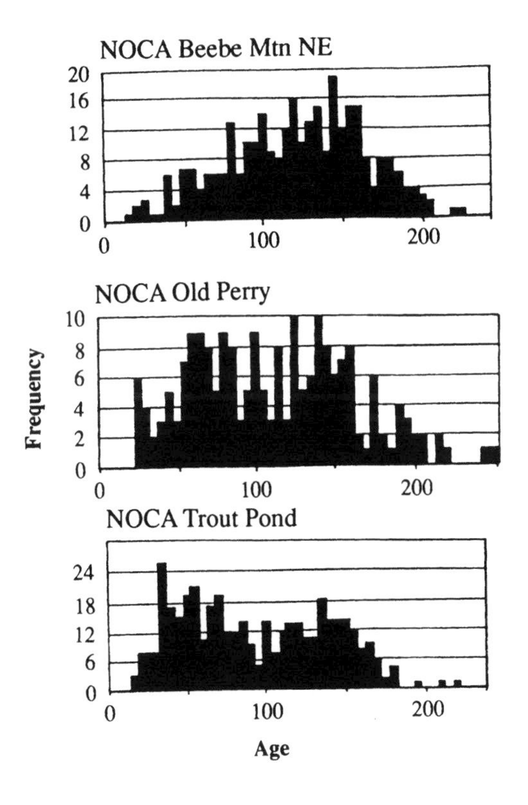

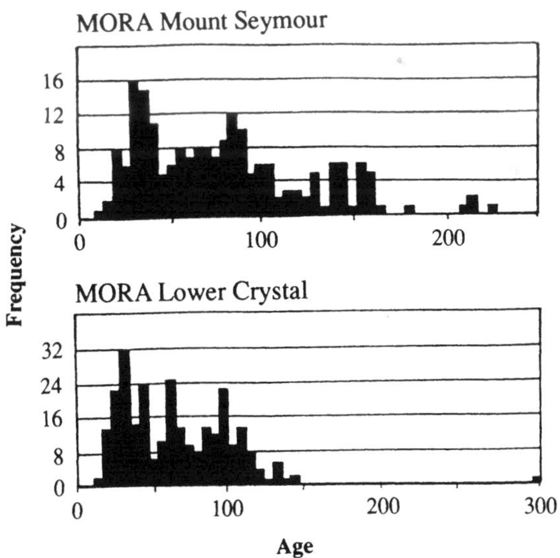

FIG. 9.5. *Age-class distributions after fire for mature (125–250-year-old) burns in* Abies lasiocarpa/ Vaccinium *forests.*
(From Huff and Agee 1991)

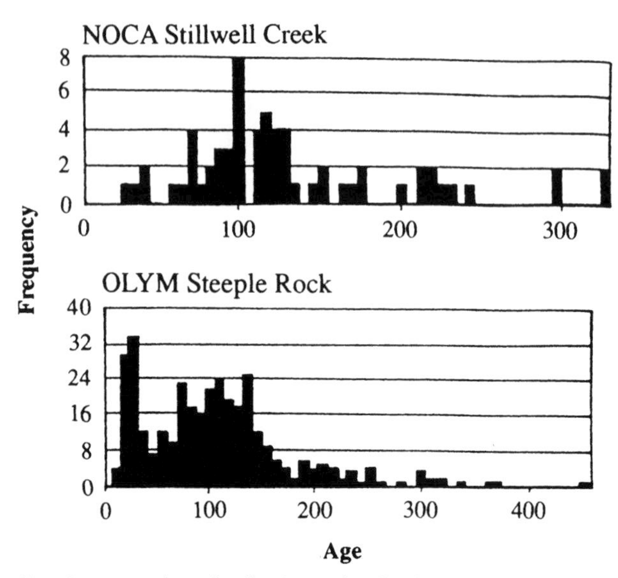

FIG. 9.6. *Age-class distributions after fire for old (>250-year-old) burns in* Abies lasiocarpa/Vaccinium *forests.* (From Huff and Agee 1991)

subalpine firs are reaching the average maximum life span (250 years; Franklin and Dyrness 1973) at about the same time that the understory reinitiation stage is under way. Clearly, the older sites are "old growth" forest, but because of the slow recruitment and short life span of the trees the classic characteristics of the old-growth stage cannot develop. The sequence defined by Oliver (1981) apparently truncates at the understory reinitiation stage.

Another way of evaluating stand development is to compare the shapes of the age-class curves when the data are smoothed by compressing age-classes into 15-year increments and fitting the rather flexible Weibull distribution to the data. Scale and shape parameters derived from the modeled age-class distributions can be compared between sites. The scale parameter (b) is associated with stand age. It increases from the young (2.83) to mature (6.73) to old (7.53) sites, and represents the increasing number of older trees on older burns. However, with the amount of recent recruitment on the old sites, and the high probability that the remaining older trees will die before long, the old sites may eventually develop a scale parameter characteristic of younger stands. In fact, these older burns will eventually be composed of primarily younger trees.

A shape parameter (c) equal to 1 is equivalent to a negative exponential distribution, while a shape parameter nearing 3 is approximately a normal distribution. The shape parameter averaged 1.854 for young stands, 2.171 for mature stands, and 1.673 for old stands. The young stands have a shape parameter indicative of a normal distribution skewed to the right; the mature stands have the closest to a bell-shaped distribution; and the old stands have the closest to a negative exponential distribution. The difference in average is not high, and there is overlap among the three groups of burns. An ecological interpretation of these shapes suggests that the young sites are entering the stem exclusion stage of stand development, the mature sites are in the stem exclusion stage, and the old sites are in the understory reinitiation stage, with many more stems in the younger age classes and a declining number in older age classes. Whether the understory reinitiation stage persists in the absence of disturbance, or whether it cycles back to the stem exclusion stage, is unknown. In the presence or absence of fire, subalpine fir eventually dominates these modal sites, but stand architecture will vary significantly by stand age.

In the eastern Cascades, where lodgepole pine is part of the stand at the time of the fire event, a tremendous amount of seed can be released by the opening of serotinous or semi-serotinous cones, and tree regeneration in the first year after the fire may be sufficient to restock the stand. Both subalpine fir and Engelmann spruce are common coclimax associates in such stands (Henderson 1981). In the Rocky Mountains, postfire tree regeneration of spruce and fir may be slow and unpredictable, depending on the presence of a seed source, a good seed year, and above-normal precipitation in the growing season (Habeck and Mutch 1973).

In the Pasayten Wilderness, Fahnestock (1976) inventoried a suite of *Abies lasiocarpa* plant associations, from recently burned to over 400 years old. Lodgepole pine is an early dominant in such stands, almost to the exclusion of Engelmann spruce and subalpine fir. The spruce and fir that do establish grow more slowly than the pine and are relegated to subordinate canopy positions. The lodgepole pine reach peak density and basal area within 50 years and may dominate the site for up to 150 years. Spruce and fir in the understory continue to increase over time (Table 9.4), and subalpine fir is the major climax dominant on these sites (Daubenmire and Daubenmire 1968, Agee et al. 1990). Occasional lodgepole pines will persist up to about 400 years (Fahnestock 1976) in these fir-spruce forests.

An early lodgepole pine stage may not occur if the fire-return interval has been so long that lodgepole seed source has disappeared from the site. This requires fire-free intervals of more than 300 years. In such cases, subalpine fir and Engelmann spruce will compose the dominant tree component, but as in *Tsuga mertensiana* forests without lodgepole pine, the tree recolonization time may be extended for decades to centuries. Conversely, if fires occur at short intervals, lodgepole pine may be the sole conifer dominant, and the sites may be difficult to identify as *Abies lasiocarpa* forests.

Whitebark pine may grow in pure stands as the climax dominant or in association with subalpine fir and Engelmann spruce as an early seral species (Steele et al. 1983, Keane et al. 1990). Whitebark pine has wingless seeds and is dependent on animals such as Clark's nutcracker (*Nucifraga columbiana*) to disperse its seed to the interior portions of burns. In climax stands, whitebark pine may sometimes grow in krummholz form because of severe winter wind exposure. Such stands burn like chaparral, although they may not be of large extent. After such a severe burn, Tomback (1986) documented a density of whitebark pine seedings of 0.2 m^2. At each germinant site there were 1–4 seedlings from a Clark's nutcracker cache, and these birds were observed caching seeds in the burned area from unburned erect whitebark pines located downslope. In large severe burns, seed dispersal by nutcrackers may give whitebark pine an advantage over other conifers that rely on wind dissemination of

TABLE 9.4. REPRODUCTION (SEEDLINGS PER HECTARE, <4 CM DBH) OF LODGEPOLE PINE, ENGELMANN SPRUCE, AND SUBALPINE FIR AFTER FIRE IN PASAYTEN WILDERNESS, WASHINGTON

Years since Burn	*Lodgepole Pine*	*Engelmann Spruce*	*Subalpine Fir*
0–50	2,746	259	116
50–100	597	476	141
100–200	2	60	244
200–400	0	96	1,394

SOURCE: Fahnestock 1976

seed (Tomback et al. 1990). In small or less severe burns, dispersal by nutcrackers may still be important. For example, 10 years after a whitebark pine clump burned at Crater Lake (Fig. 9.2), seven whitebark pine seedlings 9–60 cm in height had replaced seven trees killed by the fire, and some were clearly cache-related (J. K. Agee, unpublished data).

Where whitebark pine is seral to subalpine fir, it may colonize the site first, along with lesser numbers of subalpine fir. Later establishment is primarily by subalpine fir (Fig. 9.7). The replacement appears to take so long that fire is likely to occur before whitebark pine has been eliminated from the site (Morgan and Bunting 1990).

Simulation of the presence/absence of fire in seral and climax whitebark pine stands (Keane et al. 1990) indicated that in climax stands the major effect of small fires at 80 year fire-return intervals was to reduce the basal area of whitebark pine and essentially eliminate subalpine fir. In seral stands, whitebark pine and subalpine fir were largely replaced by lodgepole pine, due to the stand-replacing fire intensity (500–1,000 kW m^{-1}) of most fires on the higher productivity sites. Under simulated natural fire frequencies, whitebark pine was able to maintain itself in seral and climax stands. However, the inclusion of white pine blister rust into the simulations reduced the importance of whitebark pine in both seral

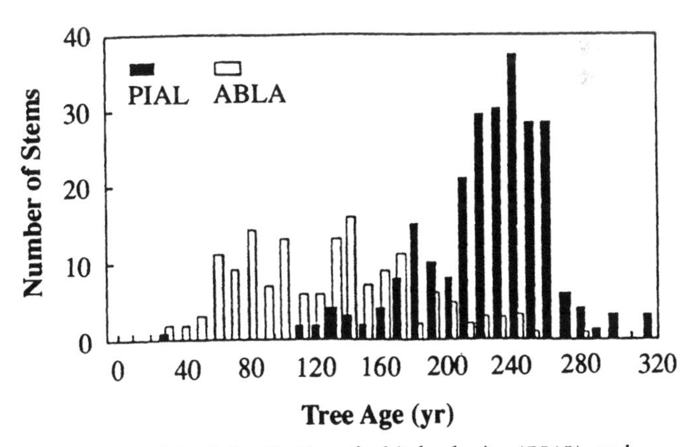

FIG. 9.7. *Combined distribution of whitebark pine (PIAL) and subalpine fir (ABLA) trees by 10-year age classes for 10 stands on the Shoshone National Forest, Wyoming.*
(From Morgan and Bunting 1990)

and climax stands; inclusion of mountain pine beetle has a larger negative effect in seral stands. Although Keane et al. (1990) did not evaluate the synergistic effects of fire and blister rust, observation of western white pine and blister rust after fire in the Olympic Mountains suggests that the future of whitebark pine is not bright in the face of these multiple disturbances.

Subalpine Nonforest Communities

All subalpine plant communities will burn, but not under all conditions. In the *Tsuga mertensiana* zone, closed or parkland forest has the highest probability of burning, because of the dead fuel loads that can be desiccated during east wind events and the presence of flammable lichens in crowns low to the ground (Agee and Smith 1984). In the 1978 Hoh fire in western Olympic National Park, only tree clumps burned; intervening red heather/huckleberry (*Phyllodoce empetriformis/Vaccinium deliciosum*) meadows did not. In the 1981 Chimney Peak fire about 20 km to the east, high winds drove the fire across heather meadows and patches of heather being invaded by small mountain hemlocks. The only plant community that escaped fire was the late snowmelt *Carex nigricans* community, growing in the bottoms of microtopographic basins.

Little is known about the response of nonforested subalpine communities to fire. Henderson (1973) suggested that fire would transform red heather meadows to either lush herbaceous (*Valeriana sitchensis/Carex spectabilis*) or pure *Vaccinium* communities. Some of this speculation was based on the North Cascades study of Douglas and Ballard (1972), who compared a 29-year-old burned area with an adjacent unburned area, assuming that the differences they found were the result of burning. They found that red heather and white heather (*Cassiope mertensiana*) were rare in the burned area, and that the sprouting huckleberry was the major postfire dominant. A postfire reconnaissance of the Chimney Peak fire in 1985 revealed that red heather had sprouted vigorously after the 1981 fire. Red heather was less likely to sprout near forest edges, where deep litter layers had probably smoldered and killed its above- and below-ground stems and roots (Potash 1989). Heather was observed to sprout after fire at several other locations in the Olympic

Mountains and Cascades (Fig. 9.8), so the type of fire, rather than its presence or absence, is important in predicting postfire response. Experimental prescribed fires in heather meadows showed that after one year, in comparison to control plots, burned plots had higher stem density and lower dry matter production (Potash 1989).

The 12 species of huckleberry (*Vaccinium* spp.) in the Pacific Northwest cover 60,000 ha (Minore 1972). Most of these fields are potentially forested sites that were transformed to brushfields by past burning and are slowly reforesting. Berry production can approach 1,000 l ha^{-1} in a high quality field, but production declines over time as tree crown cover increases. Native Americans used these fields extensively to collect berries over the summer months (Minore et al. 1979). Management techniques to maintain thin-leaved (big) huckleberry (*Vaccinium membranaceum*) fields have included burning, slashing invading trees and burning them, grazing, borax application, and herbicide application. Herbicide applied to frilled cuts on trees was recommended as a preferred treatment (Minore et al. 1979), since it had the least short-term impact on berry production. Fire was used under such conservative weather conditions that diesel oil (up to 1 l per 2 m^2) was necessary to carry the flame. Free-burning fire would have been a more natural experimental treatment, but it will usually occur only under conditions that threaten to allow fire escape. With the diesel oil/burn treatment, berry production was delayed for more than 5 years, when the control plot produced 35 kg ha^{-1} of berries, 7 times the berry mass on burned plots and 300 times the mass on slash/burn plots

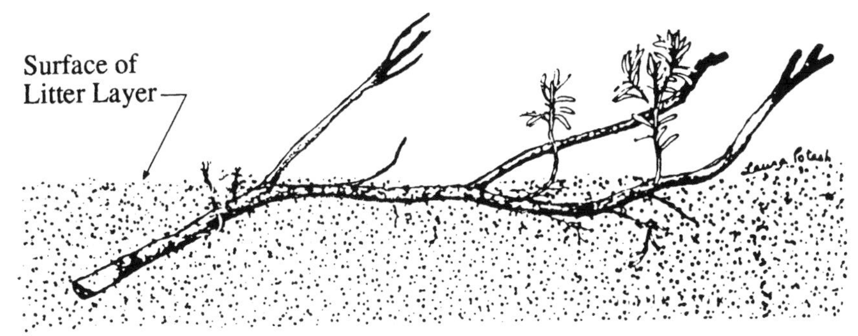

FIG. 9.8. *Sprouting pattern of red heather after topkill by fire.*
(From Potash 1989)

(Minore et al. 1979). The slashing treatment apparently created enough heat to significantly damage the huckleberry rhizomes in the soil.

Burning can increase stem density for some huckleberry species. In Montana, Miller (1977, 1978) found that spring burning on lower elevation sites increased density of blue huckleberry (*Vaccinium globulare*), probably because of topkill without significant soil heating. However, late summer burns resulted in fewer stems. In the Pacific Northwest longer time periods after burning need to be monitored, and if possible free-burning fires should be studied more intensively. Such fires may pose a threat of escape or may be wildfires, but they are likely to produce conditions more similar to those that created the historic productive huckleberry fields of the region.

The ecotone, or boundary, between subalpine meadows and forest is quite dynamic but changes slowly. As climate has changed, there has been a lag time between those changes and the adjustment of the ecotone. During warming/drying episodes, tree establishment may move upslope into what were meadow environments. During cooling episodes, mature forest may be affected most by changes in growth rates. If disturbance occurs, however, killing mature forest, a new ecotone between forest and meadow may be created.

Regional warming during the twentieth century has been associated with substantial tree invasion into heather meadows in the Cascades and Olympics (Fig. 9.9). This type of invasion into cool, normally snow-dominated meadows with short growing seasons has occurred with a pattern almost opposite to that of burned subalpine forests, which tend to occur on relatively warm, dry subalpine sites, on steep south aspects. Along a gradient from wet/cold (e.g., heather meadows) to warm/dry (typical steep south aspect forest sites that burn), a bell-shaped curve for tree regeneration potential is hypothesized to exist (Fig. 9.10). Potential for tree regeneration is not high in heather meadow or burned forest under "normal weather." A climatic shift to wet/cold weather moves the curve to the right (curve B) and makes the heather site even worse for tree recruitment, while providing better conditions on the burned site. A shift to drier/warmer weather moves the curve to the left (curve C), allowing tree invasion into meadows but little tree recruitment onto burned sites. The potential for tree establishment

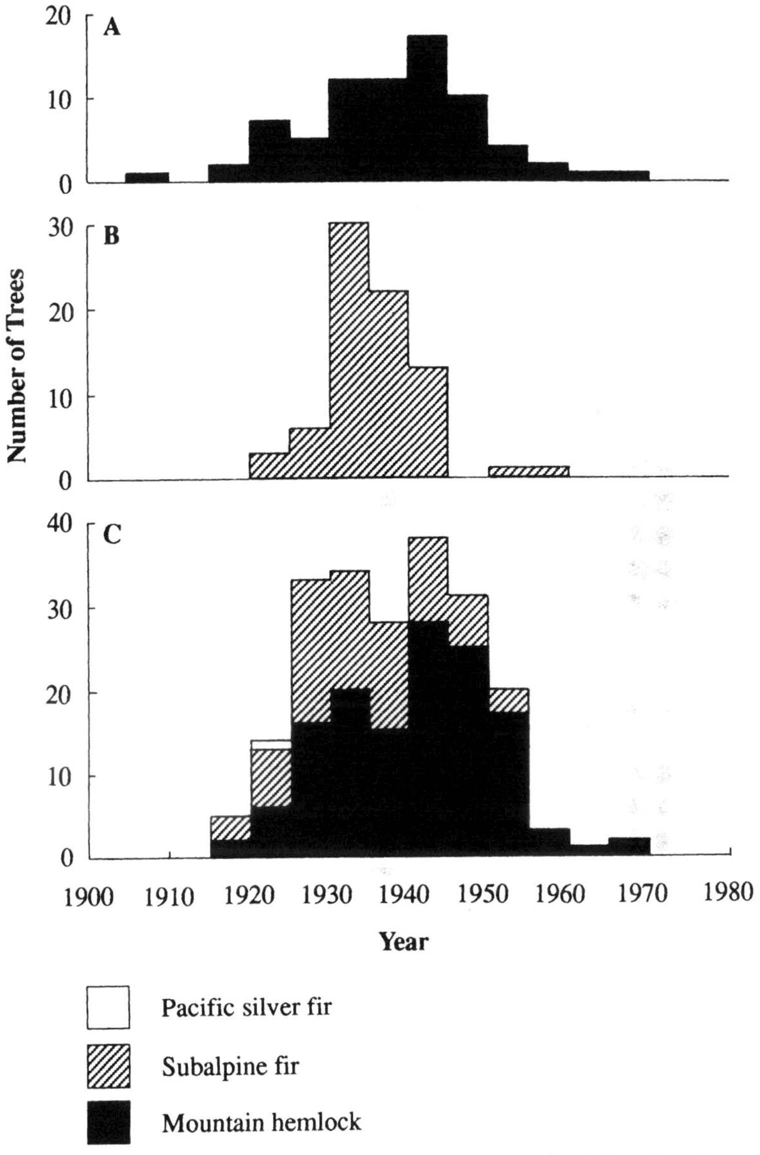

FIG. 9.9. A: *Tree invasion into heather meadows near High Divide, Olympic National Park.* B: *Invasion of trees into meadows at Mount Rainier.* C: *Invasion of trees into meadows at Mount Baker. The major invasion period at all sites is 1925–45.*
(From Agee and Smith 1984 [A], Franklin et al. 1971 [B], and Lowery 1972 [C])

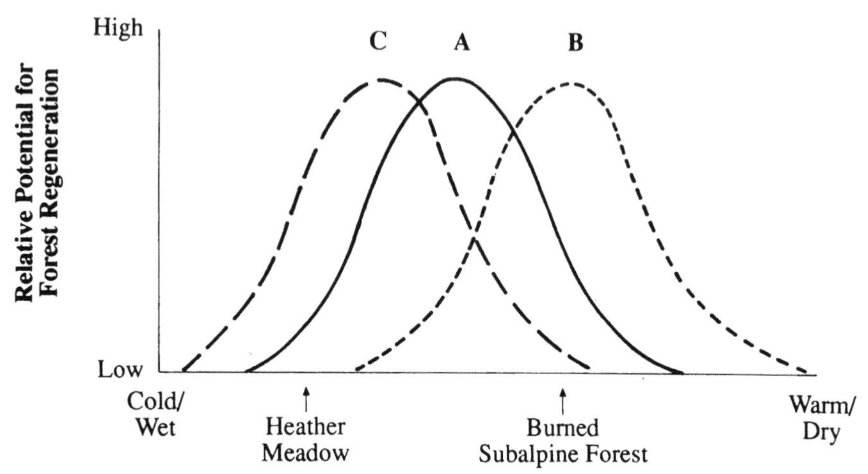

FIG. 9.10. *Relative potential for tree establishment along an environmental gradient of cold/ wet to warm/dry, with a heather meadow and burned subalpine forest site located along the gradient. In "normal" climate (A), tree recruitment potential is low for both heather meadow sites (cold/wet) and burn sites (warm/dry). A climate shift to "colder/wetter"* (B) *shifts curve to the right, making tree recruitment potential better on burns but worse for heather meadows. A climate shift to "warmer/drier"* (C) *shifts curve to the left, making tree recruitment potential worse on burns but better in meadows (see Fig. 9.9).*

in these various environments after fire (Fig. 9.11) is therefore a function of site as well as changes in climate.

Drier subalpine meadows, such as those dominated by *Festuca* spp., have burned frequently (Kuramoto and Bliss 1970), and Bockheim (1972) found charcoal in all the meadow soils excavated in the North Cascades. East of the Cascades, the role of fire in communities dominated by fescue (*Festuca viridula*) is confounded by extensive overgrazing for the last half-century, reducing potential fine fuels that carried fire and significantly changing species composition (Franklin and Dyrness 1973). These meadows probably burned along with the intermixed forest types, and if typical of other *Festuca* grasslands, regained their prefire species composition and cover within several years (Wright and Bailey 1982). Little is known about fire effects in wet subalpine meadows. In the Sierra Nevada, normally wet sedge-tallgrass meadows may burn during drought (DeBenedetti and Parsons 1979). Invading lodgepole pine may be killed as the fire smolders along the organic layers, leaving ash layers 10–25 cm thick.

FIG. 9.11. *Subalpine fire (Chimney Peak, Olympic National Park, 1981 [photographed in 1987]) burned not only mature forest but also meadows being invaded by trees. The potential for tree establishment depends on the site as well as on postfire climate.*

FUELS

After a crown fire in subalpine fir forest, coarse woody debris may exhibit a U-shaped pattern of abundance (Harmon et al. 1986) in the first few centuries after the fire. In a series of *Abies lasiocarpa/Vaccinium membranaceum* plant associations, Huff et al. (1989) found differences between size classes of fuel over a stand age range of 300 years (Table 9.5). Young stands had lower amounts of 1 hr and 10 hr timelag fuels in comparison to stands of other ages. High levels of 100 hr timelag fuels on early sites represent branches falling from fire-killed trees. The low mass of larger (>1,000 hr timelag) fuels on mature sites suggests that a "trough" occurs in stands at 100–200 years postfire, after snags have fallen and decayed but before new trees from the postfire stand have matured,

TABLE 9.5. DEAD AND DOWN FUEL LOADS (T HA^{-1}) IN CHRONOSEQUENCE OF SUBALPINE FIR STANDS

Stand Age[a]	*Number of Stands*	*1 hr Timelag*	*10 hr Timelag*	*100 hr Timelag*	*1,000 hr Timelag*	*>1,000 hr Timelag*	*Total Fuel Load*
Early	2	0.3	1.6	11.0	11.7	53.6	78.2
Young	10	0.1	0.7	3.0	12.0	62.0	77.9
Mature	9	0.5	1.6	2.7	8.4	12.2	25.4
Old	2	0.4	2.7	8.3	17.4	68.8	97.6

SOURCE: Huff et al. 1989.

[a] Early = 0–25 years; young = 40–100 years; mature = 125–250 years; old = >300 years.

died, and fallen to create a new source of coarse woody debris. Similar troughs are evident at the same stand ages in the eastern Cascades (Fahnestock 1976) and at Yellowstone (Romme 1982).

In the eastern Cascades, multiple fires in a few decades are possible (Woodard 1977). If two fires occur in a period of several decades, the coarse woody debris generated by the first fire will be consumed before the new stand has a chance to generate any, leaving a relatively debris-free ground (Fahnestock 1976).

MANAGEMENT IMPLICATIONS

Subalpine fires tend to be erratic and unpredictable. Although they are infrequent in most Pacific Northwest subalpine forests, fires have been important in shaping the landscapes we see today. Many subalpine meadows bordering forest were created by fire; today they provide diverse habitats for wildlife and recreational sites for people. Fire has been important in the distribution and abundance of lodgepole pine in eastern Cascade forests and in the maintenance of habitat for whitebark pine at treeline. The fire suppression period during the twentieth century so far has not had much impact on landscape structure in subalpine zones because of the fairly long fire-return intervals. A continued policy of fire exclusion should not have major impacts in this zone for continued decades. Subtle yet significant shifts may occur in species composition over time, and exclusion of a natural process such as fire may be at odds with park and wilderness mandates.

In many subalpine areas, the isolation of forests and meadows by snowfields or mountain masses makes them relatively safe for prescribed natural fire policies. Fires are likely to be confined to a particular slope or valley. Where the terrain becomes more uniform and plateaulike (e.g., Yellowstone) natural fire barriers are few, and sustained fire runs over long periods of time are possible.

One option contained in some early wilderness plans was to allow fires below a certain flame length to burn, suppressing them when predicted fire weather became more severe. This is not a preferred management option because once fires become large, immediate suppression may be impossible and attempts may create severe suppression impacts (bulldozer lines and so forth). Even if it were possible to develop a flame length criterion, it might be associated with changes in species composition that are not "natural" (Marsden 1983).

Another option is prescribed fire, which has not been applied at a large scale in subalpine ecosystems. It is possible to ignite such fires in early spring in some areas (Woodard et al. 1983) or in late summer (Woodard 1977). Natural fires in early spring are an unusual occurrence, and prescribed fire at that time would not substitute for natural fire on a large scale. Prescribed fire ignitions have potential to substitute for natural fires but have some of the same drawbacks: Under many conditions they will not burn, while in severe fire weather they will be unpredictable.

A third option is to continue a fire suppression policy but apply "light on the land" suppression techniques that use natural barriers wherever possible. In these areas, the highest values are often natural, with much lower values in timber or threats to resources or people. In subalpine areas in the western Cascades, fire might be allowed to play almost a natural role under such circumstances. In the more broadly distributed subalpine forests of the eastern Cascades, this might work for the highest elevation forest, but prescribed fire might be a better option for lower elevation, more contiguous forests, if a natural fire policy were not possible.

CHAPTER 10

MIXED-CONIFER/ MIXED-EVERGREEN FORESTS

THE MOST COMPLEX SET of forest types in the Pacific Northwest includes those called mixed-conifer or mixed-evergreen forests. These types have in common a wide variety of coniferous, and sometimes hardwood, tree species. They differ in their specific mix of species, their fire regime, and the successional patterns likely after disturbance. Ponderosa pine (*Pinus ponderosa*) as a seral species and Douglas-fir (*Pseudotsuga menziesii*) as a seral or climax species can be found in each type, although not at every site. Four types have been identified across the region: *Pseudotsuga*/hardwood, *Pseudotsuga menziesii*, *Abies concolor*, and *Abies grandis* (Fig. 10.1). *Pseudotsuga menziesii* is a climax or coclimax species with hardwoods across much of the Mixed Evergreen zone of southwestern Oregon and northern California, as identified by Franklin and Dyrness (1973) and Sawyer et al. (1977). A *Pseudotsuga menziesii* zone is sometimes present along the eastern Cascades, separating the *Pinus ponderosa*

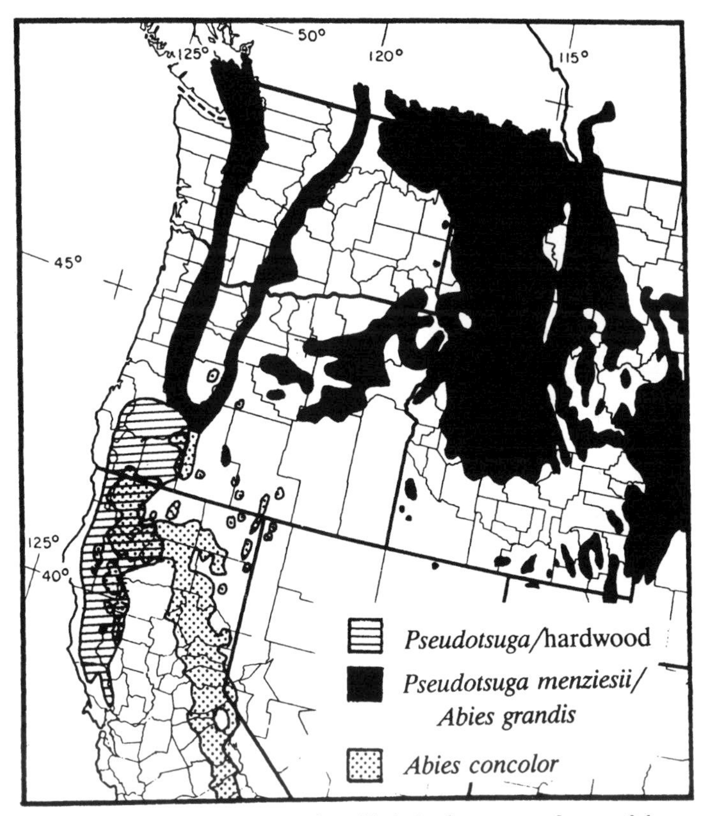

FIG. 10.1. *Location of the mixed-conifer/mixed-evergreen forests of the Pacific Northwest. The* Pseudotsuga menziesii *and* Abies grandis *forests are shown as one pattern because they are so intermixed across much of their range.*
(Adapted from Little 1971)

and *Abies grandis* zones (Franklin and Dyrness 1973), although it or the *Abies grandis* zone may be locally missing (Hall 1967, Franklin and Dyrness 1973). True firs are either absent or minor components of the *Pseudotsuga menziesii* zone. On more mesic, higher elevation sites in southwestern Oregon, and extending from the southern Cascades south along the western slope of the Sierra Nevada, the *Abies concolor* zone is found. Along the eastern Cascades north into Washington, the *Abies grandis* zone, where present, separates the lower elevation and subalpine forests. The major tree species within each of the zones are listed in Table 10.1.

TABLE 10.1. STATUS OF MAJOR TREE SPECIES OF BROAD FOREST ZONES IN MIXED-CONIFER/MIXED-EVERGREEN FORESTS

Zone and Major Species	*NW Calif., SW Oregon*	*E. Oregon Cascades*	*Ochoco/ Blue Mtns.*	*E. Wash. Cascades*
Pseudotsuga/hardwood zone				
Douglas-fir	M	—	—	—
Ponderosa pine	M	—	—	—
Pacific madrone	M	—	—	—
Tanoak	M	—	—	—
Incense-cedar	m	—	—	—
Chinquapin	m	—	—	—
Port Orford–cedar	m	—	—	—
Knobcone pine	m	—	—	—
Sugar pine	m	—	—	—
Canyon live oak	m	—	—	—
Pseudotsuga menziesii zone				
Douglas-fir	M	M	M	M
Ponderosa pine	M	M	M	M
Lodgepole pine	—	m	m	M
Western larch	—	m	m	m
Jeffrey pine	m	—	—	—
Incense-cedar	M	—	—	—
Abies concolor zone				
White fir	M	M[a]	—	—
Ponderosa pine	M	M	—	—
Douglas-fir	M	m	—	—
Incense-cedar	m	m	—	—
Lodgepole pine	m	m	—	—
Sugar pine	m	m	—	—
Abies grandis zone				
Grand fir	—	M[a]	M	M
Ponderosa pine	—	M	M	M
Douglas-fir	—	M	M	M
Lodgepole pine	—	M	M	M
Western larch	—	m	M	M
Western white pine	—	m	m	m

NOTE: M = major species; m = minor species.

[a] white fir/grand fir complex in some places.

PSEUDOTSUGA/HARDWOOD FORESTS

These *Pseudotsuga* forests are rich in tree species; commonly, three to five major conifers, including Douglas-fir, will be found in association with three to five hardwood species—rather unimpressive by eastern U.S. standards but very impressive for the West. Whittaker (1960) noted this to be one of the most diverse and complex forest regions of western North America, and the *Pseudotsuga*/hardwood type is the most widespread forest type within this diverse region. Subdivisions of this type into an inland Klamath phase and a coastal North Coast phase were made by Sawyer et al. (1977), and the map accompanying that book shows a more hardwood-codominant mixed-evergreen phase (Kuchler 1977). As effective moisture decreases, the dominant hardwood component gradually shifts from tanoak (*Lithocarpus densiflorus*) to Pacific madrone (*Arbutus menziesii*) and canyon live oak (*Quercus chrysolepis*), with chinquapin (*Castanopsis chrysophylla*) on cooler sites.

THE FIRE REGIME

Most of the drier forest types in the western United States had a significant component of Native American burning, although the use of fire was often restricted to certain locales. Burning of tanoak was done to "kill diseases and pests" and to clean the ground under the trees so that the acorns could be picked up more easily (Lewis 1973). Interior *Pseudotsuga*/hardwood forests with less tanoak were probably also frequently burned along ridgetops to maintain travel corridors and openings for production of hazel (*Corylus* sp.) and beargrass (*Xerophyllum tenax*), both of which were used for basketry material, one or two years after a site was burned (Boyd 1986).

Lightning is also common in these forests. The patterns of highest lightning occurrence in the Pacific Northwest (Fig. 2.1) are somewhat coincident with these forests. The fire cycle model of Agee and Flewelling (1983) shows the Siskiyou Mountain area to have more than twice as many ignitions as McKenzie River at the southern end of the *Tsuga heterophylla* series (Agee 1991b). The differences are largely due to increased lightning frequency and decreasing summer precipitation patterns to the south. July and August are the months of greatest ignition activity, but September has a higher proportion of total ignitions for southerly stations such as the Siskiyou Mountains. The model, which was initially designed for the Olympic Mountains, does not include October ignitions, which would likely increase the total ignitions for the southern Oregon area and northern California.

The variable fire history of this area, together with complex geology, land use history, and "steep" environmental gradients, has prevented generalizations about fire and its ecological effects. The Native American activities before 1850; the activities of miners, settlers, and trappers in the mid-nineteenth century; and twentieth-century fire suppression policies have all influenced fire history in *Pseudotsuga*/hardwood forests (Atzet and Wheeler 1982, Atzet et al. 1988).

On the slopes of the Salmon River in northern California, Wills and Stuart (in prep.) found presettlement mean fire-return intervals of 11–17 years, but with a large range (3–71 years on one site), using composite stump samples from a 5–8 ha area. Fire-return intervals averaging 18 years over 1 ha areas (Agee 1991a) to 20 years (Atzet et al. 1988) have been found in the eastern Siskiyou Mountains of Oregon. At the high elevation end of the *Pseudotsuga*/ hardwood type in the Siskiyou Mountains, transitional to *Abies concolor* forest, Agee (1991a) found a mean fire-return interval of 37 years between 1650 and 1930, using conservative natural fire rotation techniques.

Fires most likely spread over periods of weeks to months in these forests. Morris (1934b) quotes the *Jacksonville Sentinel* in September 1864: "during the past few weeks . . . the fire [in the Siskiyous] has been raging with increasing fury." During the large wildfire events of 1987 in southwest Oregon–northern California, so many lightning-caused fires started that control actions were ineffective in stopping them. The fires, which began on 30 August, burned into November 1987 (Helgerson 1988). The low probability of precipitation probably allows fires to "die down" but not be extinguished during periods of low winds or moderate weather, and to remain capable of renewed spread under patterns of windy or warmer weather. Once started, fires in these Douglas-fir forests can continue to burn until autumn rains come, so they can often cover large areas. The 1987 fires of southern Oregon, when 39 fires burned over 70,000 ha, did not set a record; their toll was surpassed twice in the recorded history (since 1907) of the Siskiyou National Forest (Helgerson 1988, Atzet et al. 1988), suggesting that 1987 was an unusual but not unprecedented year. The same year more than 70,000 ha also burned on the Klamath National Forest in northern California.

The intensity of presettlement fires encompassed a wide range of fire severity with many fires, or large portions of them, burning at low to moderate severity. Evidence of low severity burns is seen in the predominance of live older trees with fire scars (many of which are healed over) or basal char in these forests; moderate severity burns are indicated by the coincidence of fire dates and regeneration cohorts, suggesting creation of some growing space by postfire tree mortality. High severity burns are revealed by the absence of conifers on some sites dominated by hardwood trees or shrubs, the dominance of species such as knobcone pine (*Pinus attentuata*), or even-aged stands of other conifer species.

Stand Development Patterns

Douglas-fir has long been characterized as a species adapted to fire (Flint 1925, Starker 1934). Recognition of its superior ability to adapt to the presence of fire is critical to an understanding of the ecological impacts of fire. Its life history strategy to fire (e.g., Rowe 1981, Wright and Bailey 1982) as a *resister* when mature enables it to maintain dominance as long as fires are not too intense. In *Pseudotsuga*/hardwood forests, it is likely to remain a dominant after moderate severity fire in a mature stand where Pacific madrone and tanoak are understory species. In the presence of a moderate fire these associated species are top-killed (*endurers*) and have to resprout from the ground. However, if a young stand with the same species mix is burned again in one or two decades, Douglas-fir as a small tree will not be fire-tolerant, and as an *avoider* it will then be killed, leaving the madrone and tanoak to resprout again and dominate the postfire stand for many decades. Eventually, a mixed forest of Douglas-fir overstory and tanoak understory will dominate these

Pseudotsuga/hardwood forests (Sawyer et al. 1977, Thornburgh 1982). All-sized forests with multiple-aged cohorts are common.

The patterns of stand development in *Pseudotsuga*/hardwood forests represent variable fire severities and the ability of the different species to take advantage of postfire conditions. A modal sequence in the *Pseudotsuga*/hardwood type is shown in Figure 10.2. Beginning after a stand replacement fire, the Douglas-fir regenerating on the site may survive several moderate severity fires that thin the

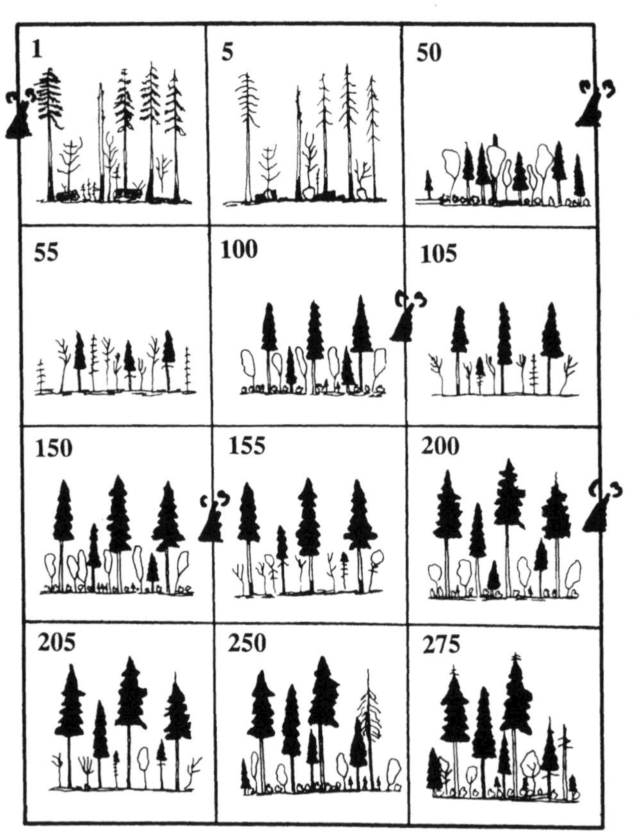

FIG. 10.2. *Stand development sequence for* Pseudotsuga menziesii*/hardwood forest. Fire-return interval is about 50 years. Fires are denoted by flame pattern with smoke puffs at edge of panels. Old-growth character is found in stands that burn with low to moderate severity over several fires, allowing the development of large residual Douglas-fir.* (From Agee 1991b)

Douglas-fir *resisters*, remove the *avoiders*, and topkill the associated hardwoods, such as madrone and tanoak (*endurers*), unless the fire is of very low intensity and duration. Several recurrences of such fires will create a stand with various age classes of Douglas-fir (some of which are large), and an age class of Douglas-fir and hardwoods representing regeneration after the last disturbance. Associated fire-resistant conifers include ponderosa pine, sugar pine (*Pinus lambertiana*), and incense-cedar (*Calocedrus decurrens*), with Port Orford–cedar (*Chamaecyparis lawsoniana*) common in draws. At age 250+, the structure of this stand may superficially appear much like a seral Douglas-fir stand in a *Tsuga heterophylla* forest, although subject to considerably more fire disturbance. Such stands will usually be intermixed with others that have experienced a stand replacement event during one of the intermediate fires, so that the stand pattern on the landscape is quite patchy.

Successional patterns after a severe fire, such as the one beginning the sequence in Figure 10.2, are summarized by Thornburgh (1982). Because most of the hardwoods sprout from an established root crown, they have a significant competitive advantage over Douglas-fir. Douglas-fir seed crops are irregular in *Pseudotsuga*/hardwood forests, with heavy crops perhaps once in seven years. The seedlings must compete not only with the hardwood trees but also with a variety of shrubs with *endurer* or *evader* strategies (manzanita [*Arctostaphylos*], snowbrush [*Ceanothus velutinus*] and deerbrush [*C. integerrimus*], gooseberries and currants [*Ribes* spp.]), as well as with a variety of herbs. The initial colonization phase by Douglas-fir can take decades, as shown by a several-decade-wide age class of 240–270-year-old dominant Douglas-fir at Oregon Caves National Monument (Agee 1991a). In areas where tanoak and Douglas-fir occur, Thornburgh (1982) suggests that either Douglas-fir eventually emerges through the tanoak canopy, or if Douglas-fir was immediately successful at establishment, tanoak develops under its canopy in a codominant role.

An example of repeated intense fire in the eastern Siskiyous is shown by the Kinney Ridge landscape west of the Applegate River, originally photographed by Hofmann (1917b) (Fig. 10.3A) and rephotographed in 1988 (Fig. 10.3B). The repeated fires have eliminated Douglas-fir across most of the landscape if it existed there, and 70 years later the area remains dominated by *endurer* species,

FIG. 10.3. A: *The south-facing slope of Kinney Creek, near the Applegate River in southwest Oregon, 1917. The numbers on the photo correspond to fire dates: 1 = 1917; 2 = 1914; 3 = 1910; 4 = 1897; 5 = 1886; 6 = 1854.* B: *The same landscape, 1988. A fire-scarred Douglas-fir in a clearcut just north of the ridgeline (arrow) showed a fire-return interval of 18 years over a 2 ha area.*
(From Hofmann 1917 [A], Agee 1991a [B])

such as manzanita (*Arctostaphylos* spp.) and oaks (*Quercus* spp.), and the *evader* species knobcone pine. Just to the left of this photo, on the north side of the ridge, a Douglas-fir stand underburned repeatedly with a fire-return interval of about 18 years between 1740 and 1860, and now has little hardwood component at all (Agee 1991a).

In a *Pseudotsuga*/hardwood stand on the Salmon River, Wills and Stuart (in prep.) reconstructed conifer age classes of Douglas-fir and sugar pine on three sites. Hardwoods were not included in the analysis, because of topkilling after most events and poor condition of current-day cross-sections. Two of the three sites (each 5–8 ha) had trees up to 500 years old, and each of three smaller plots within each site showed unique regeneration patterns due to variable fire severity (Fig. 10.4). The majority of conifers on plot 1 established themselves after an apparently severe fire in 1802. Almost continuous recruitment after that time suggests that later fires were of low severity. Plot 2 apparently experienced a high severity event in the early 1700s, with about 50 years of continuous conifer recruitment. Little additional recruitment occurred until after 1910. Plot 3 had a similar large gap in conifer recruitment. Some of these gaps may be explained by the omitted hardwood component of these stands, but fire must have thinned the smaller trees many times. On this site, 20 fires occurred between 1740 and 1987.

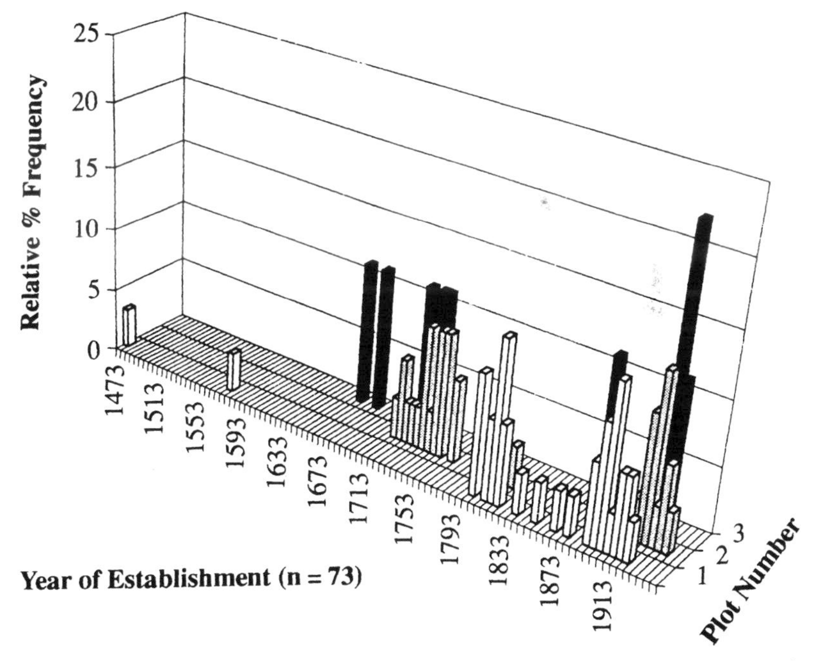

FIG. 10.4. *Conifer establishment over time in a* Pseudotsuga menziesii/*hardwood forest, Salmon River, California. Species include Douglas-fir and sugar pine, not differentiated by species.*
(From Wills and Stuart, in prep.)

The *Pseudotsuga*/hardwood forests of Oregon Caves, at about 1,300 m elevation, are at the transition to *Abies concolor* forests. They have white fir mixed with a variety of conifers and hardwoods across the age range sampled (Agee 1991a; Fig. 10.5). Douglas-fir in this 45 ha site (of 197 ha studied) are as much as 270 years old; scattered trees of the same age across higher elevation sites suggest that a high severity event covered the area sometime before 1730. Many of the trees that likely became established during that time have been killed by later fires; at least eight fires burned portions of the *Pseudotsuga*/hardwood type between 1730 and 1921. The multiple peaks in recent (<125-year-old) age classes represent patchy fire spread over the area; not every fire covered the whole area, so that tree establishment appears almost continuous.

PSEUDOTSUGA MENZIESII FORESTS

Pseudotsuga menziesii plant associations are found through northwestern California and southwest Oregon, the eastern Cascades of Washington and Oregon, and the Blue Mountains (Table 10.1). In the northwestern California–southwest Oregon area, these associations are mixed with *Pseudotsuga*/hardwood forests and may sometimes represent occurrences of ultrabasic parent material, such as serpentine, particularly if Jeffrey pine (*Pinus jeffreyi*) is present. The Sacramento and McCloud River drainages in northern California have *Pseudotsuga menziesii* forests where Douglas-fir and incense-cedar are codominant (J. D. Stuart, personal communication, 1992). In the eastern Cascades and Blue Mountains, *Pseudotsuga menziesii* forests are found as a transitional type between nonforest areas or *Pinus ponderosa* forests at lower elevation and *Abies grandis* or *Abies lasiocarpa* forests at higher elevation. Its typical elevation range in northeastern Washington is 600–1,300 m (Franklin and Dyrness 1973).

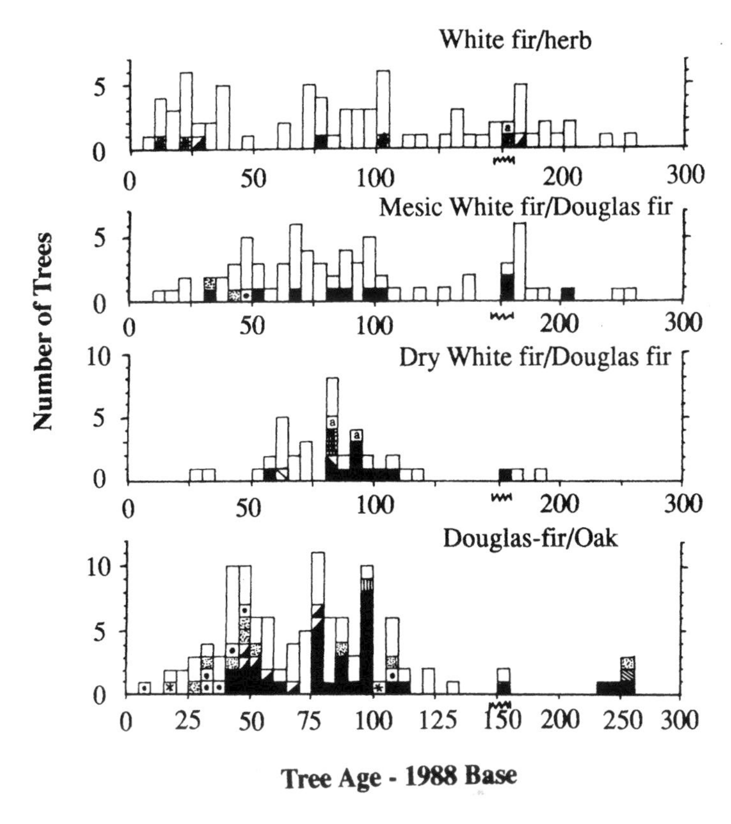

FIG. 10.5. *Age-class information from an elevational gradient of forest types at Oregon Caves National Monument. Lower elevation types are near the bottom of the figure. The Douglas-fir/oak community is discussed under* Pseudotsuga*/hardwood; the others are discussed under* Abies concolor *forests.*
(From J. K. Agee, "Fire History along an Elevational Gradient in the Siskiyou Mountains, Oregon," Northwest Science 65 [1991]: 197.

THE FIRE REGIME

As with almost all drier forest types in the West, Native American and lightning ignitions combined to produce relatively short fire-return intervals in *Pseudotsuga menziesii* forests. In the eastern Siskiyou Mountains in September 1841, this account was recorded separately by three members of the Wilkes Expedition (in Boyd 1986, p. 73):

> On the flank of a hill stretching to the E.S.E., in passing thro the woods we suddenly came on to an Indian woman who was blowing a brand to set fire to the woods probably, we stoped to speak with her, but she was sullen, & dogged, & made no reply, & we passed.
>
> Saw a squaw who was so busy setting fire to the prairie & mountain ravines that she seemed to disregard us.
>
> We found the woods on fire in several places & at times had some difficulty in passing some of the gullys where it was burning.

This pattern of human interaction with fire was common throughout *Pseudotsuga menziesii* forests of the Cascades and northern Rocky Mountains (Barrett and Arno 1982). Lightning fires added to the human ignitions: in the Montana-Idaho region, lightning fire frequency in the *Pinus ponderosa* and *Pseudotsuga menziesii* series (766 fires/million ha/yr) is more than 10 times the average for forests in the region (67 fires/million ha/yr) (Bevins 1980).

Fire-return intervals in *Pseudotsuga menziesii* forests of the eastern Cascades appear to be somewhat similar to those of *Pseudotsuga*/hardwood forests. In the Wenatchee Valley of Washington, Wischnofske and Anderson (1983) found a mean fire-return interval of 7–11 years in *Pseudotsuga menziesii* forests by combining fire records over 15–20 ha areas. In the Blue Mountains, Hall (1976) found a 10 year fire-return interval using single stump samples. On the Okanogan National Forest in Washington, Finch (1984) found individual trees to have fire-return intervals of 10–24 years. Over roughly 40 ha areas, he found a fire-return interval of 14 years for *Pseudotsuga menziesii*/*Vaccinium* spp. and *Pseudotsuga menziesii*/*Calamagrostis rubescens* plant associations. Fires that scarred >50 percent of the stumps occurred at a 22 year return interval. In a dry western Cascades area, Agee et al. (1990) found a 52–76 year mean fire-return interval for Douglas-fir forests with either ponderosa pine (*Pinus ponderosa*) or lodgepole pine (*Pinus contorta*) as codominants. These apparently long intervals reflect the mix of these forest types with more typical westside forest types that have longer fire-return intervals.

In the Mud Creek and Mills Canyon area of the lower Entiat Valley, Washington, fire-return intervals of 8–18 years on individual stumps were counted (J. K. Agee, unpublished data). In *Pseudotsuga menziesii* plant associations in the northern Rocky Mountains, Arno (1980) summarized a range of 13–26 years, with longer fire-return intervals in areas where this series was of minor extent.

Historical fire intensities spanned the ranges of the forest types at higher and lower elevations. Many of the drier *Pseudotsuga* types have a substantial component of ponderosa pine with 15–20 fire scars, suggesting frequent, low intensity fires. Often the Douglas-firs in these types appear to be unscarred, or the scars may have grown over. Plant associations in the eastern Washington Cascades with this fire regime of low to moderate severity include *Pseudotsuga menziesii/Carex geyeri, Pseudotsuga menziesii/Calamagrostis rubescens-Carex geyeri,* and *Pseudotsuga menziesii/Arctostaphylos uva-ursi* (Williams et al. 1990).

Higher severity fires also occurred. In the Stehekin Valley, Washington, a *Pseudotsuga menziesii* forest at the High Bridge campground is stocked with ponderosa pine and Douglas-fir that all germinated between 1715 and 1745, an indication of a stand replacement fire sometime before 1715. Moderate to high severity fire regimes may be found in *Pseudotsuga menziesii/Physocarpus malvaceus, Pseudotsuga menziesii/Symphoricarpos albus,* and *Pseudotsuga menziesii/Vaccinium* spp. (Williams et al. 1990), representing at least occasional fires of high severity.

Stand Development Patterns

Most *Pseudotsuga menziesii* plant associations were once more open in appearance than they are today, and many were dominated by ponderosa pine, with western larch (*Larix occidentalis*) in selected areas. Western larch is more frequent in *Abies grandis* plant associations in the eastern Cascades, but it is more common in *Pseudotsuga menziesii* plant associations in the Ochoco and Blue Mountains and east into Idaho and Montana (Daubenmire and Daubenmire 1968, Schmidt et al. 1976). An example of forest change with fire exclusion is shown in Figure 10.6 (Gruell et al. 1982). Frequent, low intensity fires kept such sites open so that they were less likely to burn intensely even under severe fire weather. The tendency for stocking to increase with protection is shown by photographs of the same scene over 40 years. Fires are likely to be more intense over time with protection.

A computer model of *Pseudotsuga menziesii* forests in the presence of disturbance has been developed by Keane et al. (1990). Their model is capable of simulating successional patterns in a variety of inland Northwest forest types, but here it is applied to a *Pseudotsuga menziesii/Physocarpus malvaceus* plant association of western Montana. The four scenarios shown (Fig. 10.7) have increasing fire-return intervals, including a no-fire scenario. They support the changes seen in Figure 10.6. In the simulation for a 10-year fire interval, fireline intensities were in the range of 50–100 kW m^{-1} because of limited time for fuel to accrue between fires; intensities increased to 80–150 and 300–1,200 kW m^{-1} at the 20- and 50-year fire intervals. The slight decline over time in basal area of western larch and ponderosa pine in most simulations was due to an artifact of the model. In most of the eastern Cascades, the western larch component would be absent, but the curves for ponderosa pine and Douglas-fir would be about the same. The "natural" scenario (Fig. 10.8A), using a stochastically selected set of fire intervals, incorporates characteristics of the first three regular intervals but appears much like the simulation for a 10-year fire-return interval; the "natural" scenario has a mean fire-return interval of 8 years but substantial range (fire-free intervals of up to 33 years).

In Washington the *Pseudotsuga/Physocarpus* plant association appears to have had occasional intense fires (Williams et al. 1990), suggesting that it fits into the moderate severity fire regime at the western end of its range. The presence of western larch in the eastern Cascades is a good indicator of a moderate severity fire regime that occasionally opens up enough growing space for western larch establishment. Protection from fire for long periods of time increases Douglas-fir in a multilayered architecture and has been associated with increased duration and intensity of western spruce budworm (*Choristoneura occidentalis*) attack.

The tendency for increasing fuel buildup over time in the Keane et al. (1990) model (Fig. 10.8B) is kept in relative equilibrium by the natural fire scenario. *Pseudotsuga menziesii* forests in Glacier National Park dominated by ponderosa or lodgepole pine and protected since 1900 had 40–84 t ha^{-1} of dead and down fuel, and one stand had over 300 t ha^{-1} (Lunan and Habeck 1973). Thus, long periods of protection are associated with fuel buildup. The longest fire-return interval (between years 35 and 70) in the Keane et al. (1990) model is associated with greatly increased scorch height. On dry sites, the scorch height of 10 m can be 50–75 percent of tree height for Douglas-fir and lodgepole pine, although not necessarily an equivalent proportion of the crown volume. Two-way contingency table analysis for trees on burns in Montana, Wyoming, and Idaho shows postfire tree survival for these two species to be a function of crown scorch, and additionally a function of insect damage for Douglas-fir and of basal scorch for lodgepole pine (Peterson and Arbaugh 1986).

Growth of residual Douglas-fir and lodgepole pine can be significantly reduced for at least four years after wildfire (Peterson et al. 1991). Although significant variation exists as a result of individual tree damage, radial stem growth was negatively affected compared to unburned stands (Fig. 10.9). It is possible that the thinning effect of fire will eventually increase growth on the residual trees (e.g., Oliver and Larson 1990).

When mortality of either species occurs, growing space is opened up for both. Densities of Douglas-fir and lodgepole pine regeneration are inversely correlated with density of residual trees (Fig. 10.10), and the proportion of the more shade-tolerant Douglas-fir increases with residual density (Finney 1986). Microsites farthest away from residuals contain the tallest regeneration and the highest proportion of lodgepole pine. Variable fire intensity affects not only the scale of the pattern but also the species most likely to dominate the next age class in these multiple-aged stands. With continued protection, Douglas-fir becomes the only understory species, and ponderosa pine and/or lodgepole pine are found as older overstory dominants (Agee et al. 1990). Over time, because of the relatively

A

B

FIG. 10.6. *Landscape change with fire exclusion in* Pseudotsuga *forests of western Montana.* A: *In 1909, the forest is wide-spaced ponderosa pine, with little Douglas-fir, that has been lightly and selectively cut. The understory is snowberry and herbs.* B: *By 1928, Douglas-fir invasion is evident.*

C

D

C: *In 1938, the pattern continues.* D: *By 1948, a major structural change is largely complete, and the herbaceous understory has been shaded out.* (From Gruell et al. 1976; photos courtesy of USDA Forest Service Archives.)

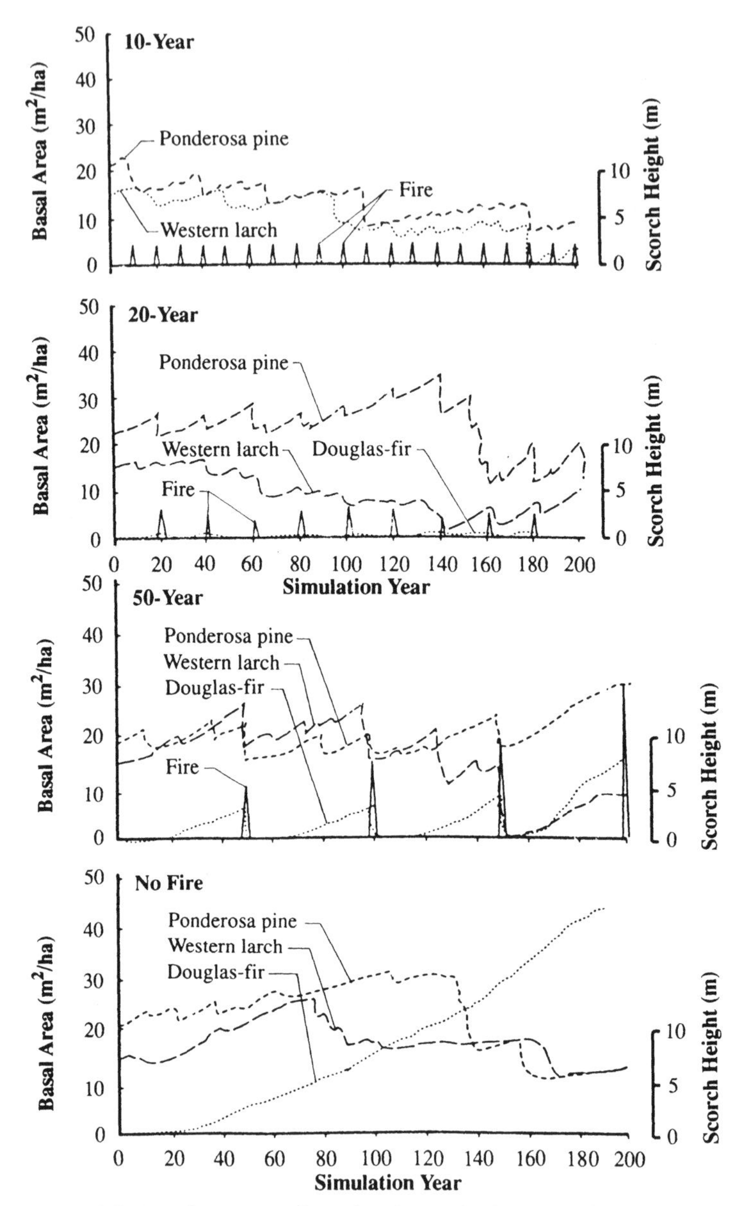

FIG. 10.7. *Basal area over time of major species in a* Pseudotsuga menziesii/Physocarpus malvaceus *plant association for different fire-return intervals. As fire-return intervals lengthen, ponderosa pine decreases in importance relative to Douglas-fir. Where grand fir is present, its response is similar to Douglas-fir, and Douglas-fir assumes an intermediate response between ponderosa pine and grand fir.* (From Keane et al. 1990)

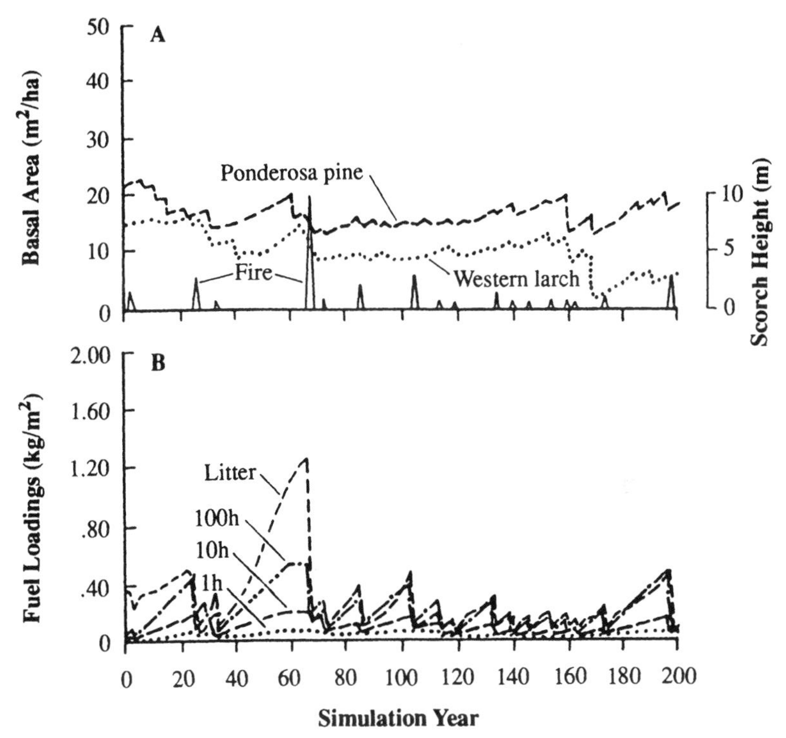

FIG. 10.8. A: *A "natural" fire scenario in the same type as Figure 10.7.* B: *Fuel loading associated with accumulation over time and consumption by periodic fire.* (From Keane et al. 1990)

short lifespan of lodgepole pine, Douglas-fir is likely to dominate all forest layers on sites where these two species are currently co-dominants; this is likely to take up to 200 years (Larson 1972). In contrast, because of its longevity, ponderosa pine will remain codominant with Douglas-fir longer. Insects are likely to be the mortality agent for both pines (Hessburg and Everett, in prep.) and may prematurely kill the pines if they are stressed by the presence of thick understories of shade-tolerant trees.

Many understory shrubs and herbs in *Pseudotsuga menziesii* forests are adapted in some way to frequent fire. Most of the dominant shrubs either sprout or have seed stimulated to germinate after fire. After the Pattee Canyon fire in a *Pseudotsuga menziesii/Physocarpus malvaceus* plant association, the ninebark (*Physocarpus malvaceus*), serviceberry (*Amelanchier alnifolia*), and Rocky Mountain maple

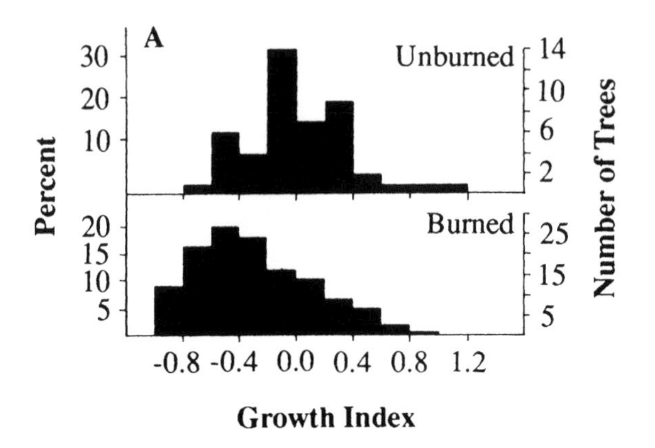

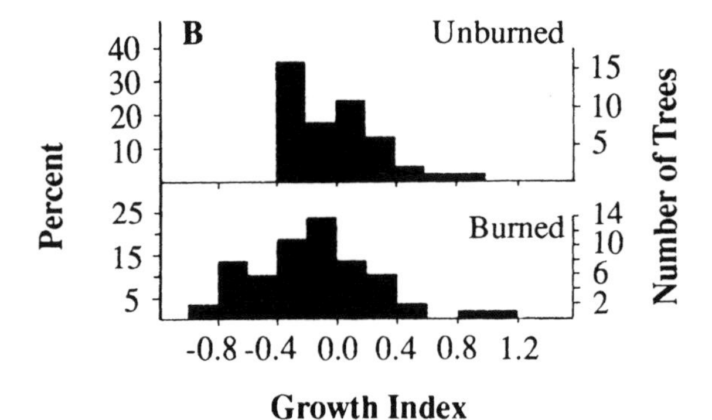

FIG. 10.9. *Radial growth of control vs. burned sites for* (A) *Douglas-fir and* (B) *lodgepole pine after four years. Growth index is defined as the ratio of preburn to postburn radial growth and shows a negative impact of burning.* (From Peterson et al. 1991)

(*Acer glabrum*) resprouted from root crowns, while the associated shrubs snowberry (*Symphoricarpos albus*), blue huckleberry (*Vaccinium globulare*), birchleaf spirea (*Spiraea betulifolia*), and thimbleberry (*Rubus parviflorus*) sprouted from rhizomes (Crane et al. 1983). After a 1970 wildfire in the Entiat River Washington drainage, snowbrush (*Ceanothus velutinus*) germinated from fire-stimulated seed and was the dominant species with up to 12 percent cover on south aspects (Tiedemann and Klock 1976). Repeated fires

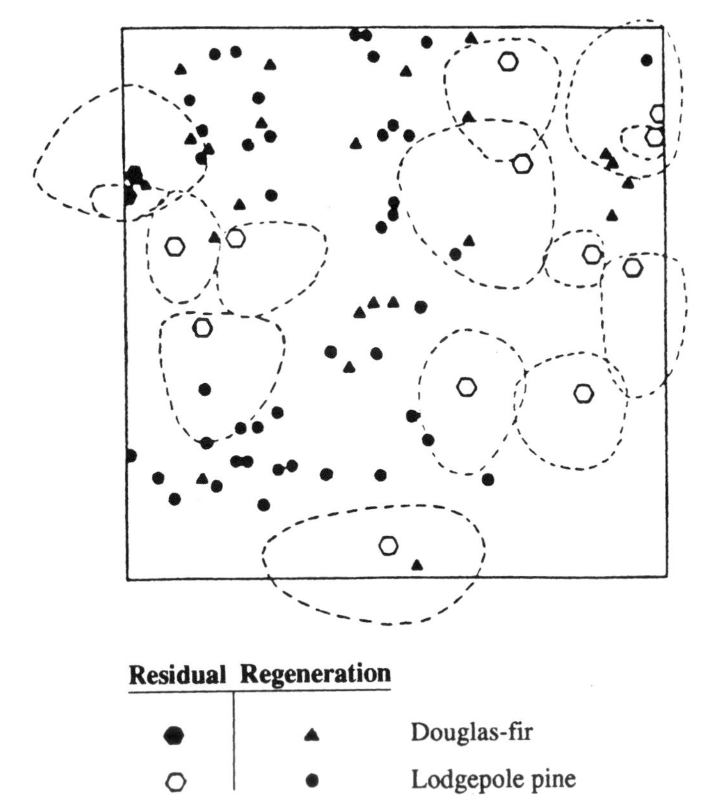

FIG. 10.10. *Spatial relationship between residual trees after a 1926 fire near Ross Lake in the North Cascades and postfire regeneration of Douglas-fir and lodgepole pine. Canopy projections of residual trees are shown by dashed lines. Few trees establish under the canopy of residuals, likely due to shading and root competition.*
(From Finney 1986)

over short intervals, however, will kill obligate seeding species and eventually exhaust the supply of seed in the soil (Biswell 1973).

Several classification systems have been developed for shrub and herb response to fire. McLean (1969) developed three categories based on root systems. The *resistant* category has species with taproots or a fibrous root system with rhizomes 5–13 cm deep: snowberry (*Symphoricarpos albus*), birchleaf spirea (*Spirea betulifolia*), and Oregon boxwood (*Pachistima myrsinites*). The *intermediate* category includes species with shallower rhizomes, such as pinegrass (*Calamagrostis rubescens*), Oregon grape (*Berberis repens*), and arnica

(*Arnica cordifolia*). The *susceptible* category includes those species with rhizomes in duff (rattlesnake plantain [*Goodyera oblongifolia*], wintergreen [*Pyrola* spp.]), fibrous roots and stolons (twinflower [*Linnaea borealis*] and bearberry [*Arctostaphylos uva-ursi*]), and fibrous roots only (trail plant [*Hieracium albiflorum*]). A thorough description of the fire response of major shrubs in *Pseudotsuga menziesii* forests of Montana and Idaho is presented by Noste and Bushey (1987).

For *endurer* shrubs, shrub canopy mortality has a significant impact on postburn growth. In a *Pseudotsuga menziesii/Physocarpus malvaceus* plant association, current annual twig growth of redstem ceanothus (*Ceanothus sanguineus*), ocean spray (*Holodiscus discolor*), ninebark (*Physocarpus malvaceus*), and Scouler's willow (*Salix scouleriana*) increased directly with the degree of shrub canopy mortality due to fire (Owens 1982).

ABIES CONCOLOR FORESTS

The *Abies concolor* forests of the Pacific Northwest are a northern extension of more widespread forest in the Sierra Nevada to the south. In the Cascades the species composition is somewhat similar to the Sierra Nevada, but in the Siskiyou Mountains white fir (*Abies concolor*) seems to occur in cool sites and often forms pure stands. Although Franklin and Dyrness (1973) distinguished an *Abies concolor* zone from the adjacent, lower elevation mixed-conifer zone dominated by Douglas-fir, they noted that the zone is not widespread in the Pacific Northwest and that its status in California is more clear. In the Siskiyou Mountains it ranges from 1,400 to 1,800 m elevation, and in the Cascades, from 1,400 to 1,700 m, where it often grades into *Abies magnifica* plant associations.

The Fire Regime

Both Native American and lightning ignitions were important sources of fire in *Abies concolor* forests. Native Americans burned these forests regularly, as indicated by these accounts from various areas of mixed-conifer forest in California (Lewis 1973, p. 73):

> It was a general practice of these Indians [Nisenan tribe, north and east of Sacramento] to clear the forests of undergrowth by setting fire to it. They seem to have been able to keep the fire under control.
>
> Like most of the Californians who inhabited forested tracts, the Maidu [Feather/American Rivers] frequently burned over the country often annually.
>
> In the spring . . . the old squaws began to look about for the little dry spots of headland and sunny valley [Yosemite], and as fast as dry spots appeared, they would be burned. In this way, fire was always the servant, never the master.

Such narratives are commonly found in the ethnographic literature (Lewis 1973). Yet it is not entirely clear whether Native American ignitions augmented or substituted for lightning ignitions. In an environment as prone to burn as the drier *Abies concolor* forests, human ignitions may only have substituted for inevitable lightning fires across the same portions of the landscape. Generally, it is believed that Native American burning augmented lighting ignitions (Kilgore and Taylor 1979), although a fire model based on long-term climatic data including lightning storm frequency suggests that much of the fire history could be explained solely by the occurrence of lightning (van Wagtendonk 1972).

Fire history of *Abies concolor* forests of Crater Lake National Park was studied by McNeil and Zobel (1980), who found an increasing fire-return interval with elevation in the southern "panhandle" section of the park. The average fire-return interval, using a composite sample of two trees per site, was 9–42 years along an elevational gradient from about 1,350 to 1,550 m elevation. Variation around the mean at each sample site was high (Fig. 10.11); the longer intervals may be associated with establishment and growth past the most susceptible stage of fire sensitive species. For example, Thomas and Agee (1986), working in the same area, found substantial white fir 180–200 years old that had established itself during one such extended fire-free period. At the northern margin of the type (Lookout Mountain near Bend), Bork (1985) used cross-dated specimens from 16 ha plots to determine a 9–25 year fire-return interval for an *Abies concolor/Ceanothus* community, equivalent to some of the lower elevation sites studied by McNeil and Zobel (1980). In higher elevation *Abies concolor* forest in the Siskiyou

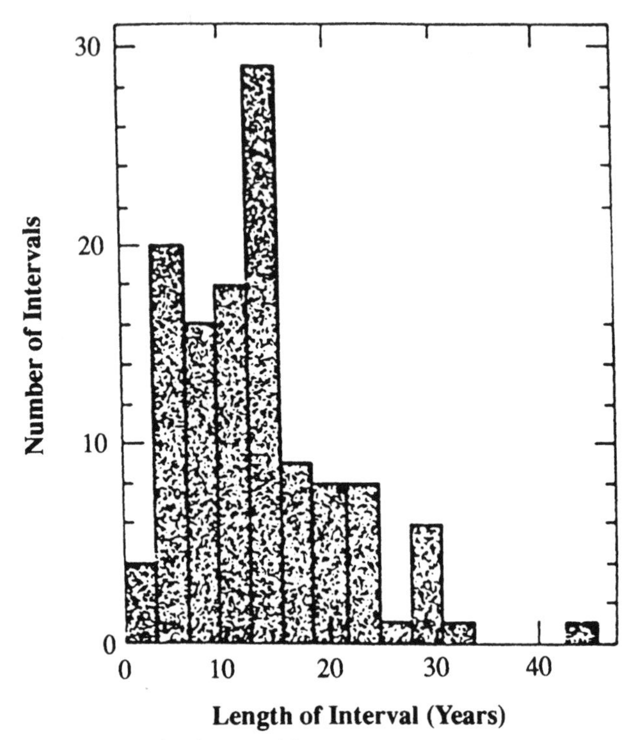

FIG. 10.11. *Distribution of fire-return intervals in an* Abies concolor *forest at Crater Lake National Park.*
(From R. McNeil and D. Zobel, "Vegetation and Fire History of a Ponderosa Pine–White Fir Forest in Crater Lake National Park," Northwest Science 54 [1980]: 42. Copyright 1980 by Washington State University Press, Pullman. All rights reserved. Reprinted by permission of the publisher.)

Mountains, Agee (1991a) found a 43–61 year fire-return interval for a white fir/Douglas-fir community and a 64 year fire-return interval for white fir/herb communities, reflecting a lengthening of the fire-return interval with increasing elevation.

Presettlement fire histories in California *Abies concolor* forests show a similar pattern. At the ecotone of higher elevation *Abies magnifica* forests at Swain Mountain near Mount Lassen, Taylor and Halpern (1991) found 1 ha plots to have fire-return intervals of 40–42 years. Farther south and at lower elevation Kilgore and Taylor (1979), using composite fire intervals from 0.4–0.8 ha areas, found mean fire-return intervals of 9–18 years.

The intensity of these historical fires was usually low, because the frequent fires removed understory ladder fuels and consumed the forest floor (Kilgore and Taylor 1979). Fires occurring after an extended fire-free period would likely have been more intense and were probably the norm in the higher elevation stands with larger proportions of white fir. Seasonality of burning has not been studied in detail, but it was probably a summer to autumn phenomenon. Fires beginning in mid-summer might slowly burn across the landscape into the autumn. At Oregon Caves, several fires seemed to stop at the lower elevation boundary of the white fir/herb plant community, possibly in response to high cover of green herbage with high fuel moisture (Agee 1991a).

Fire extent is a difficult parameter to characterize, as it requires cross-dated samples from a wide area of the landscape. At Crater Lake, McNeil and Zobel (1980) found few fires that burned their entire 7.5 km^2 study area. Areas on opposite sides of a steep, narrow canyon usually did not burn in the same year (although recent prescribed fires have spotted across it), and occurrence of a widespread fire usually reduced the size (or intensity as judged from lack of fire scars) over the next decade. To the north Bork (1985) found no single fire that scarred trees across each of five 16 ha plots scattered across a 125 ha area at Pringle Falls. In the Sierra Nevada, Kilgore and Taylor (1979) concluded that most fires burned small, irregularly shaped areas. These data suggest either that fires were of too low an intensity to leave sufficient scarred trees, even though they covered substantial areas, or that fires were indeed small, scattered events over the landscape.

Stand Development Patterns

Abies concolor forests have a gradient of stand development patterns associated with the fire regime gradient. As fire-return intervals lengthen, likely due to cooler, wetter climate, there is a tendency to have higher proportions of white fir in the overstory. The mixed-conifer successional model of Kercher and Axelrod (1984) predicts that as elevation increases from 1,500 to 1,800 m, ponderosa pine is reduced from 60 percent to 5 percent of the basal area, while white fir, which covers only 5 percent of the basal area at 1,500 m, increases to over 60 percent at 1,800 m. These simulations, which included fire as a disturbance, suggest that white fir is a much better competitor in the cooler environments.

An example of the stand development patterns in the cooler, high elevation *Abies concolor* type is Oregon Caves (Agee 1991a). Three white fir communities occur with increasing elevation: a dry Douglas-fir/white fir type, a mesic Douglas-fir/white fir type, and a white fir/herb type. The proportion of Douglas-fir basal area declines from 63 percent to 45 percent to 3 percent along this gradient as fire-return interval gradually increases. There is a tendency toward an all-aged stand as the proportion of white fir increases, due in part to an increasing role of wind disturbance (blowdown) in the stand (also found by Taylor and Halpern 1991).

The species composition and structure of lower elevation stands of *Abies concolor* forests are significantly affected by fire. The frequent fires have the largest impact at the seedling and sapling stage. Ponderosa pine and sugar pine are more resistant to fire when small than are white fir or incense-cedar (van Wagtendonk 1983). Frequent fires with low flame lengths therefore determine the potential future canopy dominants by selectively favoring the pines over the white fir and incense-cedar.

A contagious pattern of forest structure is statistically apparent in these lower elevation *Abies concolor* forests. In the Sierra Nevada, "aggregations" of age classes and species exist at a small scale: 135–1,600 m^2 (Bonnicksen and Stone 1981). At Crater Lake, similar species contagion was found for ponderosa pine, sugar pine, and white fir, in presettlement-era forest (Thomas and Agee 1986). Decades of fire suppression have not erased that pattern in unlogged areas, as most of the presettlement tree dominants are still alive on these sites. The source of the clumping by individual species is not clear, but it probably involves good seed years, presence or absence of fire in a young patch, and competitive relationships between the species able to colonize the patches.

The dynamics of fire in *Abies concolor* forests have been modeled extensively (van Wagtendonk 1972, Kercher and Axelrod 1984). Periodic, low intensity fire maintains a cyclic stability in fuel loads (van Wagtendonk 1985) and understory plant biomass, and maintains dominance of pines in the overstory. Starting with a bare patch, Kercher and Axelrod (1984) followed the development of a mixed-species stand over a 500 year period in the presence of an average fire-return interval of eight years with variation around that mean. Ponderosa pine becomes an early dominant and maintains dominance over the half-millennium (Fig. 10.12). The growth rate, growth form, and thicker bark per unit diameter of ponderosa pine help it to maintain dominance. Over 50 percent of the stand basal area is composed of either ponderosa or sugar pine. Some of the dips in the graphs are associated with death of individual trees of that species (e.g, Douglas-fir) and are not a part of the general pattern. This model suggests that ponderosa pine has a competitive advantage in the presence of fire that it loses in a fire-free environment.

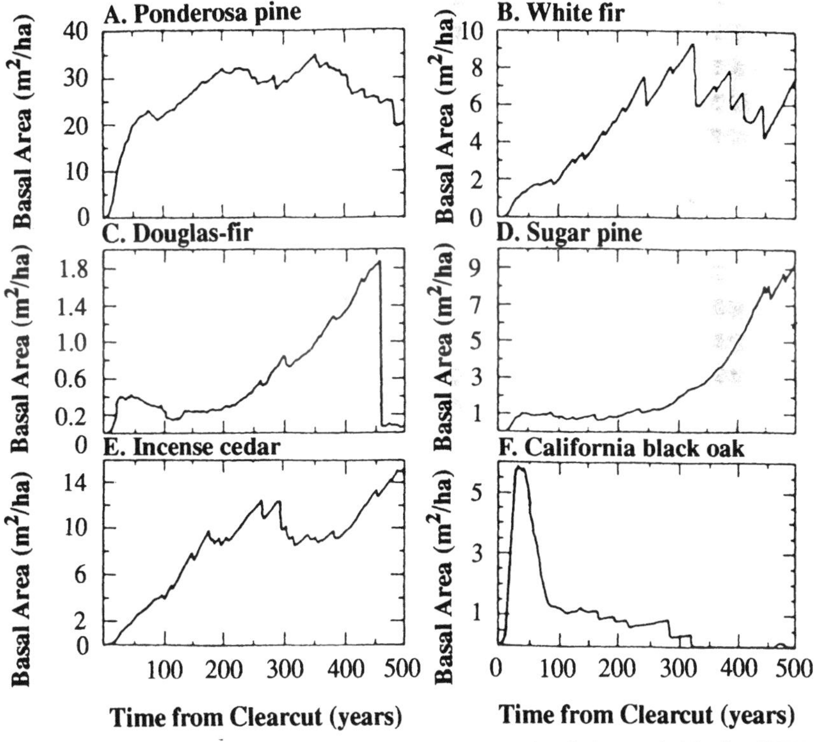

FIG. 10.12. *Basal area of major species over a 500 year simulation period in the SILVA model with an 8 year fire-return interval. Note the different scales on the Y axis.* (From J. R. Kercher and M. C. Axelrod, "A Process Model of Fire Ecology and Succession in a Mixed-Conifer Forest," Ecology 65 [1984]: 1725–42. Copyright 1984 by Ecological Society of America. Reprinted by permission.)

In presettlement times, the spatial pattern of these forests may have been visually as well as statistically apparent. Decreases in fire activity have allowed massive invasion of the understory by trees (Parsons and DeBenedetti 1979, McNeil and Zobel 1980). Most of the trees are less than a century old (Fig. 10.13) and became so dense that regeneration of even the shade-tolerant species essen-

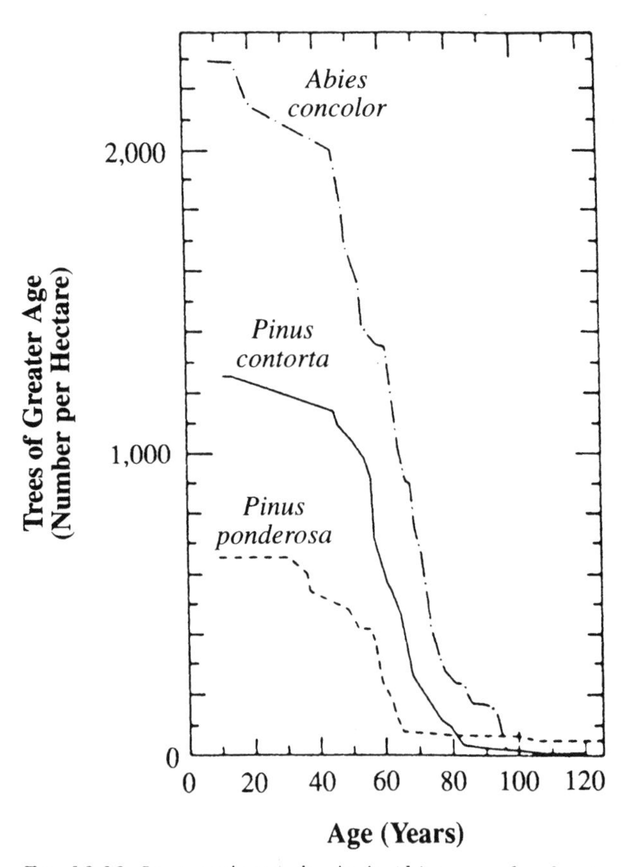

FIG. 10.13. *Increases in tree density in* Abies concolor *forest, largely a twentieth-century phenomenon. Tree establishment is indicated by steeper slopes, while flatter slopes indicate a lack of establishment.*

(From R. McNeil and D. Zobel, "Vegetation and Fire History of a Ponderosa Pine–White Fir Forest in Crater Lake National Park," Northwest Science 54 [1980]: 44. Copyright 1980 by Washington State University Press, Pullman. All rights reserved. Reprinted by permission of the publisher.)

tially ceased after about 50 years of fire protection. The pines stopped regenerating soonest (as shown by the flattening slope of the curve), and the white fir continued for several more decades. Even so, much of the stunted white fir of 2–3 m height in 1980 was established by 1955 (Agee 1982).

Reintroduction of fire into these altered systems has met with mixed results. Fuel objectives have largely been achieved. Low intensity fires have been successfully introduced to northern California *Abies concolor* forests with 50–175 t ha^{-1} of dead and down fuel, with consumption of up to 90 percent (van Wagtendonk 1974, Kauffman and Martin 1989). A similar total fuel load at Crater Lake was reduced 67 percent (Table 10.2); near future additions from live material killed but not consumed by the fire totaled about half the fuel consumed in the first fire (Thomas and Agee 1986). The fuel "ladder" from the surface into the crown of the trees has generally been broken by the first fire (Kilgore and Sando 1975), so that a second fire is generally easier to control, 5–10 years after the first. These fires burn across forest floors protected for many decades, and due to low decomposition rates (Agee et al. 1978, Stohlgren 1988) forest floor depth is probably greater than any levels ever experienced by the mature tree component.

Most researchers suggest that foliar bud phenology best explains seasonal differences in tree mortality (Dieterich 1979, Wyant et al. 1983), and this is likely the case where crown scorch is the major

TABLE 10.2. FUEL REDUCTION (KG/HA) BY PRESCRIBED BURNING IN A WHITE FIR FOREST AND POTENTIAL ADDITIONS DUE TO TREE MORTALITY

Fuel Time Lag Category	*Before Fire*	*After Fire (1 yr)*	*Near Future Additions*	*Near Future Totals*
1 hr	3,470	1,455	4,615	6,070
10 hr	3,290	2,277	4,761	7,038
100 hr	6,160	2,940	12,683	15,623
>100 hr	67,285	38,067	43,190	81,257
Litter-duff	92,130	12,525	—[a]	12,525
Total	172,335	57,264	65,249	122,513

SOURCE: Thomas and Agee 1986.

[a] Litter additions from trees killed by fire are included in the 1 hr timelag category. Duff additions are assumed to be negligible in the short run. Annual litter additions from the residual stand are not included.

ecological effect of fire. Where forest floors are deep, smoldering and long duration heating are an important associated cause of tree damage and susceptibility to later bark beetle attack. Smoldering of the duff layer long after the passing of the fire front is common in stands that have not burned for many decades (Agee 1973). Duff combustion temperatures exceed 500°C, and both the cambial tissues insulated by the bark (Ryan and Frandsen 1991) and fine roots insulated by the soil (Grier 1989, Swezy and Agee 1991) can be killed by smoldering that can persist for 6–10 hours. The cambial effects are a function of bark thickness and fire duration, while root effects are a function of rooting depth and root size (Fig. 10.14). Although autumn burning may be associated with higher fuel consumption (Kauffman and Martin 1989) there appears to be less pine mortality after such burns (Swezy and Agee 1991). This may be due to autumn fires preceding the period of new root growth, allowing trees to recover over the next winter and spring, in contrast to spring fires that occur at the onset of the dry season.

The lower elevation *Abies concolor* forests, with a large component of ponderosa pine, were open and parklike in the mid-nineteenth century. The Leopold Report (1963), which brought about a radical shift in natural area fire policy, noted:

> When the forty-niners poured over the Sierra Nevada into California, those that kept diaries spoke almost to a man of the wide-spaced columns of mature trees that grew on the lower western slope in gigantic magnificence. The ground was a grass parkland, in springtime carpeted with wildflowers. Deer and bears were abundant. Today much of the west slope is a doghair thicket of young pines, white fir, incense-cedar, and mature brush—a direct function of overprotection from natural ground fires.

This description resembles climax ponderosa pine communities (see chapter 11). The filling in of the understory by trees has reduced species richness and cover of shrubs and herbs.

Most of the shrubs in the *Abies concolor* zone are adapted to burning either through an *endurer* or *evader* strategy. Between 80 percent and 90 percent of the shrubs will be top-killed after a fire (Kauffman and Martin 1985). Those capable of sprouting will do so based on season of burn and degree of duff consumption, related to

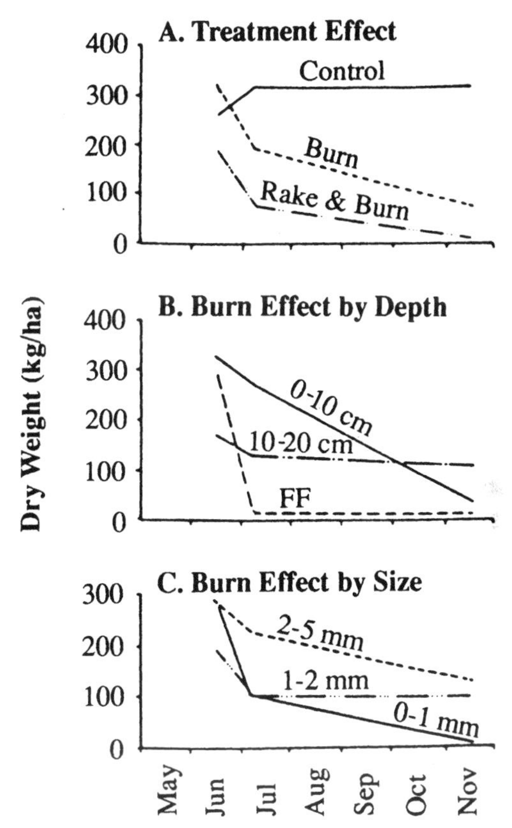

FIG. 10.14. *Effects of prescribed fire on fine-root biomass in a pumice soil at Crater Lake.* A: *Fine-root mass* (*<5 mm diameter*) *for control, spring burn, and rake/spring burn* (*where debris has been moved away from the tree base*) *treatments.* B: *For burned trees, deeper roots are less affected than surficial roots. Layers are: FF = forest floor; 0–10 cm, 10–20 cm = soil horizons.* C: *The smallest fine roots* (*0–1 mm*) *are more affected than larger fine roots* (*2–5 mm*). (From Swezy and Agee 1991)

heat duration at the root crown. Late spring is the time when sprouting shrubs are most susceptible to mortality: Root reserves are at a low point, and the fresh leaves are sensitive to heat. About two-thirds of root crown resprouting shrubs were killed by late spring fires, compared to half for late fall fires and about 15 percent for very early spring fires (Kauffman and Martin 1985). Early fall fires are also associated with high fuel consumption and more shrub mortality. Tanoak and shrubby black oak (*Quercus kelloggii*) were more sensitive than mountain whitethorn (*Ceanothus cordulatus*) or chinquapin (Kauffman and Martin 1990).

Deerbrush (*Ceanothus integerrimus*) seeds were effectively scarified by heat treatment, with wet heat being more effective than dry heat below 100°C (Kauffman and Martin 1991). Germination was less than 2 percent for nonscarified seeds, and less than 5 percent for nonstratified seeds. Early spring fires permit later stratification of scarified seeds, so that seeds needing stratification can germinate the same year. In California areas burned before April met those requirements (Biswell 1973, Kauffman and Martin 1984), while areas burned after April experienced fire-stimulated seed germination the next spring. Germination soon after a spring fire may not be advantageous to the shrub component of an ecosystem, as grasses generally have established root systems and may compete with young shrubs for moisture. Squaw carpet (*Ceanothus prostratus*), deerbrush (*Ceanothus integerrimus*), and whitethorn ceanothus (*Ceanothus cordulatus*) increased from no seedlings to 1,500–2,000 ha^{-1} after early fall burns, and to about 15,000 ha^{-1} after late fall burns (Kauffman and Martin 1984, Kilgore and Biswell 1971). Early fall burns tend to have higher duff consumption and consume many of the potential germinants (Weatherspoon 1988). Spring burns were associated with little successful fire-stimulated seed germination from shrubs (Kauffman and Martin 1984). Repeated short-interval burns in areas with nonsprouting shrubs will eventually deplete the nonsprouters from the shrub understory (Biswell 1974).

In some forests, thinning and pile burning may be an option to reduce fuels. Around the edges of piles, where heat has stimulated but not destroyed seed, circles of shrubs may result from such activities if the fire is induced in the proper season (early spring or late fall). The *Ceanothus* species, as nitrogen fixers, tend to be favored browse species for deer and may be a tradeoff for reduction of hiding cover on spring-fall transitional ranges.

ABIES GRANDIS FORESTS

Abies grandis forests have the most moderate environment of the forest zones of the eastern Cascades area. They are normally bounded at upper elevation by *Abies lasiocarpa* forest and at lower elevation by *Pseudotsuga menziesii* or *Pinus ponderosa* forest. The *Abies grandis* zone occurs from 1,100 to 1,500 m in the eastern Cascades (Franklin and Dyrness 1973), from 1,500 to 2,000 m in the Ochoco and Blue Mountains (Hall 1967), and from 1,200 to 1,500 m where it occurs in the eastern Washington Cascades. As a transitional forest zone, it tends to have a fire regime intermediate between those of the upper and lower elevation forests.

In *Abies grandis* forests, associated species of grand fir may often be the primary dominants: western larch, lodgepole pine, Douglas-fir, or ponderosa pine. A fire regime of variable severity helps maintain the dominance of these species. For example, western larch is found in 11 of 21 habitat types in eastern Washington and northern Idaho (Daubenmire and Daubenmire 1968) but is a dominant in only about 3 and is a seral species in all the habitat types in which it is found.

THE FIRE REGIME

On the Warm Springs Reservation, Weaver (1959) counted one stump with a 47 year fire-return interval in this mixed-conifer type. In the eastern Cascades of Washington, Wishnofske and Anderson (1983) found 33–100 year fire-return intervals over 30 ha sites. Trees that contain fire scars represent sites that have burned at low to moderate severity; however, fires of higher severity also occur in *Abies grandis* forests. In an *Abies grandis* forest in Montana, the intermediate interval was about 17 years (Arno 1976), while stand replacement fires occurred every 100–200 years (Antos and Habeck 1981). At higher elevation, where *Abies lasiocarpa* is the climax dominant, the intermediate interval was 25–75 years, and the stand replacement interval was 140–340 years (Barrett et al. 1991).

Unusual stand conditions or fire weather result in the stand replacement type of fire. In the Rocky Mountains, a ridge-aloft pattern produces warm, dry weather, while short-wave troughs of dry air moving from northwest to southeast create windy weather. Abundant lightning can strike from systems carrying moist air from the Gulf of Mexico into the northern Rockies (Schroeder and Buck 1970). The Great Fires of 1910 (Pyne 1982) and the 1967 Sundance fire (Stickney 1986) occurred under such unusual synoptic weather patterns. In the eastern Cascades and Blue Mountains, strong north winds are often associated with severe fires, such as the 1970 Entiat fires.

Stand Development Patterns

The successional dynamics of *Abies grandis* forests depend on the intensity of the fire as well as the species composition and structure of the vegetation at the time of the fire. Low intensity surface fires encourage canopy dominance by ponderosa pine, western larch, and Douglas-fir (Arno et al. 1985), as these species are *resisters*. The thick bark of the first two insulates the cambium against damage, and in mature stands they are usually the taller trees, which helps them avoid crown scorch. The deciduous habit of larch makes crown scorch less important, especially for late season fires (Davis et al. 1980).

A good example of the effects of moderate severity fire is provided by Peterson and Ryan (1986). The fire was under moderate fuel moisture conditions with a scorch height of 10 m (Table 5.7). Mature western larch and ponderosa pine suffered no mortality in the simulation, while the basal area of other species declined by 75–100 percent. After such fires, radial growth of western larch will often increase (Reinhart and Ryan 1988), while residual Douglas-fir may show no change in radial growth increment.

Fires of low to moderate intensity in these stands increase the relative dominance of Douglas-fir or western larch relative to lodgepole pine or grand fir. The thinner barked, and usually smaller, true firs will be killed, while a new cohort of regeneration will appear. Four years after the 1988 Dinkelman fire north of Wenatchee, Washington, regeneration of western larch, Douglas-fir, lodgepole pine, and grand fir was abundant on sites with a scattered residual western larch overstory. Over time density is expected to decline, and a shift in species composition may occur. On sites with substantial overstory mortality, some individuals of all species may survive, while if the overstory is relatively intact, the new cohort will develop as predominantly shade-tolerant grand fir (Larson 1982). Substantial tree understory development, due to this multiaged structure created after low to moderate severity fire or perhaps after the breakup of lodgepole pine in the canopy, will encourage an understory reinitiation stage in stand development (Oliver 1981), which may be associated with increased crown fire potential and another set of multiple successional postfire pathways.

A crown fire will kill all the trees in the stand. Herbs and shrubs may dominate the floristics of early succession, with some herbs (e.g., fireweed [*Epilobium angustifolium*]) peaking and declining within the same decade (Stickney 1986). Western larch has light-winged seeds, which can blow onto a burned site from adjacent stands or from lightly scorched cones in the fire-killed stand (Haig et al. 1941). Lodgepole pine, if present on the site, will establish from serotinous cones in the area. If the crown fires have been spaced more than 150 years apart, western larch with some western white pine (*Pinus monticola*) are the most probable tree dominants at the time of disturbance, because lodgepole pine, the other early seral species with fast growth, is short-lived and may have been killed by mountain pine beetles (Haig et al. 1941). Where the crown fire interval is shorter than 150 years, lodgepole pine will at least share dominance in the postfire tree cohort (Gabriel 1976, Antos 1977, Antos and Habeck 1981).

If two crown fires occur in quick succession, the site may revert to a brushfield (Antos and Habeck 1981), as neither western larch nor lodgepole pine will survive such fires in the pole stage (Davis et al. 1980). Where a single crown fire occurs, stand establishment will usually include individuals of other species as well as lodgepole pine and larch. On dry sites, ponderosa pine may be included. On average sites, Douglas-fir and grand fir will normally be present, and on moist sites western white pine may be a codominant. None of the associated species typically grows as fast as western larch and lodgepole pine (Haig et al. 1941, Cobb 1988). Individual species will create different strata over time, although all may establish in the same time period. On a series of stands in eastern Washington that regenerated after crown fires, Cobb (1988; Fig. 10.15) found western larch to be a consistent dominant, with lodgepole pine sharing dominance on some sites. Douglas-fir and grand fir showed a variety of stratification patterns but always in intermediate or suppressed crown positions.

Fuel conditions can also affect the probability of crown fire be-

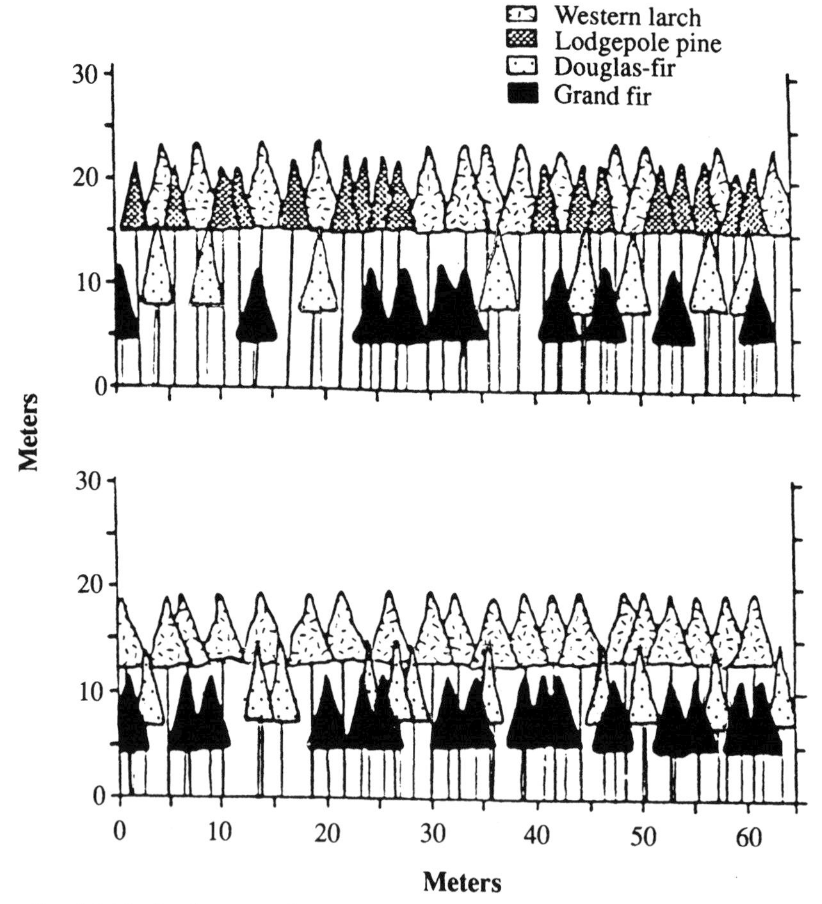

FIG. 10.15. *Development of even-aged stands on two sites after stand-replacing wildfire in the* Abies grandis *zone. Western larch is the clear dominant where it exists, and lodgepole pine will be a codominant where it exists. Douglas-fir, and to a greater extent grand fir, will be subordinate to the other two species.*
(From Cobb 1988)

havior. Pole-sized, heavily stocked stands have a high crown fire potential (Davis et al. 1980). If a young stand survives an initial light burn, perhaps due to its burning under average to moist weather conditions, later underburns act as a negative feedback mechanism for crown fires by reducing fuels that might encourage crown fire spread. A long interval between underburns will allow a tall understory to develop, which has a higher probability of crowning (Davis et al. 1980). Larch, as a deciduous conifer, has lower foliage flammability during the growing season than its associated evergreen conifers, because all foliage is young (<1 yr) and thus more succulent (Schmidt et al. 1976). By the end of the growing season, however, the foliar moisture of western larch is probably similar to or even lower than the other conifers on the site as the tree prepares to drop its needles.

MANAGEMENT IMPLICATIONS

The four mixed-conifer and mixed-evergreen forests described here have several features in common: major tree dominants, overlapping ranges in historical fire-return intervals, significant changes in forest structure since the fire protection era began, and similar challenges to living with fire in the future.

The two most constant tree species are Douglas-fir and ponderosa pine, which provide a clue that fire has been important in historical stand development. However, as important as low intensity fire was in each of these types, there is evidence that higher intensity fires were at times important too, particularly in the *Pseudotsuga*/hardwood and *Abies grandis* forests. As a group, in terms of historical fire activity, these forests fall between the low and moderate severity fire regimes (Fig. 1.5).

Fire protection has created homogeneous high fuel conditions across the entire landscape, a condition that was once associated only with protected refugia, such as canyon bottoms, some north aspects, and small pockets of these forests surrounded by cooler, moister types (Hessburg and Everett, in prep.). Selective logging of pine and Douglas-fir has accelerated successional processes, so that reintroduction of fire alone may not restore historical conditions

(Thomas and Agee 1986). Insects, such as Douglas-fir tussock moth (Williams et al. 1980) and spruce budworm (Hessburg and Everett, in prep.), are at epidemic levels and are less limited in time and space. Diseases, including foliar, stem, and root pathogens, are becoming more important (Agee and Edmonds 1992). Wildfires are likely to be larger and on the high end of the intensity scale because of higher total and dead fuel loads and greater horizontal and vertical fuel continuity. We have essentially been "saving" biomass on these sites, and although the savings account has a large balance, bankruptcy is inevitable. We cannot continue to be successful in excluding fire from these fire environments (Brown and Arno 1991), but reintroducing fire as a cure is not painless.

The scale of the problem is tremendous and will be costly. The issues of area treatment versus corridor treatment, or treatment versus no treatment, will be vigorously debated. Should we try to treat every slope or maintain fuel-limited corridors of variable width in an attempt to segregate the problem into manageable units?

In parks and wilderness areas, two significant management problems exist. The first is that prescribed fire under moist conditions is essential for restoration of low fuel conditions in the drier forest types where fire-return intervals averaged 20 years or less. These restoration burns need to be done under conditions where important structural elements, such as the larger, old trees, can be protected (Agee and Huff 1986a). Once one or two restoration burns are complete, continued prescribed burning or natural fires can maintain these mosaics. Prescriptions are generally available for prescribed burning (van Wagtendonk 1974, Kilgore and Curtis 1987).

The second challenge is the sporadic, patchy nature of fires where fire-return intervals exceed about 25 years, particularly in the cooler *Pseudotsuga menziesii* and *Abies grandis* forests. Prescribed burning is difficult because the "window" for adequate burning (from too moist to too dry) is narrow. Natural fires in these systems can be unpredictable, burning slowly one day and crowning the next. The management of such fires, as in *Abies magnifica* forests, can be quite difficult unless there are well-defined natural or human-created fuelbreaks. These forests tend to be transitional to cooler-moister or warmer-drier types, so forests are often continuous with these adjacent types.

The mixed-conifer and mixed-evergreen forests of the Pacific Northwest provide the most significant fire management challenges of all the Northwest forests. The solutions are not likely to be the same between forest types or within a particular type. Solutions will likely be driven by specific management objectives and by the failure of current fire suppression policy as larger wildfires prompt changes in our relationship with fire.

CHAPTER 11

PONDEROSA PINE AND LODGEPOLE PINE FORESTS

PONDEROSA PINE and lodgepole pine are the two conifers with the widest geographical range in the Pacific Northwest. For lodgepole pine this is partly due to its ecological range; it inhabits coastal bogs, a variety of montane forests, and subalpine forest. Ponderosa pine is found as an early seral dominant in many plant associations and as a climax dominant in several others. This chapter focuses on communities where these species are the potential (climax) vegetation—a small part of their wide geographical distribution. Where they are seral to other species, ponderosa and lodgepole pines are discussed in the chapters devoted to those other species or vegetation types.

Pinus ponderosa plant associations are found as a band of vegetation separating closed forest communities, often *Pseudotsuga menziesii* plant associations at higher elevation, from grassland communities in drier or lower elevation locales (Fig. 11.1). Historically, these forests had substantial understory vegetation. In south-central Oregon, this zone also borders *Pinus contorta* forests. These latter forests are largely a topoedaphic climax, using polyclimax

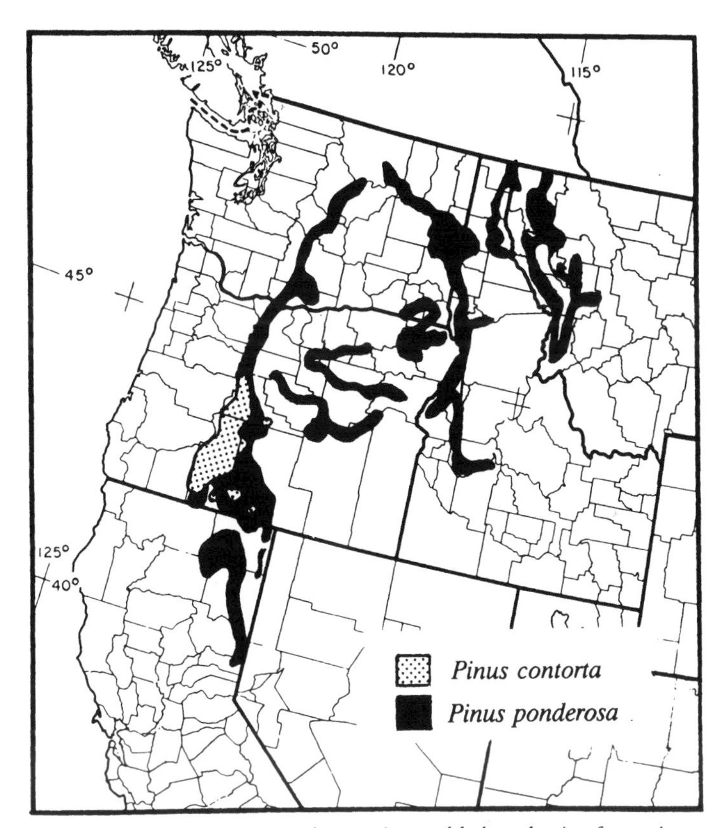

FIG. 11.1. *Map of climax ponderosa pine and lodgepole pine forests in the Pacific Northwest. Small pockets of both forests are scattered in other locations across the ranges of both species.*
(Adapted from Little 1971)

terminology: They are found on coarse-textured pumice deposits north, east, and to some extent south of Crater Lake, as a result of the eruption of Mount Mazama about 6,600 years ago, and on associated flats where extended frost pockets occur (Fig. 11.1). Some forests dominated by lodgepole pine are part of the *Pinus ponderosa* series and may eventually succeed to ponderosa pine. Most stands are between 1,200 and 1,800 m elevation (Franklin and Dyrness 1973, Zeigler 1978).

PINUS PONDEROSA FORESTS

Pinus ponderosa forests are not widely distributed in the Pacific Northwest. Ponderosa pine is more widely distributed as a dominant seral species in plant associations named for Douglas-fir, grand fir, or white fir. This section focuses on *Pinus ponderosa* plant associations, in which ponderosa pine is generally the only conifer dominant. Common plant associations include shrub-dominated *Pinus ponderosa/Purshia tridentata, Pinus ponderosa/Physocarpus malvaceus,* and *Pinus ponderosa/Symphoricarpos albus,* and herb-dominated *Pinus ponderosa/Stipa comata, Pinus ponderosa/Festuca idahoensis,* and *Pinus ponderosa/Agropyron spicatum* (Daubenmire and Daubenmire 1968). The latter three were once lumped as *Pinus ponderosa/Stipa comata* (Daubenmire 1952). In some areas, codominance of the understory layer by associated shrubs or herbs has allowed more finely identified plant associations to be named (Volland 1976, Williams et al. 1990, Johnson and Clausnitzer 1992). Overgrazed, dry *Pseudotsuga menziesii* plant associations will sometimes appear floristically much like *Pinus ponderosa* plant associations (Peek et al. 1978). In most presettlement *Pinus ponderosa* plant associations, there was a sizable perennial herbaceous component, which together with pine needles created a flashy fuel that encouraged frequent, widespread burning.

THE FIRE REGIME

Considerable information exists for the fire history of *Pinus ponderosa* communities in the southwestern United States. Weaver (1951a) did some of the early work in the Southwest; based on individual point samples, he found fire-return intervals of 4–12 years. Because individual fires do not often scar every tree that has an existing fire scar, this technique is relatively conservative as a point frequency estimate of fire. In Arizona, Dieterich (1980b) found a mean fire-return interval of 1.8 years on a *Pinus ponderosa* site by combining the cross-dated records of fire-scarred trees over a 40 ha area; the lowest mean fire-return interval for a single tree was

4.1 years. More recently, Savage and Swetnam (1990) summarized data from similar studies in the U.S. Southwest and found mean fire-return intervals (scarring >10 percent of trees) from 1.9 to 6.4 years. In comparison, relatively little work has been done in climax *Pinus ponderosa* forests in the Pacific Northwest.

Individual trees in a *Pinus ponderosa* forest on the Warm Springs Indian Reservation in Oregon were found to have mean fire-return intervals of 11–16 years (Weaver 1959). The one sample showing a 47 year fire-return interval was taken from mixed-conifer forest at higher elevation. In the same area over the following decade, Soeriaatmadja (1966) found mean fire-return intervals of 3–36 years, again using individual fire-scarred trees and some from mixed-conifer locations. Farther south in the Oregon Cascades, Bork (1985) completed the most rigorous, cross-dated fire history so far for ponderosa pine forests of the Pacific Northwest (Table 11.1). The Cabin Lake site is a *Pinus ponderosa/Purshia tridentata–Artemisia tridentata/Festuca idahoensis* plant association, on the edge of the shrub-steppe. The Pringle Butte site is a slightly more mesic *Pinus ponderosa/Purshia tridentata/Stipa occidentalis* plant association. Fire appears to be slightly more common on the Pringle Butte site, which at about 30 m^2 ha^{-1} has about 50 percent more basal area than the Cabin Lake site. The low composite mean fire-return interval (MFRI) of 4–7 years across an entire site is somewhat misleading as it covers up to 165 ha and is an area frequency. Bork recommends the 16 ha MFRI as a meaningful stand-level MFRI.

Pacific Northwest *Pinus ponderosa* forests are at the edge of grasslands that became heavily grazed by domestic livestock, primarily cattle, in the late nineteenth century. In the American Southwest, Savage and Swetnam (1990) documented an unusual decline in fire activity in the Chuska Mountains in northeastern Arizona beginning in the 1830s, which they associated with an increase in Navajo sheep populations in the early 1800s. This decline in fire activity was decades ahead of similar declines in fire activity between 1870 and 1900 in other areas of the Southwest. The hypothesized cause was reduction in grass that fueled episodic light surface fires. In the Pacific Northwest, livestock populations increased tremendously in the mid- to late 1800s; most were cattle until the late 1890s (Galbraith and Anderson 1991). Weaver (1959) suggested a decline in fire activity around Simnasho Butte and Butte Creek after 1870,

TABLE 11.1. VARIATION IN MEAN FIRE-RETURN INTERVAL (MFRI) AS A FUNCTION OF AREA SAMPLED

Variable	*Cabin Lake*	*Pringle Butte*
Number of trees sampled	31	35
Oldest tree at site (yr)	539	619
Oldest recorded fire (yr A.D.)	1460	1391
Individual tree MFRI	4–100	13–74
16 ha plot MFRI	16–38	7–20
Average MFRI across plots	24	11
Composite MFRI by site	7	4

SOURCE: Bork 1985.

which he attributed to grazing activity near native settlements. Bork's (1985) data do not show similar fire declines, but she notes that the Cabin Lake site is water-limited for cattle, so it likely was spared the heavy use received by areas closer to water.

The low severity of these fires is inferred from the pattern of scarring but not killing residual trees, and from early accounts of free-ranging fire in this forest type. Fires tended to be of low intensity, rarely scorching the crowns of the older, mature trees. The first federal forest rangers in the southern Cascades yellow-pine (ponderosa pine) woods noted (Munger 1917, p. 37): "Ordinarily, a fire in yellow-pine woods is comparatively easy to check. Its advance under usual conditions may be stopped by a patrolman on a fire line a foot or so wide, either with or without backfiring. The open character of the woods makes the construction of fire lines relatively easy, and in many cases horses may be used to plow them."

Even in low severity fire regimes, however, intense fires may sometimes occur, possibly due to longer than normal fire-return intervals that allow litter and understory fuels to build up, or due to very unusual fire weather, such as strong north winds. Even-aged stands of several hundred hectares have been identified in the Pringle Butte area (D. Frewing, cited in Bork 1985), implying a stand replacement fire. As noted below, such large single-age classes may also be related to insect epidemics. Large blocks of even-aged stands are unusual in *Pinus ponderosa* forest.

Not much is known about the extent of individual fires in ponderosa pine forests. Although regional patterns of extensive fires are well documented in the American Southwest, there is less evidence to show such widespread fire activity in ponderosa pine forests of the Pacific Northwest. Weaver's (1959) two closest sites at Warm Springs were separated by about 10 km, and of 32 different fires, only 2 were recorded in the same year, although these specimens were not cross-dated. Bork's (1985) sites showed extent of fires better because of a sampling density designed to do so. Most of the fires in her study did not scar all trees within 16 ha plots, except for an 1865 fire at Pringle Butte and an 1835 fire at Cabin Lake. This suggests that fires were either small, scattered, and frequent, or consumed too little fuel to scar trees. Once such a pattern was established, it might tend to perpetuate itself by maintaining a scattered pattern of fuels sufficient to carry fire. More work is clearly needed on the extent and seasonality of fires in ponderosa pine forests.

Fire is linked with other disturbance factors in ponderosa pine forests, most notably postfire insect attack. Scorched trees are more likely to be successfully attacked by western pine beetle (*Dendroctonus brevicomis*), mountain pine beetle (*D. ponderosae*), red turpentine beetle (*D. valens*), or pine engraver beetles (*Ips* spp.) (Fischer 1980, Furniss and Carolin 1977). Crown scorch at levels above 50 percent is associated with 20 percent or more mortality by western pine beetle in mature trees (Miller and Keen 1960). Younger trees can survive more than 75 percent scorch with about 25 percent mortality (Dieterich 1979). The crown classification work of Keen (1943) identified four age classes (1–4) and four vigor classes (A–D) (Fig. 11.2); age classes 3–4 and vigor classes C–D are mostly likely to be attacked by insects when growth rates decline due to fire damage. Weaver (1959) noted increased growth of ponderosa pine of crown classes 2-B (66 percent) and 3-B (23 percent) for 20 years following a wildfire. Reduction in tree vigor during drought is also associated with insect attack, and major losses in mature and old-growth ponderosa pine occurred during the drought of the 1920s and 1930s in the Pacific Northwest (Keen 1937).

Defoliation by the pine butterfly (*Neophasia menapia*) is described by Weaver (1961) as being responsible for even-aged stands of ponderosa pine in the Klickitat River valley of eastern Washington. The epidemic occurred in 1893–95, with natural control by a wasp-like parasite (*Theronia atalantae*). During the epidemic, butterflies were so common that streams were choked by them, and travelers

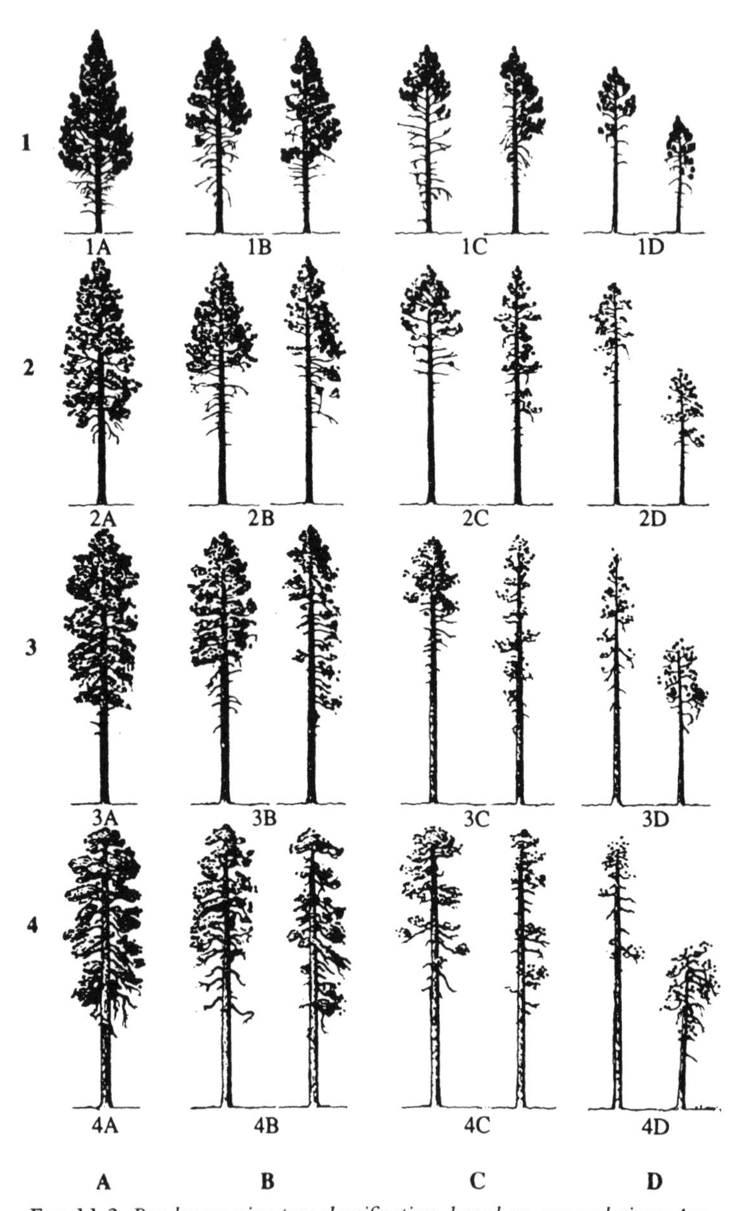

FIG. 11.2. *Ponderosa pine tree classification, based on age and vigor. Age increases from 1 to 4, and vigor decreases from A to D.*
(From F. P. Keen, "Ponderosa Pine Tree Classes Redefined," Journal of Forestry 41(4) (1943): 249–53. Copyright 1943 by the Society of American Foresters, 5400 Grosvenor Lane, Bethesda, MD 20814-2198. Reprinted by permission. Not for further reproduction.)

were covered head to foot with larval webs. Ponderosa pines were killed or reduced in vigor, encouraging attack by western pine beetles. The result was a widespread age class of ponderosa pine on these sites, which were heavily grazed by sheep at the time. Some of these young stands were thinned by fires in the early 1900s, and many residuals were later logged in the 1920–30 period, resulting in a forest structure that differed greatly from that of the 1800s.

Fire may help control dwarf mistletoe infestation by pruning dead branches and consuming tree crowns that have low-hanging witches' brooms. Near the Grand Canyon, Harrington and Hawksworth (1990) found that the largest difference in mortality between trees with or without mistletoe in uneven-aged stands is in the crown scorch range of 40–75 percent. Little mortality occurs below that amount of scorch, while above 75 percent scorch there is high mortality in both classes of trees. Similar findings in young even-aged stands were made in Oregon (Koonce and Roth 1980).

There is not much direct evidence of interactions between fire and various pathogens of ponderosa pine. Little decay may be present around fire scars, which may exist for the life of the tree (Morris and Mowat 1958). Fire might scar roots and encourage root rot formation, but this has not been documented in detail. Conversely, fire often burns out old stumps that harbor decay organisms, and in presettlement ponderosa pine forests it likely had a net beneficial effect for live trees by locally sterilizing the soil around stumps and logs that burned. Laboratory tests on armillaria root rot (*Armillaria mellea*) showed slight decreases in growth and development of armillaria when exposed to moderate concentrations of ash leachate, but increased growth occurred at higher concentrations (Reaves et al. 1984). It is difficult to infer the effects that ash leachate from burned forest floors may have on this fungus as the leachate moves through the soil and contacts rhizomorphs.

Stand Development Patterns

In ponderosa pine forests soil moisture is usually more important than light for the establishment of seedlings. Available soil moisture, particularly in spring and early summer, is critical to the survival of ponderosa pine seedlings (Barrett 1979). Competition for soil water between trees and other understory plants is greatest in the 0–20 cm soil zone (Riegel et al. 1992), within which ponderosa pine will be rooting in its first year. Presence of other understory vegetation, such as perennial grasses (Larson and Schubert 1969) or mature trees, can reduce height growth of established ponderosa pine compared to open-grown trees (Barrett 1973), making the small pines more susceptible to thinning by periodic fire. Establishment on bare soil, historically provided by periodic fire, is higher than on microsites covered by pine needles (Schultz and Biswell 1959). The roots more quickly penetrate into the soil, and moisture availability may be enhanced on bare, open patches across the landscape. Shading of the seedlings can be important for protection from heat and frost (Cochran 1970) by reducing incoming short-wave radiation during the day and long-wave radiation loss at night. This can be accomplished either through maintenance of a partial canopy cover or by small regeneration patches sheltered by adjacent clumps. Nevertheless, mortality of ponderosa pine seedlings may occur from later burns, frost heaving in large openings, drought, or animal damage (Sackett 1984).

Most of the research on stand development in *Pinus ponderosa* forests has been conducted in the southwestern United States, where these forests are more widespread than in the Pacific Northwest. The classic stand development paper was published by Cooper (1960), who studied stands in the White Mountains of eastern Arizona. Cooper was struck by the apparent small-group, even-aged nature of the open ponderosa pine forests (Fig. 11.3) and quantitatively documented the pattern and age structure of these old-growth forests.

Cooper found a wide age-class range across 2–4 ha plots, suggesting a pattern of periodic tree establishment at a small scale. He then performed a contiguous quadrat analysis (Grieg-Smith 1952) to determine the size of the apparently clumped pattern. The number of pines in each cell of a 256 cell matrix are counted, and then adjacent cells are combined; Cooper used cell sizes of 1/250 ha. An analysis of variance is conducted on the various block sizes to determine the mean square variance associated with each block size. If clumping occurs, then the combination of blocks at the scale of the clumping will show a high variance, because of the high number of pines in some blocks (within clumps) and low numbers

FIG. 11.3. *An old-growth ponderosa pine forest in central Oregon. Note the clumping of similar-sized trees at a very small scale.*

elsewhere (between clumps). Cooper found stands of mature trees to have variable clump sizes, with the clumping usually at 0.06–0.13 ha, equivalent to clumps 25–35 m on a side (Fig. 11.4). He interpreted a similar pattern for younger "reproduction" stands, although his data show a wider range of clump sizes (0.06–0.26 ha).

The process of stand development in *Pinus ponderosa* forests is a result of the shade intolerance of ponderosa pine, periodic good years for seedling establishment associated with years of above-normal precipitation, and frequent fire (Cooper 1960). Gaps in the forest, created by mortality of an existing small, even-aged group, allow the shade-intolerant pine to become established when a good seed year and appropriate climate coincide. In this opening, the stand of young trees will be protected from fire because of lack of fuel on the forest floor, while the fire will burn under mature stands and eliminate any reproduction there. As the trees in the opening continue to grow, they provide enough fuel to carry the fire and thin the stand. Within a group, relatively uniform spacing is the

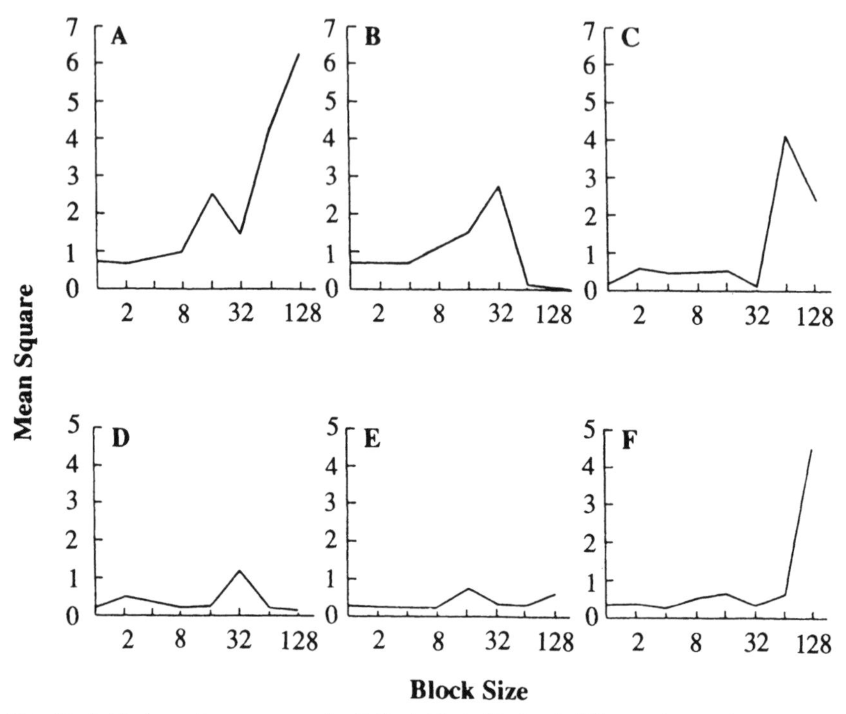

FIG. 11.4. *Variance (mean square) attributable to blocks of different sizes, using a contiguous quadrat technique.* A, B: *Mature stands at Malay Gap.* C–F: *Mature stands at Maverick. Peaks indicate increased variance due to a clustering effect at that quadrat size, with peaks to the right (e.g.,* F) *being less interpretable due to lack of degrees of freedom. To convert to block size in ha, divide by 247.*
(From C. F. Cooper, "Changes in Vegetation, Structure, and Growth of Southwestern Ponderosa Pine Forests since White Settlement," Ecological Monographs 30 [1960]: 129–64. Copyright 1960 by Ecological Society of America. Reprinted by permission.)

result of moisture competition and a tendency for closely spaced trees to be selectively killed by fire (Cooper 1961a).

In current-day thickets, fire tends to thin nonuniformly (Morris and Mowat 1958, Wooldridge and Weaver 1965), leaving open patches and unthinned clumps as well. This pattern of nonuniform thinning is due to altered forest structure and differs from what occurred in presettlement forests. In presettlement forests, as a young group and adjacent clumps continued to grow and underburn at frequent intervals, fire maintained the even-aged, clumped pattern over time (Fig. 11.5; Cooper 1961b). When a clump became old, insects, possibly associated with one or more drought years,

FIG. 11.5. *Parklike forests of ponderosa pine, showing a mosaic of small groups.* A: *Mature stand is in the middle distance with a young stand behind it.* B, C: *Frequent fires kept the stands open and consumed debris, creating sites for seedlings.* D: *As saplings grew, they shed needles that allowed subsequent fires to kill new seedlings, maintaining the mosaic character of the forest.*
(From C. F. Cooper, "The Ecology of Fire," Scientific American 204[4] [1961]: 150–60.)

likely attacked one tree and then another in the clump. Disease may also have attacked members of a clump, especially if their roots were grafted to one another. Cooper (1960) suggested a relatively rapid breakup (20 years or less) for an old clump. One implication of this pattern is that it is likely to remain spatially stable over time.

A more recent study has suggested that some elements of Cooper's hypothesis do not fit all ponderosa pine stands (White 1985).

In support of Cooper's findings, White found an all-aged distribution over a 7.3 ha area in northern Arizona, and stems older than 106 years were strongly aggregated. The size range of clumps of mature trees, determined by a nearest-neighbor analysis (Clark and Evans 1954), was more variable, with clumps from 0.02–0.29 ha (15–55 m on a side). Within a group of 3–44 stems, at densities from 62–285 trees ha^{-1}, trees were randomly distributed, with nonsignificant tendencies toward both uniform and clumped patterns. These are but slight variations on Cooper's findings. The most significant difference concerned Cooper's hypothesis of rapid clump breakup and regeneration: White found age variations within mature tree clumps of 33–268 yrs. White suggested that rather than the whole group dying simultaneously, one or two trees within a group might die and contribute additional fuel to the forest floor. When they burned, limited sites for regeneration were created, and this pattern, repeated over many decades for an old-growth group, might result in the large within-group age ranges he found. Cooper's hypothesis was supported primarily by observation of one clump that had broken up rapidly, with later regeneration; he did not conduct extensive within-group age-class analysis. White's findings may represent a general pattern rather than an exception. If true, his results would imply that the spatial nature of clumps may vary over time on the landscape.

Pattern in Pacific Northwest *Pinus ponderosa* forests has been documented by West (1969) and Morrow (1985) in the eastern Oregon Cascades. West used the same technique as Cooper (1960) along a gradient of increasing moisture, from western juniper communities to mixed-conifer communities within the Warm Springs Indian Reservation where Weaver's (1959) fire scars were sampled. The middle section of the gradient, where *Pinus ponderosa* plant associations occurred, showed the highest degree of aggregation, at a block size of 0.25 ha—larger than shown by Cooper (1960) but within the range of White's (1985) data. Trees within a clump tended toward regular, uniform spacing, probably for the same reasons noted by Cooper (1960). Sapling clusters in the Oregon study were described as elliptical and oriented east-west, averaging 55 x 67 m. West (1969) hypothesized that this was due to prevailing west winds blowing dead trees down in an east-west direction, with later fires scarifying the soil along this axis and providing acceptable conditions for ponderosa pine establishment. Most ponderosa pine seed falls within about 30 m of the source (Barrett 1966), so patches under most conditions would be completely within the range of surrounding mature seed-source trees.

Morrow's (1985) study of pattern in ponderosa pine was conducted at the Pringle Falls Experimental Forest, near where Bork (1985) developed composite fire-return intervals. He found a wide age range across 1 ha plots and a similar clustering of three age cohorts of ponderosa pine >10 cm diameter: >230, 110–230, and <110 years old, with clusters from 0.025–0.35 ha in size (Fig. 11.6). Clustering in old clumps was less pronounced than in intermediate-aged groups. Tree distribution within clumps was also clustered, in contrast to the Arizona stands, but no elliptical reproduction clumps were noted as West (1969) observed. The temporal pattern of regeneration suggested that longer fire-free intervals were associated with surviving cohorts of trees, implying that growth into the sapling stage without frequent fire helped regeneration to persist into the mature stage. Periods with apparently more frequent fire had fewer surviving cohorts to the present. Morrow concluded that continuous fire protection, such as occurred in the twentieth century, would erode both the spatial and temporal pattern of these ponderosa pine forests.

Early ponderosa pine forests had considerable understory vegetation. In the Southwest, mountain muhly (*Muhlenbergia montana*) was a dominant grass; in the Pacific Northwest, Idaho fescue (*Festuca idahoensis*), bluebunch wheatgrass (*Agropyron spicatum*), and Sandberg's bluegrass (*Poa sandbergii*) were common, with long-stolon sedge (*Carex pensylvanica*) and pinegrass (*Calamagrostis rubescens*) on more mesic sites. Total understory production is inversely related to tree crown cover (Pase 1958). In eastern Washington, McConnell and Smith (1970) found production below 200 kg ha^{-1} above 50 percent crown cover, with forbs and shrubs the dominant components. They found production above 300 kg ha^{-1} below 50 percent crown cover, with grasses more important. Increasing tree spacing to 5.6 m in second-growth stands resulted in 665 kg ha^{-1} of understory production after seven years.

Several factors have been responsible for the decline of perennial grasses in *Pinus ponderosa* plant associations over time. Livestock grazing since the late 1800s has reduced perennial grasses while

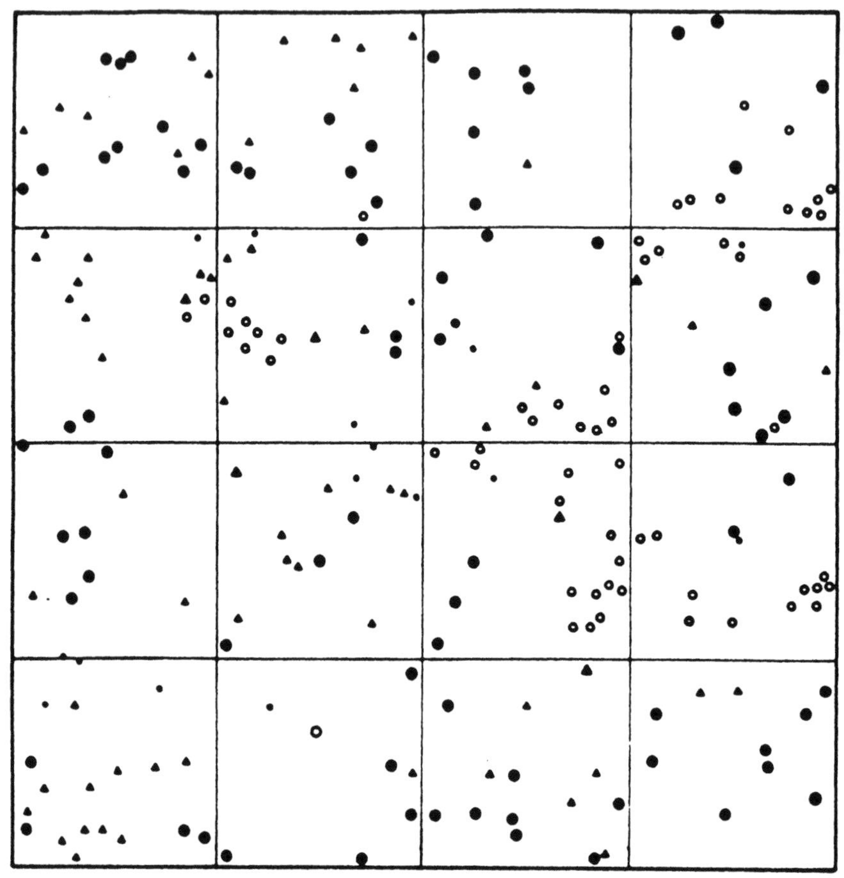

FIG. 11.6. *Stem map of a 1 ha stand at Pringle Falls Experimental Forest, Oregon. Two species (ponderosa pine [circles] and lodgepole pine [triangles]) and three age cohorts above 10 cm diameter are represented: >230 years (large black circles, ponderosa pine only), 110–230 years (open circles and triangles), and <110 years (small dark circles and triangles). Clustering is present in all classes but most obvious in the 110–230-year-old class.*
(From Morrow 1985)

encouraging annual grasses, such as cheatgrass (*Bromus tectorum*), and has allowed shrubs to invade. Even before a fire exclusion policy was enacted, some areas apparently experienced declines in fire activity, thought to be due to absence of sufficient grass fuels (Savage and Swetnam 1990, Weaver 1959). As a result of fire exclusion and logging, many stands now have higher tree density and generally smaller, more uniformly dispersed stems, creating dense overstory canopies (Laudenslayer et al. 1989). Fire exclusion has not allowed thinning or pruning of the trees or topkill of shrubs by frequent fires. The observations of Dutton (1881, pp. 136–37) in the Grand Canyon country of the 1880s are likely applicable to most of the *Pinus ponderosa* forests of the presettlement West:

> The trees are large and noble in aspect and stand widely apart. . . . Instead of dense thickets where we are shut in by inpenetrable foliage, we can look far beyond and see the tree trunks vanishing away like an infinite colonnade. The ground is unobstructed and inviting . . . from June until September there is a display of wildflowers which is quite beyond description. The valley sides and platforms above are resplendent with dense masses of scarlet, white, purple, and yellow. It is noteworthy that while the trees exhibit but few species the humbler plants present a very great number, both of species and genera.

The typical fire in ponderosa pine forests had little effect on the herbaceous component besides removing the cured material above the ground. Removing the accumulated needles and topkilling shrubs generally aided the grasses and forbs (Weaver 1951b, Biswell 1973) and stimulated flowering. Bunchgrasses normally recover in one to three years, and most forbs recover quickly (Wright et al. 1979). Bulb and tuberous species are rarely harmed by fire. Competition from grasses and forbs in areas protected from domestic livestock (Rummell 1951) limits tree regeneration to areas scarified by fire. In areas dominated by cheatgrass, fire cannot be used to restore native conditions, but it may not further degrade the area. Merrill et al. (1980) found no increase in bluebunch wheatgrass under ponderosa pine after a fall wildfire, but initial increases in cheatgrass height returned to preburn conditions after four years.

Bitterbrush was likely much less common across the landscape than today. Although bitterbrush can sprout after light spring burning (Martin and Driver 1983), fire often kills it. About 35 percent of spring-burned bitterbrush will sprout, with about 15 percent surviving for more than a year, while fall-burned bitterbrush is almost totally killed (Clark et al. 1982). Bitterbrush is capable of fixing atmospheric nitrogen (Youngberg 1966), but repeated burning favors grasses over bitterbrush (Nord 1965, Wright et al. 1979). Bitterbrush seedlings were found in profusion after burning in central Oregon and northern California (Martin 1983). Bitterbrush densities of 3,500–19,000 ha^{-1} averaged 15–40 cm height after seven years. Wider tree spacing is associated with increased bitterbrush: for each 1 m increase in spacing, bitterbrush production increases 8 kg ha^{-1} (McConnell and Smith 1970).

Fuels

Fuels in *Pinus ponderosa* forests were rarely at high levels because of the frequent fires that consumed forest floor fuels and pruned residual trees. Fine fuels are produced either by needlefall or by understory vegetation. Biswell (1973) described needlefall in California pine forests to be about 3,650 kg ha^{-1} at basal area of 38 m^2 ha^{-1}, while needlefall was 2,000 kg ha^{-1} in stands thinned to 15 m^2 ha^{-1}. Herbage increased after thinning from 250 to 1,600 kg ha^{-1}, so the total fine-fuel production on an annual basis was essentially the same at both sites: 3,600–3,900 kg ha^{-1}. However, the grasses are less likely to carry fire until mid-summer when they are cured and react like other dead fuels to changes in atmospheric moisture. In openings in Arizona ponderosa pine forests, Cooper (1960) estimated that regeneration patches were about eight years old before a typical fire would carry through the patch.

Branches and logs can also contribute to fire behavior. In related forest types, Norum (1977) and Agee et al. (1978) found consumption of larger fuels (up to 7.6 cm diameter) increased consumption of duff. Branches tend to fall more sporadically than needles; in a California pine stand, no branch fuels fell in thinned stands over a 10-year period, while about 6,500 kg ha^{-1} fell in an adjacent unthinned stand (Agee and Biswell 1970). In presettlement stands, downed logs were probably clumped at the same scale as the live tree components from which they were created, and such clumps contributed to local increases in fire behavior. It is doubtful that logs remained long on the forest floor to provide wildlife habitat, rooting media for seedlings, or sites for nitrogen fixation by microorganisms, as they were probably consumed by the next several frequent fires on the site. Seedlings often had better survival in ash of the burned logs, because of reduction in competition and the soil-sterilizing effect from the heat of the fire.

MANAGEMENT IMPLICATIONS

Ponderosa pine forests have experienced significant ecological change since fire suppression began. In most forests, selective removal of the large "yellow belly" pines has eliminated the visually obvious clumping pattern of the older forests and replaced it with more uniformly spaced, smaller "black bark" forests. In unharvested areas, establishment and growth of ponderosa pine in the understory of the larger pines have also reduced spatial diversity and created fuel ladders to the larger trees. Where these larger trees are old and of low vigor, they might have been removed in past decades by insect attack. They are vulnerable to significant increases in insect attack either from prescribed fire or wildfire.

The increased damage due to wildfires is effectively summarized by Biswell et al. (1973). This team evaluated the effects of wildfires in stands of increasing "rough," or time of protection. Not surprisingly, they found that fires in roughs 17 years or older tended to be more intense and create some stand replacement patches; they recommended prescribed burning at intervals of six to seven years to keep fuels at low levels.

Prescribed burning, particularly in stands protected from fire for considerable time, must be carefully done. Biswell (1960) recommended a reintroduction over several cycles, beginning with fire under very moist conditions, followed by fire under successively drier conditions, to avoid excessive fuel consumption in one burn. In Arizona, Allen et al. (1968) and Harrington (1987) provided some site-specific prescriptions for fuel reduction based on 10 hr timelag fuel moisture, relative humidity, wind speed, forest-floor moisture content, preburn fuel loading, larger woody fuel loading, and overstory density. If stand conditions are patchy, fire behavior is likely to vary (Harrington 1982).

In the eastern Cascades of Oregon, Landsberg et al. (1984) studied second-growth ponderosa pine, which had been protected from fire since its establishment, after a single prescribed fire. They found decreases in stand volume increment, height growth, and basal area increment after eight years (Fig. 11.7). Cochran and Hopkins (1991) extrapolated from these results to suggest that height growth, from which productivity estimates are made, is reduced at least 4 percent by repeated, moderate underburning. They recognize that firm conclusions require more experience with repeated spring and fall burns under different conditions, as later fires may have less severe effects, and long-term reductions in the risk of high severity wildfire are ignored in such extrapolations.

Reintroduction of fires into *Pinus ponderosa* forests of park and wilderness areas needs careful prescription. First, it is unlikely that any such forests can absorb the shock of naturally occurring fires during the summer months, given the altered fuel and forest structure due to fire exclusion. The fires will generally be far too intense, consume too much fuel, and cause the death of many of the older, presettlement forest trees. Although some researchers argue that prescribed fire in wilderness is unnatural (Worf 1984), a natural fire in such significantly altered forest structure is even less "natural." Prescribed fire is essential to restore such sites and can be done through a series of low intensity fires. Attempts to complete the restoration in a single fire are not realistic. Autumn fires appear to harm low vigor ponderosa pines less than spring fires (Swezy and Agee 1991) in *Abies concolor* forests, and it is assumed that the same conclusions hold for *Pinus ponderosa* plant associations. Once a series of prescribed fires has reduced surface fuel loads, fuel ladders to the overstory, and total tree density, either continued prescribed fires or natural fires can maintain the role of fire into the future for these forest types. The choice at that point depends on factors other than forest biology: nearness to the boundary of the unit, likelihood that fires that once moved into the unit from outside will continue to be able to do so, air quality constraints, and other factors considered in management planning.

PINUS CONTORTA FORESTS

Most lodgepole pine forests consist of early seral cohorts of lodgepole pine, which will eventually be replaced by another species: often Douglas-fir, grand fir, or subalpine fir. The lodgepole pine

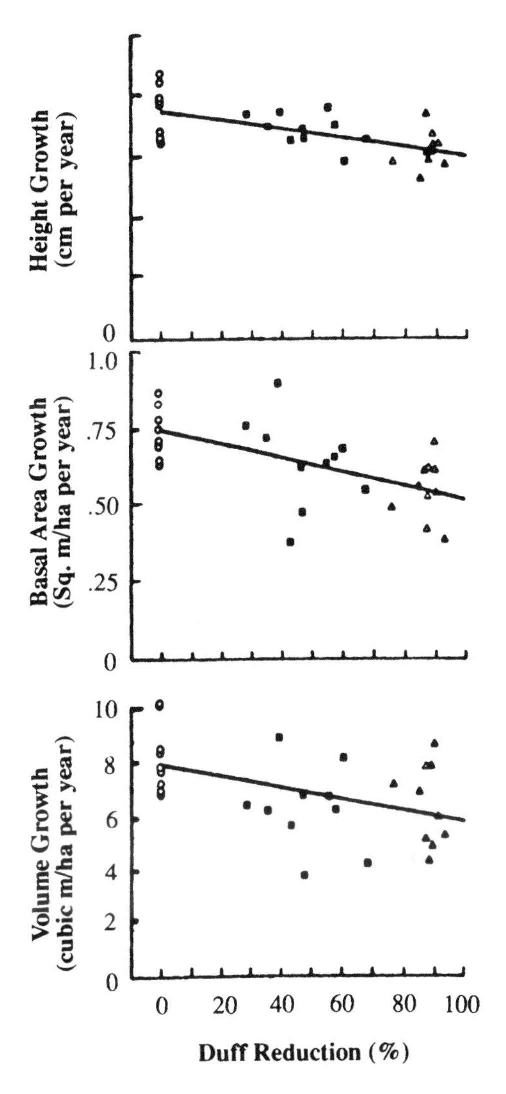

FIG. 11.7. *Reduction of periodic annual growth in height, basal area, and volume for ponderosa pine four growing seasons after burning as duff reduction increased.*
(From Landsberg et al. 1984)

forests discussed in this chapter are those that replace themselves, in which lodgepole pine is the sole tree dominant. The *Pinus contorta* series is found scattered across the western United States, occurring in Colorado (Moir 1969) and Yellowstone (Despain 1983); Figure 11.1 maps only the most concentrated area of climax lodgepole pine forest. These forests are often confused with seral lodgepole pine stands, where mature lodgepole pines have dominated for a century or more and later successional species have not yet invaded (a relay floristics pattern). The lodgepole pine series in Oregon, also called climax lodgepole forest, can be separated from seral lodgepole pine forests by the absence of these other species, a general lack of understory shrubs and herbs, and the relative dominance of a few shrub or herb species. Bitterbrush (*Purshia tridentata*) is a common associate of lodgepole pine on transitional forests, while understories with long-stolon sedge (*Carex pensylvanica*), western needlegrass (*Stipa occidentalis*), and pussypaws (*Spraguea umbellata*) indicate the lowest productivity basins of climax lodgepole pine forest. These stands are shallow-rooted forests of very low productivity: 1–2 m^3 ha^{-1} yr^{-1} of volume increment, with 1–2 m^2 m^{-2} of leaf area (Volland 1976, Stuart 1983). The pumice soils have low thermal conductivities, which result in a wide range of maxima-minima temperatures near the ground surface (Cochran et al. 1967).

The Fire Regime

Climax lodgepole pine forests have a moderate severity fire regime. A combination of low, moderate, and high severity fires occurs in space and time. Fire frequency is not well documented for these forests. Within Crater Lake National Park, Agee (1981b) estimated a fire-return interval of about 60 years based on scanty data. In a nearby area, Chappell (1991) estimated a fire-return interval of 39 years for a red fir stand adjacent to a flat dominated by climax lodgepole pine forest. To the east on Fremont National Forest, Stuart (1984) documented one fire-free interval of 60 years and located at least one stand that had been fire-free for about 350 years. The average fire-return interval is probably in the range of 60–80 years, with areas surrounded by higher productivity forests at the lower end of the range.

An important related disturbance is insect attack by mountain pine beetles (*Dendroctonus ponderosae*). In these low vigor stands, insect attack is often only partially successful, scarring rather than killing trees (Stuart et al. 1983). The scars can be mistaken for fire scars (see chapter 4); the three most common types of fire wounds can also be mimicked by mountain pine beetle scars (Gara et al. 1986).

The magnitude of natural fires ranges from slowly burning logs across the forest floor to crown fires. A low intensity fire was identified in 1899 at Wickiup Springs on the Fremont National Forest (Stuart 1983). At Crater Lake in 1980 a prescribed fire burning in ponderosa pine forest spread into climax lodgepole pine forest by means of down and decayed logs, with little fire spread outside of log corridors (Agee 1981b). A fire roughly 60 years before was also of low severity. During some fire events in adjacent forest, even during summer, fire has not spread into climax lodgepole pine stands (Gara et al. 1985), suggesting that these stands may be a natural fire barrier except under unusual fire weather.

Stand replacement fires were noted as a probable origin of a stand in 1840 near Wickiup Springs (Stuart 1983). The 1988 Prophecy fire at Crater Lake, a prescribed natural fire that started in red fir forest, was pushed by a strong westerly wind through a 100 ha patch of lodgepole pine forest as a crown fire, even though the crowns were not touching or continuous across much of the area. Strong winds are probably required for fires to crown through these climax lodgepole pine flats.

Most climax lodgepole pine stands have scattered charcoal near or at the surface, an indication of a time when a log or tree was partially consumed. Zeigler (1978) used the absence of charcoal from other species to help identify climax lodgepole pine stands. About three-quarters of his seral lodgepole stands (n = 37) had charcoal of other species mixed with lodgepole pine charcoal, while none of the climax lodgepole pine stands (n = 19) had charcoal of other species. The presence of charcoal from other species indicates that at one time these species existed on the site, and all are later successional species than lodgepole pine. The absence of such charcoal is no guarantee that the site is a climax lodgepole pine site, however, as 25 percent of the seral lodgepole sites contained only lodgepole pine charcoal.

STAND DEVELOPMENT PATTERNS

Most stands that have been surveyed have multiple age classes (Fig. 11.8; Agee 1981b, Stuart et al. 1989), either from forest fires or mountain pine beetle attacks that have removed some but not all of the mature stems. A typical disturbance scenario includes interactions of insects, fire, and fungal pathogens with the lodgepole pine stands. The scenario described here begins with a mountain pine beetle attack in a mature stand.

Mountain pine beetles in these low productivity stands tend to strike in epidemic proportions after trees are large enough to sustain brood populations and after a period of sustained low radial growth (Mason 1915, Waring and Pitman 1980, Stuart 1984). This pattern is similar to that in seral lodgepole pine stands in the Rocky Mountains, where beetle attack may be linked to stand age (>80 years), tree size (dbh >20 cm), and thick phloem (>2.5 mm) (Cole and Amman 1980). Tree size and stress in climax lodgepole pine stands may be correlated, as trees increase in size and compete for limited soil moisture on the site (Smith 1981).

The beetles tend to land preferentially on trees that have been

FIG. 11.8. *Lodgepole pine ages on a 0.04 ha plot at Crater Lake National Park, showing pulses of tree establishment after disturbances at roughly 60 year intervals.*
(From Agee 1981)

previously fire-scarred and those in a state of advanced decay (Gara et al. 1984), with a similar pattern for successful attacks (Geiszler et al. 1980). These "focus" trees, once attacked, will continue to attract beetles, which will then attack the largest nearby trees—the "switching mechanism" described by Geiszler and Gara (1978). If these trees have sufficiently thick phloem, the tree will likely be killed, but in many of these stands the bark and phloem of mature trees is so thin that attacks may be successful only on one side of the tree. The north side of the tree, protected from temperature extremes, where the attack may be successful will usually be the scarred side. Anywhere from one-third to two-thirds of the stand dominants may be attacked by beetles, and about 75 percent of those attacked are killed (Stuart 1983). Basal area and leaf area may be reduced by 50 percent or more. At Agee's (1981b) Crater Lake site, beetle and fire kill reduced average stand age from 124 years to 70 years; residual stems averaged 8.65 cm dbh while dead stems averaged 18 cm.

Stems killed by the beetles will normally fall within a 5–10 year period, at which time existing or incipient decay will continue to decompose the log. On the site studied by Agee (1981b), 85 percent of stems killed either by fire or by later beetle attack had fallen by year 10. The thin bark exfoliates and small branches break off, so that within a decade or two the log is in direct contact with the soil.

Lodgepole pine in this area has nonserotinous cones (Mowat 1960). Annual seed production may be 1.5 million–1.8 million seeds ha^{-1}, with high viability and relatively uniform seed dispersal (Stuart et al. 1989). Seedlings first emerge on the north side of snags, and soil moisture availability is associated with seedling establishment. Therefore, the areas where groups of fire- or insect-killed lodgepole pines occur will be major regeneration sites within a year or two after disturbance. Frost heaving occurs in areas farthest from existing live trees or snags, and soil moisture competition from herbaceous plants and tree roots of adjacent lodgepole pines limit successful regeneration in other seemingly open gaps. With the microclimate buffering associated with snags and logs, a new age class of lodgepole pine becomes established on the site. Just as soil moisture affects mature tree growth (Smith 1981), it also affects seedling establishment trends. On the Fremont National Forest, Stuart (1983) showed the clear association between seedling groups and insect-killed trees (Fig. 11.9). After a 1980 fire at Crater Lake, seedlings increased from 100 ha^{-1} in the year of the fire to 350 ha^{-1} by year 2 and to 6,800 ha^{-1} by year 10 (J. K. Agee, unpublished data).

Over the next few decades, the downed logs continue to decay. After 50 years, they commonly have a hard sapwood rind but decayed heartwood interior and become prime vectors for fire spread. Unless the fire is blown by exceptional winds, it will move along the narrow corridors of available logs, which may be continuously linked across the landscape (Fig. 11.10). Where discontinuities occur in log contact, fires often extinguish themselves. The flaming logs may be close enough to a sapling or mature tree to fatally scorch it or to scar it in one of several patterns (Gara et al. 1986). The heat from the burning log will penetrate into the soil and damage roots of trees that are otherwise unaffected by the fire.

These fires of low to moderate severity encourage secondary disturbance by insects and pathogens. Mountain pine beetles will attack trees damaged by the fire just as they attack other low vigor trees (Gara et al. 1985). Fungal pathogens enter the damaged root sections and create a decay column that slowly moves up the bole of the tree (Fig. 11.11; Littke and Gara 1986). In 50 years, the decay column will be about 40 cm high, while after 100 years it will be 1.5 m or more up the bole (Gara et al. 1985). Such decay has in the past been thought to enter directly through fire wounds on the bole (Nordin 1958). The infections begin with white-rot Basidiomycetes, which are later outcompeted by brown-rot Basidiomycetes (Littke and Gara 1986).

The trees killed by either fire or insects form the next generation of logs, which will become ideal fire vectors in another 50–75 years. At that time the disturbance may be another insect attack, another fire of low to moderate severity, or perhaps a crown fire if conditions are right.

Crown fires can consume all the understory vegetation beneath climax lodgepole pine stands. Bitterbrush is usually killed, although it can sprout after low intensity spring fires. The sedges and grasses normally sprout back the next season, and by the time fires burn in the summer, most of the plants have completed their aboveground growth and seed production. In the log-to-log type of fire, the only plants affected will be those along the roughly 50 cm wide log

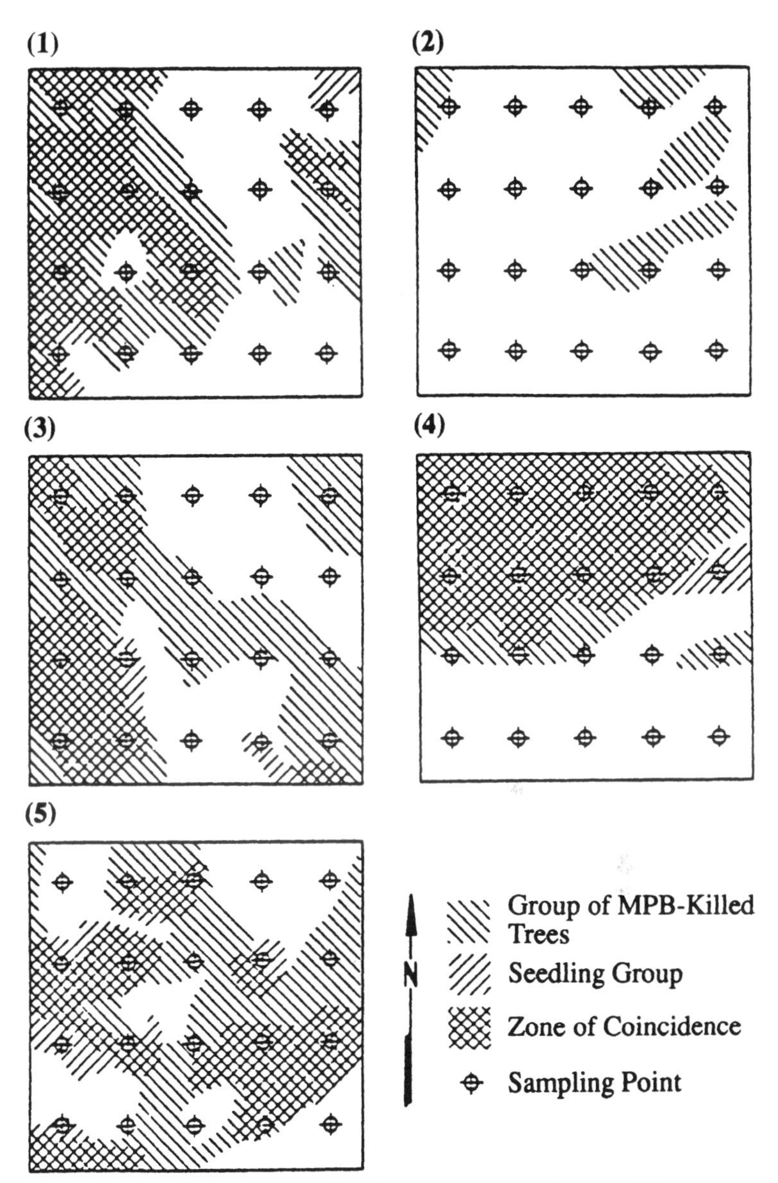

FIG. 11.9. *Coincidence of trees killed by mountain pine beetle (MPB) and seedling groups of lodgepole pine on five plots. Seedling groups are highly associated with MPB-killed groups.*
(From Stuart 1983)

FIG. 11.10. *A log-to-log burning pattern in climax lodgepole pine forest.*

FIG. 11.11. *Excavation of damaged root from 80-year-old fire (left) on a 140-year-old lodgepole pine, with little aboveground evidence of decay. Decay column in bole (right) can be traced back to the damaged root.*
(From Littke and Gara 1986)

corridors in areas of white ash, about 10–15 percent of the surface area (Agee 1981).

A synthesis of these interactions (Fig. 11.12) at Lookout Point on the Fremont National Forest shows one set of possible combinations of these interacting disturbances for climax lodgepole pine forests. An example of another disturbance scenario is an area not burned for over 200 years in the same general area. In this area, trees are being killed individually by mountain pine beetles, and small patches of regeneration are associated with each killed tree. The pattern of continual but minor disturbance differentiates this area from the more wavelike disturbances across the other areas studied (Gara et al. 1985).

Fuels

Litter fuels in climax lodgepole pine stands are almost nonexistent. Stands that have not burned for 60 years have substantial bare soil exposed, except in areas of high tree density where litter layers may be 1–2 cm thick. Grass and shrub fuels are too scattered to be effective fire carriers. Crown fuels are also limited, although crown fires have been noted to burn through the crowns under high wind conditions.

The most continuous fire vector is logs. Partially decayed logs, remnants of disturbances several decades earlier, carry most fires, so that fire behavior is more a function of coarse woody debris than of fine-fuel dynamics. Logs often have higher fuel moistures than other fuel classes. At a prescribed fire at Crater Lake in 1982, fine-fuel moisture was 4–10 percent, but these fuels were too discontinuous to carry fire. Instead, the fires smoldered along logs with relatively high moisture content (average 42 percent).

These moist, partially decayed logs burn for two reasons. First, the decayed heartwood is surrounded by a less decayed rind of sapwood. This allows reasonable air circulation while the rind helps conserve heat by reducing convective heat loss from the smoldering interior. As the interior portion burns, the sapwood rind may collapse and be consumed, or it may be left as charred residue when the fire moves down the log. Second, rotten wood generally requires less energy to produce the combustible gases required for

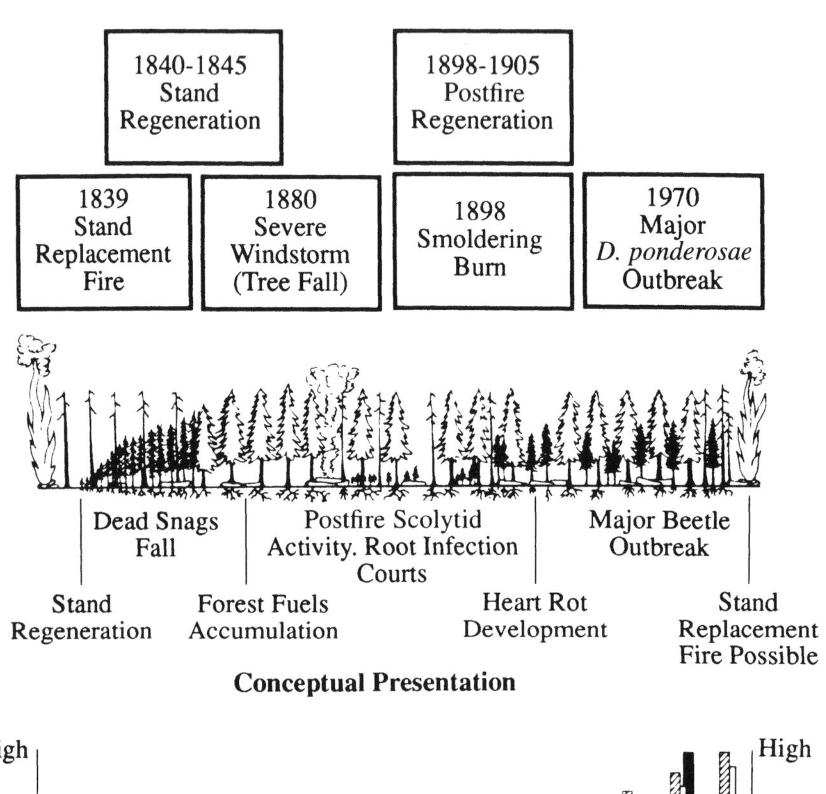

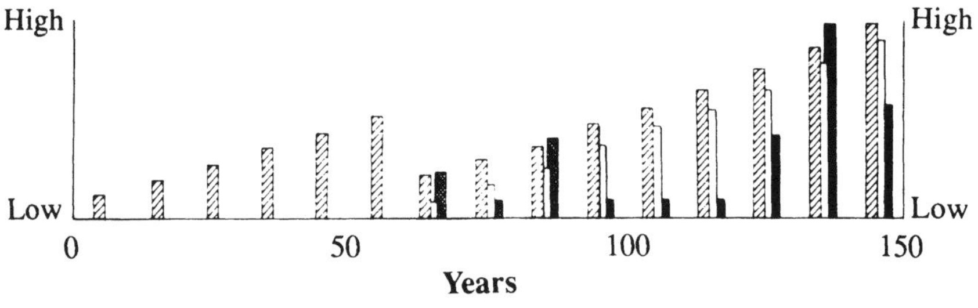

FIG. 11.12. *Interactions among fires, fungi, and mountain pine beetles observed at Lookout Point, Oregon.*
(From Gara et al. 1985)

ignition (Susott 1982). These physical and chemical characteristics favor partially decayed logs as fire corridors under most conditions. Less decayed logs would be more likely to burn soon after they have fallen, when they have attached needles and branches, or later in years of extended drought.

Management Implications

Climax lodgepole pine forests rarely grow for a century without a major disturbance by fire or insects. In the two areas where these stands have been most intensively studied (Crater Lake and Fremont National Forest), both log-to-log (or "cigarette burn") and crown fire activity has been observed or reconstructed from age data for forest stands. In areas managed for natural values, a prescribed natural fire policy may be implemented with the knowledge that human impact on such stands appears to have been minimal. The low productivity of these climax lodgepole pine forests has limited fuel buildup and other changes due to past fire exclusion policies.

These stands should be expected under most conditions to have slow-moving fires burning along logs created decades earlier by a past fire or insect disturbance. Fires moving into such stands from adjacent forest types will normally be extinguished on their own as fuels dissipate in the interior of the stand. Nevertheless, the occurrence of past crown fires in many of these stands suggests that climax lodgepole pine stands cannot be considered fire barriers under all fire weather conditions.

If fuel treatment is to be considered to manage such fuels, perhaps at the edge of a natural fire zone, total canopy removal is not recommended. This will encourage rapid regeneration of young lodgepole pine, which if dense enough may carry a low crown fire. Instead, crown thinning to about 20–25 percent cover for perhaps 200 m width and removal of continuous log fuels, followed by maintenance every 10 years or so to remove understory fuels (particularly small trees), is the preferred way to create and maintain a fuelbreak where one is desired. The existence of a canopy will help keep competition for soil moisture high, limiting understory establishment and growth. This will be a zone where natural fires will generally stop or can be suppressed. The 1988 Prophecy fire at Crater Lake crowned through a very open stand of climax lodgepole pine that appeared to be such a barrier, again suggesting that what appear to be fuelbreaks are not firebreaks.

Prescribed fire in these stands, such as underburning, is not a possibility on any widespread basis because of the lack of surface fuels. When fires spread under moderate fire weather, they move very slowly and consume little but logs. Any proposed manipulations will have to be either manual or mechanical, with pile burning a possible means of reducing the fuels generated by those activities.

CHAPTER 12

NORTHWEST WOODLANDS

THE WOODLANDS OF THE PACIFIC NORTHWEST are transitional between forest and shrub/grassland communities. These sparsely treed areas are commonly called savannas if they have less than 30 percent tree cover and fewer than 50 trees ha^{-1} (15 m spacing). They are called woodland if they have more than 30 percent cover and 150–300 trees ha^{-1} (6–8 m spacing) (Griffin 1977). In the Pacific Northwest, outside of riparian corridors, oak and juniper communities (Fig. 12.1) may fit the definitions of savannas or woodlands. Both occur in areas that are physiologically marginal for tree establishment and growth, usually because of water limitations caused by thin or coarse-textured soil or low rainfall (Minore 1979).

Northwest woodlands have in common not only some climatic and edaphic factors but also a poorly understood fire regime and ecological interactions with fire. Overstory and understory cover are usually inversely related, and alien species are now often dominant after disturbance by heavy grazing or fire.

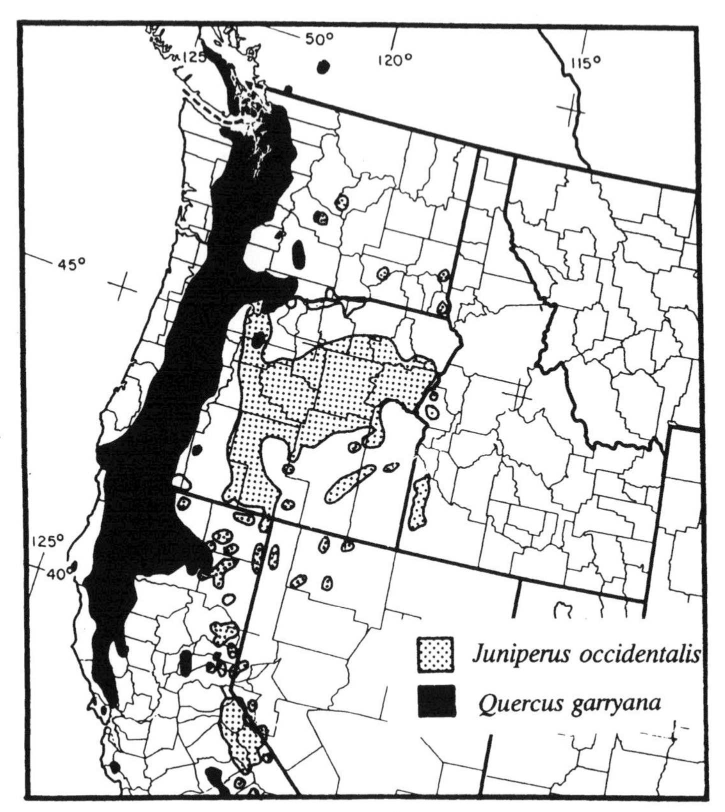

FIG. 12.1. *Location of Oregon white oak woodlands and western juniper woodlands in the Pacific Northwest.*
(Adapted from Little 1971)

QUERCUS GARRYANA WOODLANDS

Oak woodlands are found throughout the West; this chapter considers only those with Oregon white oak, also called Garry oak (*Quercus garryana*). This oak ranges from just south of 50° latitude in British Columbia to about 34° latitude in Los Angeles County, California. In California three types of oak woodland have been recognized, and Oregon white oak is a dominant only in the northernmost one (Griffin 1977). In the southern oak woodland,

Engelmann oak (*Q. engelmannii*) and coast live oak (*Q. agrifolia*) are dominant, while in the foothill woodland surrounding the large interior valleys of California, blue oak (*Q. douglasii*), valley oak (*Q. lobata*), and interior live oak (*Q. wislizenii*) are dominant. In the northern oak woodland, which Griffin believed is the least satisfactory subdivision of the California oak woodlands, the presence of Oregon white oak is the major differentiating factor. Otherwise these oak woodlands are similar to blue oak woodlands farther inland: similar grasses, the same manzanita (*Arctostaphylos* spp.) and ceanothus (*Ceanothus* spp.) shrubs, and digger pine (*Pinus sabiniana*). In northeastern California, there is some mixing of Oregon white oak into juniper woodlands, but generally these woodland types are distinct from one another.

In the Bald Hills of coastal northwest California, Sugihara et al. (1987) identified four plant communities that had Oregon white oak as a dominant, two with shrubs as the dominants but with some oak present, and a grass community with no mature oak present. These areas were classified within several years after cessation of grazing (L. Reed, personal communication, 1992). The oak-dominated communities included the xeric oak/hedgehog dogtail (*Quercus/Cynosurus*), mesic oak/snowberry (*Quercus/Symphoricarpos*), mesic oak/orchardgrass (*Quercus/Dactylis*), and a slightly mesic oak/larkspur (*Quercus/Delphinium*) type found in concave topography in a matrix of oak/hedgehog dogtail. The hedgehog dogtail appears to be replaced by wildrye (*Elymus glaucus*) and oatgrass (*Arrhenatherum elatius*) with protection from grazing (L. Reed, personal communication, 1992). Rather than the savannalike mixture of oaks and grassland elsewhere in California, these communities tended to have either well or poorly developed overstories of oaks.

In Oregon oak woodlands tend to differ slightly from those in California. The broadest classification of oak woodlands is the division of the type into "oak opening" and "oak forest" (Habeck 1961), depending on whether the trees were more or less than about 15 m apart, similar to the savanna/woodland distinction. More specific classifications seem to depend largely on local understory floristics, which may be dependent on disturbance history. In the Rogue and Umpqua valleys of Oregon, Riegel et al. (1992) identified five oak woodland communities: a xeric oak/hedgehog dogtail type, intermediate oak/brome (*Quercus/Bromus carinatus*) and oak–Douglas-fir/fescue (*Quercus-Pseudotsuga/Festuca ovina*) types, and a mesic oak–Douglas-fir/wildrye (*Quercus-Pseudotsuga/Elymus glaucus*) type. In the North Umpqua River area, Smith (1985) found similar types but with an additional poison-oak layer. In the Willamette Valley, Thilenius (1968) identified four communities: oak/poison-oak (*Quercus/Rhus*), oak/cherry/snowberry (*Quercus/Prunus/Symphoricarpos*), oak/serviceberry/snowberry (*Quercus/Amelanchier/Symphoricarpos*), and oak/hazelnut/sword fern (*Quercus/Corylus/Polystichum*). His xeric oak/poison-oak community is similar to the oak/hedgehog dogtail communities in southwest Oregon (Riegel et al. 1992) and coastal northern California (Sugihara and Reed 1987a), but Thilenius's other types appear to be lacking farther south. A mesic Oregon white oak/snowberry type is recognized from northern California (Sugihara and Reed 1987a) through Oregon (Thilenius 1968, Riegel et al. 1992) to western Washington (Taylor and Boss 1975, Kertis 1986) and seems to be a widely distributed community type.

The Fire Regime

Precise reconstruction of the fire regime in oak woodlands is difficult. The fires were apparently carried by grasses and forbs, and they tended to be flashy and of low duration. Evidence of such fires from fire scars on trees is often lacking, and the shrub and herbaceous components leave at most scant evidence of the last fire after a few years.

Native American burning is thought to have been very important in maintaining and/or expanding oak woodlands in the Pacific Northwest (Habeck 1961, Taylor and Boss 1975, White 1980). Although there are large areas of oak woodland where reliable witnesses did not record details about Native American burning (Griffin 1977), most early explorers in Oregon and Washington noted that the oak woodland areas they walked or rode through had been burned that year, which suggests frequent burning of most oak woodlands. Lightning fires must have augmented the human ignitions, but there is a clear association between oak woodland and human occupation patterns in the Pacific Northwest (Taylor and Boss 1975).

The myriad reasons why the Native Americans burned, and the sophistication of their practices, are matters that lie beyond the scope of this book. A variety of plant foodstuffs, such as camas (*Camassia* spp.), bracken fern (*Pteridium aquilinum*), tarweed (*Madia* spp.), and of course acorns, were either favored by burning or collected more easily after a fire (Lewis 1973, Boyd 1986, Norton 1979). Insects were caught as the fires burned, and roasted grasshoppers were commonly ground into a rough crust for bread (Boyd 1986). Deer and other game were herded by means of fire into ambush (Lewis 1973). Fire was clearly a major resources management tool of early peoples around the world (e.g., Stewart 1951, Hallam 1975).

When David Douglas traveled up the Willamette Valley in the 1820s, he observed (Davies 1980, p. 94): "Country undulating; soil rich, light, with beautiful solitary oaks and pines interspersed through it . . . but being all burned. . . . Camped on the side of a low woody stream in the centre of a small plain—which, like the whole of the country I have passed through, is burned."

Although Native American burning in the oak woodlands appears to have been frequent (Lewis 1973, Boyd 1986), annual burning is not well substantiated. Fire scars (Fig. 12.2A) are not testimony of frequent burning; rather, they establish either intense or long duration fire. In California, Sugihara and Reed (1987a) found evidence of burning only since 1875 but documented eight fires of unknown origin and two prescribed fires since 1917 on one prairie in the Bald Hills. In Washington near Hood Canal, Kertis (1986) found fire-return intervals of 11–34 years between the mid-1700s and 1900 over an area of several square kilometers. This evidence must underestimate the true fire-return intervals, since frequent light fires would not normally be recorded on trees. Most of the evidence of frequent burning of oak woodlands comes from accounts of Native American subsistence patterns.

The intensity of most of these fires must have been low, with flashy, short-duration fires moving quickly through the woodlands. In the 1840s, Wilkes (1845, p. 222) traveled through northwestern Oregon and noted extensive fires that "destroyed all the vegetation, except the oak trees, which appeared not to be injured." The bark of mature trees is several centimeters thick, so it would be protected from such fires, and the crown of the trees, at about 20 m height, would likely avoid being scorched.

FIG. 12.2. A: *Fire-scarred oak, the result of intense fires, particularly in restoration burns.* B: *Crown sprouting in scorched oaks.* C: *Root suckering, common from damaged oak trees.*

The season of burning is again established primarily from ethnographic accounts. In the Willamette Valley, Wilkes in 1841 reported fires "generally lighted in September," and oak woodlands in the vicinity of the Mary's River were burning in early July 1840 and near Newberg, Oregon, in August and September 1841 (Boyd 1986). In the Yamhill, Oregon, country Jesse Applegate noted in 1843 that "the Indian were wont to baptise the whole country with fire at the close of every summer" (Boyd 1986, p. 73). These reports included not just mention of burned ground but also descriptions of fire, flames, or smoke, so that the fires appear to have been burning at the time of the diary entry. Henry Eld of the Wilkes Expedition described the oaks of this hilly prairie country: "The whole country [is] sprinkled with oaks, so regularly dispersed as to have the appearance of a continued orchard of oak trees" (Boyd 1986, p. 71). The burning continued through September in the accounts of members of the Wilkes Expedition. In northern California along the Shasta River, Native Americans would hunt deer by encircling them with fire (Lewis 1973, p. 55). "The . . . method was used on the more open hills on the north side of the river, where the white oaks grew. When the oak leaves began to fall fires were set on the hills." This reference might be to blue oak, which is also a deciduous member of the white oak group and is more common in the Shasta region than Oregon white oak. However, the same patterns were noted in the Bald Hills nearer the coast (Lewis 1973), where Oregon white oak is the principal woodland tree.

The continuous nature of the grassy surface fuels allowed these fires to cover large areas. Jesse Applegate noted a fire in 1843 that was blown by a sea breeze in the Yamhill country through his farm on the Rickreal River (Boyd 1986, p. 73):

> This season the fire was started somewhere on the south Yamhill, and came sweeping up through the Salt Creek gap. The sea breeze being quite strong that evening, the flames leaped over the creek and came down on us like an army with banners. . . . The flames swept by on both sides of the grove; then quickly closing ranks, made a clean sweep of the country to the south and east of us. As the shades of night deepened, long lines of flame and smoke could be seen retreating before the breeze across the hills and valleys.

From the description, the fire must have traveled at least 20 km, and if it was half as wide as long, given that it was wind-driven, it must have covered at least 20,000 ha that night and probably burned to the wide riparian corridor of the Willamette River. Over a single season, 20 such fires could cover all the oak woodland area in the Willamette Valley, and the magnitude of such fires from descriptions provided by early explorers certainly lends credence to such a hypothesis. Boyd suspects that Applegate's account of the fire's intensity may be somewhat exaggerated, but the existence of the fire is supported through other contemporaneous accounts. Annual burning of large contiguous oak woodlands is likely—assuming that ignition is present—because of the potential for rapid fire spread. Discontinuous patches of oak, unless individually ignited each year, may have burned less frequently.

Stand Development Patterns

Establishment of Oregon white oak under natural conditions may occur by acorn germination or by root suckering (Fig. 12.2C). Acorns of the white oaks generally do not require stratification, so they can germinate immediately after seedfall (Olson 1974). However, many acorns are damaged or eaten before they have a chance to germinate. Larval stages of the filbert weevil (*Curculio occidentalis*) and white moth (*Melissopus latiferreanus*) infect acorns and render them sterile (Scheffer 1959, Furniss and Carolin 1977), while numerous mammals and birds eat the fallen acorns (Silen 1958). Oregon white oak produces 1,000–2,000 acorns per tree (3–9 m tall trees) and up to 560 kg ha^{-1} (Stein 1990), making a significant contribution to the diet of animals. Small mammals ate 61 percent of the acorns from savanna oaks and 91 percent of the smaller crop from closed-canopy forest (Coblentz 1980). Gophers eat small seedlings, and Native Americans also used acorns as a common food source (Gunther 1973, Norton 1979). Although germination of protected acorns ranges from 65 percent to 100 percent (Olson 1974, L. Reed, personal communication, 1992), few acorns or small seedlings survive to be affected by periodic burning.

Sprouting of the coastal white oaks (valley oak, blue oak, and Oregon white oak) is apparently less vigorous than for other western oaks (Griffin 1980). Oregon white oak was observed to weakly stump-sprout (Jepson 1910), yet sprouting appears to be one of its main forms of reproduction. At Redwood National Park, oaks in the Bald Hills were found to sprout from rhizomes and produce either clusters or scattered individuals (Sugihara and Reed 1987a). A similar sprouting pattern producing scattered individuals has been found in other oak species (Muller 1951). It has also been observed in Oregon white oak from 1930s logging at Oak Patch Natural Area Preserve in Mason County, Washington (Kertis 1986), and after wildfire at Fort Lewis, Washington (J. K. Agee, personal observation, 1985). In the latter cases, existing tree crowns were either removed or heavily scorched (Fig. 12.2B), encouraging a sprouting response. A more frequent and lower intensity fire regime would likely not stimulate sprouting on mature trees to the same degree. As with many tree species, substantial regeneration is not likely and may not be desirable after frequent disturbance that has little impact on the overstory.

The effect of fire on the relative importance of acorn and sprout regeneration for Oregon white oak is not well understood. A light burn under an oak canopy has little effect on germination of acorns that are in the canopy at the time of the burn, but those on the ground during the fire show a 50–60 percent decrease in germination (Sugihara and Reed 1987a). In Bald Hills prairies, large numbers of oak seedlings are found under some oak canopies burned with low or high fire intensities, while other burned areas show little recruitment (L. Reed, personal communication, 1992). Acorns that fall to the ground after a fire may have a better chance of survival if they land in more heavily burned microsites. At Oak Patch Natural Area Preserve, 10 seedlings and 10 sprouts were randomly chosen for analysis two years after a 1989 prescribed fire (J. K. Agee, unpublished data). Each of the sprout clumps was associated with a burned stem or was an isolated root sucker, and each of the seedlings was identified by the presence of the decaying acorn just beneath the soil surface. Cover of major plant species, bare ground, and wood fragments (associated with logs that burned) was recorded, and soil samples were taken at each location.

The seedlings, at about 7 cm average height, were associated with plants that normally invade heavily burned areas: fireweed (*Epilobium angustifolium*), tansy ragwort (*Senecio jacobea*), St. John's wort (*Hypericum perforatum*), velvetgrass (*Holcus lanatus*), and a disturbance-associated moss, *Leptobyrum pyriforme*. The sprouts, averaging 45 cm height, had a higher proportion of associated, sprouting residual species adjacent to them: blackberry (*Rubus ursinus*), wild rose (*Rosa gymnocarpa*), serviceberry (*Amelanchier alnifolia*), and salal (*Gaultheria shallon*). Chemical analysis of the soil showed two significant differences between samples: lower total nitrogen and lower total carbon on the seedling plots, which once again suggests higher fire severity that volatilized both elements.

These results indicate that heavily burned sites are associated with seedling establishment but do not explain why such sites are amenable to seedlings. There are several possible explanations: acorns falling in ash and being buried from predators, such as deer, bear, or small mammals; a sterilizing effect of the hot fire, thus preventing mold on the acorn; and less moisture competition during the early establishment phase of the seedling. One hypothesis is that seedlings may have been associated historically with local spots of high severity fire, where a branch or tree trunk had fallen, and may have been uncommon throughout the rest of the landscape matrix. Such sites might also be less likely to burn the next year and might give the seedlings an additional year or two to grow before the next fire came through the area, increasing the probability of survival into the more fire-resistant sapling stage. Roughly half of the 1–9-year-old seedlings burned in a 1987 prescribed fire resprouted and were alive three years later (L. Reed, personal communication, 1992). Some oak seedlings appear to undergo a prolonged seedling stage unrelated to grazing (Hibbs and Yoder 1993), which delays their transformation into a more rapidly growing sapling. Any hypotheses must remain speculative for now; it is clear that our understanding of the role of fire in the regeneration of Oregon white oak is poor, and that this will be a fruitful area for future research (Griffin 1977).

The effect of fire exclusion on oak woodlands is different in savannas and in woodlands. In savannas, it leads to an increasing density of shrubs and oaks, transforming the savanna into woodland. In woodlands, it results in an increase in shrubs and other tree species at the expense of oaks. These postsettlement impacts are intricately associated with grazing histories; many oak woodlands have been continuously grazed by livestock since the time when fire exclusion became effective. Much of the history of stand development in oak woodlands is hypothesis rather than fact.

As an example, the oak woodland at Young Hill on San Juan Island, Washington, shows a pattern of oak recruitment between 1836 and 1866, when 10 of 15 sampled mature oaks became established (Agee 1987). This is also a period associated with major declines in Native American populations. Smallpox came to Puget Sound before Vancouver's visit in 1793; a disease similar to malaria struck in 1836, and a measles epidemic arrived in 1847, along with tuberculosis; smallpox struck again in 1854 (White 1980). The native population was cut in half by 1840, and visitors to Young Hill in 1858 found a huge but empty longhouse at its base, which they assumed might hold 1,000 people. The association between oak establishment and the decline of native populations might not be cause and effect, but if Native Americans were burning the woodland frequently and then ceased, the oak establishment is at least consistent with that time period and the hypothesized lack of frequent burning. Declining tree recruitment after 1870 is probably associated with heavy grazing, first by the livestock of occupying British soldiers (Murray 1968) and then of American settlers. Determination of the time when conditions deviated from "natural" is hardly possible in oak woodlands with such long human histories. What we consider to be a pristine canopy of older oaks may have an anthropogenic origin.

At Oak Patch, much of the mature tree component is associated with sprouts from 1930s-era logging, although there are older scattered oaks across the preserve (Kertis 1986). In the Bald Hills of northern California, three patterns of mature woodland stands are found (Sugihara and Reed 1987a): (1) a "young dense" type and (2) a "clustered" type, both associated with relatively even-aged trees that appear to have become established after significant disturbances, and (3) an "all aged" type with a variety of size and age classes. These types are different stages of oak stand development (Fig. 12.3).

Quercus garryana woodlands have declined in area over time. Some have been urbanized, some have been converted to pasture, and yet others have experienced massive conifer invasion. Prairie and oak savanna in Oregon's Willamette Valley have been replaced by oak forest (Habeck 1962), and a great deal of woodland has

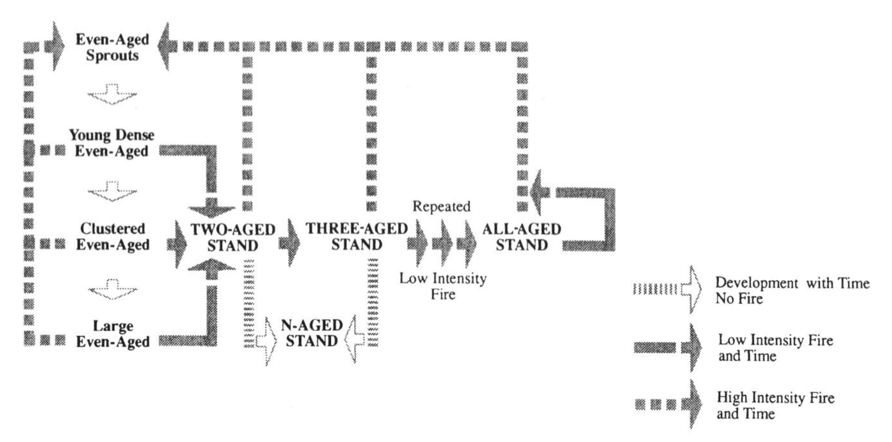

FIG. 12.3. *The stages of stand development in oak woodlands.* (From Sugihara and Reed 1987b)

disappeared over the last century through forest encroachment. Prairies with oak-dominated margins have declined by 26 percent in the Bald Hills of Redwood National Park (Sugihara and Reed 1987a; Fig. 12.4). Older scattered oaks are now found in a two-storied forest with younger oaks (Thilenius 1964). In many areas, oaks are being replaced by Douglas-fir forest (Sprague and Hansen 1946, Franklin and Dyrness 1973).

Recent tree invasion into oak woodlands has been considerable. At Young Hill, Agee (1987) found a veritable jungle of conifer reproduction under oak woodland that had begun to invade the woodland in the early 1960s (Fig. 12.5), similar to what was documented by Agee and Dunwiddie (1984) on nearby Yellow Island. The Yellow Island invasion began in a period of slightly increased summer precipitation in a summer-dry climate. The Young Hill invasion did not correlate well with the precipitation pattern, and is thought to reflect cessation of continuous grazing by sheep when the land became a national historical park. In the first five years after establishment, height growth of Douglas-fir seedlings may total only 20 cm, but it soon increases markedly. Later invaders appear to grow faster than the early invaders (Fig. 12.6), suggesting a change in microclimate (probably reduction in windspeed and less transpirational stress) beneficial to Douglas-fir growth (e.g., Agee and Dunwiddie 1984). Over time, Douglas-fir will overtop Oregon white oak (Fig. 12.7), and the shade-intolerant mature oaks will die.

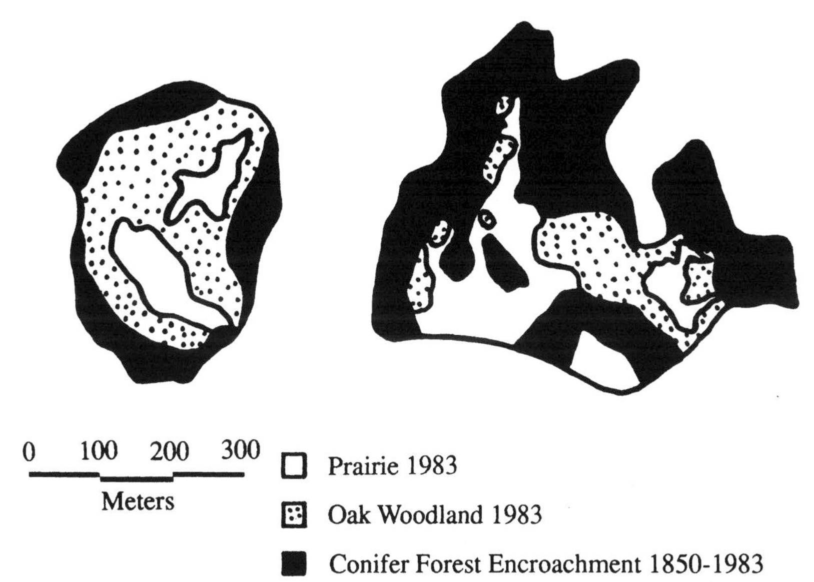

FIG. 12.4. *Changes in prairie and oak woodland area in the Bald Hills section of northern California. These two prairies are not adjacent. They show a pattern of prairie and woodland shrinkage common to most oak woodlands.* (From Sugihara and Reed 1987a)

In some locations normal variations in drought may slow the invasion process, as the oaks are more drought-resistant than the associated conifers (Minore 1979). Drought in the late 1980s in the San Juan Islands was associated with death of invading Douglas-fir at the margins of oak woodlands and prairies, and the death of small grand fir in Douglas-fir forests (J. K. Agee, personal observation, 1989). Such limited events only forestall the inevitable disappearance of many oak woodlands from the landscape.

Most oak woodlands and northwestern grasslands were historically dominated by perennial species. Understories heavily grazed by livestock tend to be grass-dominated and composed of the highest proportion of alien species (Sugihara et al. 1987). Longer seasons of grazing tend to be associated with introduced annual grasses and reduced species richness (Saenz and Sawyer 1986). Knowledge of the effect of fire in these altered communities is largely experimental at present. Not only have most of these communities been grazed, but also tree and shrub growth have shaded out some native species and altered competitive relationships for others.

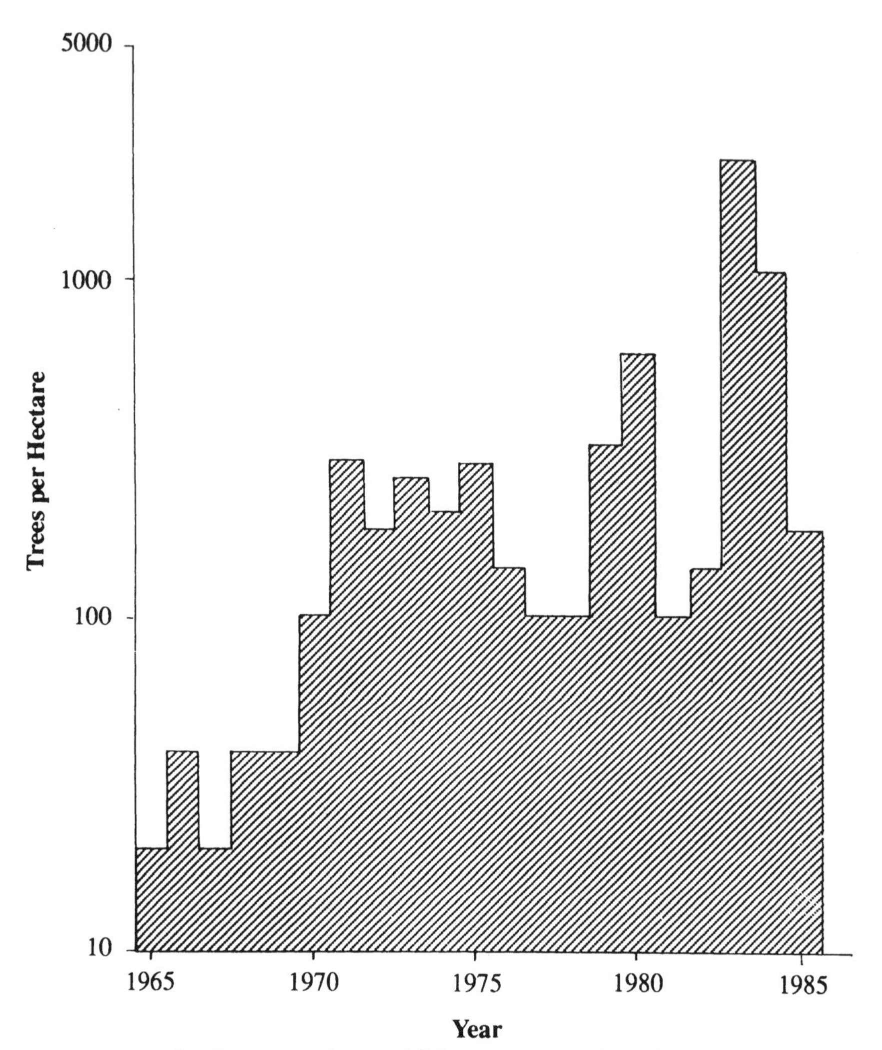

FIG. 12.5. *Douglas-fir regeneration (establishment per year) from the 1960s to 1986 in San Juan Island oak woodlands. Douglas-fir regeneration threatens the existence of* Quercus garryana *woodlands; similar invasion occurs throughout much of the range of Oregon white oak.*
(From Agee 1987)

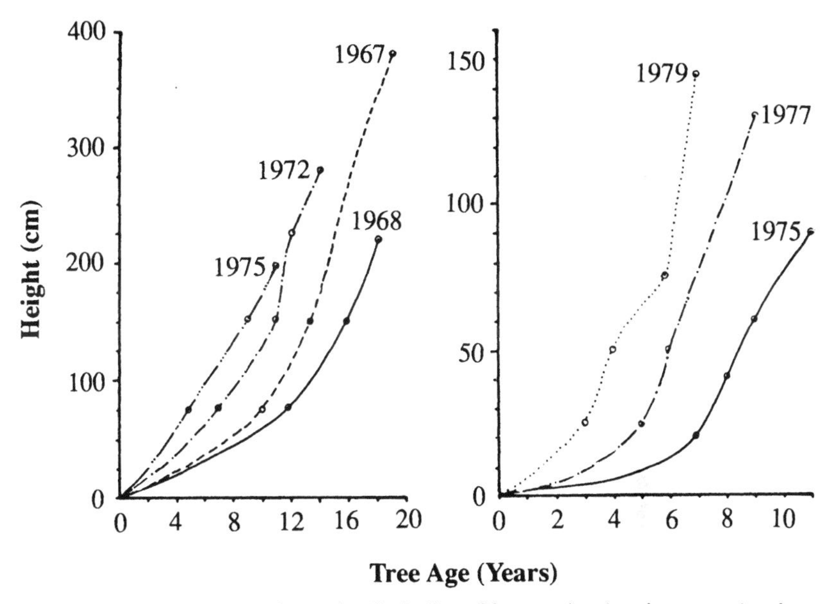

FIG. 12.6. *Later invaders of Douglas-fir, indicated by germination dates associated with each growth curve, showing significantly faster growth than the early invaders.* (From Agee 1987)

Native and introduced shrubs tend to decline after burning. Rhizomatous species, such as snowberry, are usually not killed, but cover may decline and spread of existing thickets may be at least temporarily stopped. On Yellow Island in the San Juan archipelago, rhizomes 20–30 cm distant from existing patches were growing to 30–40 cm height in the first year in the early 1980s (Agee and Dunwiddie 1984). After a 1987 prescribed fire in August, patches that were severely burned recovered more slowly than lightly burned patches. In severely burned patches, snowberry cover declined from about 20 percent to less than 1 percent after two years and remained at that level through postfire year 4 (P. Dunwiddie, unpublished data).

The alien shrub, scotch broom (*Cytisus scoparius*), has been an aggressive invader into northwestern grasslands, primarily grassland or savanna areas with less shade than woodlands (Gill and Pogge 1974). It is a nitrogen-fixing species, so it is able to establish and grow well in areas bared by overgrazing or cultivation. Scotch broom does sprout, but the response varies by season. Plants cut low

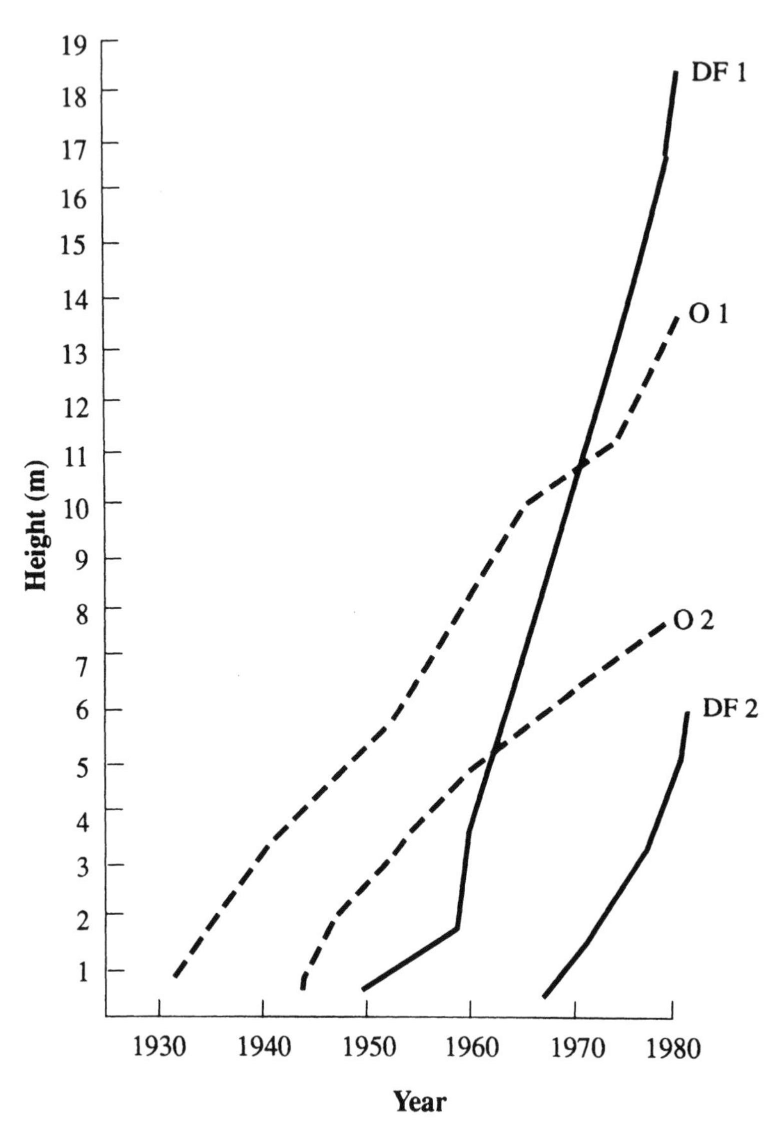

Fig. 12.7. *Height-growth curves of Oregon white oak (O) and later established Douglas-fir (DF), showing the inevitable domination of Douglas-fir if it is not removed.*
(From Kertis 1986)

to the ground in November to May had 30–100 percent sprouting, while those cut in August had a 0–10 percent response (Bossard 1990). The seeds of scotch broom are stimulated to germinate after burning, so that a single fire, even in August, can result in higher density of scotch broom than before the fire (Williams 1981). The young plants normally require two years to flower and set seed, so that a second fire one or two years after the first will eliminate much of the scotch broom. Some of the residual seed heated in the first fire will die from fungal infection (Bossard 1990). This frequency of disturbance may be necessary to reduce the scotch broom, but it may encourage other alien species. At Fort Lewis, Washington, burned fields of scotch broom were dominated by the alien forb St. John's wort; after a prescribed fire at Oak Patch, the aliens tansy ragwort, St. John's wort, oxeye daisy, and velvetgrass were dominant species.

A quite different response has been observed in the Bald Hills of northern California, where prescribed burning has increased native forbs, while introduced grasses and forbs decreased. Perennial plants with bulbs and legumes appeared to do well after the fire, prompting a conclusion that frequent burning would favor native perennials over introduced annual species (Sugihara and Reed 1987a).

A more neutral response was obtained in relatively pristine grassland on Yellow Island after a 1987 prescribed fire (P. Dunwiddie, unpublished data). Mosses and lichens made up about 38 percent of the preburn cover and recovered to about 10 percent in four years. Cover of native species recovered within two years, but cover of aliens increased severalfold, likely filling the space previously occupied by mosses and lichens, and was still twice the preburn cover at year 4. The cover of bulb-forming camas (*Camassia leichtlinii*) doubled from 14 percent to 27 percent, and since this was a staple of Native American diet, this might be one reason why they burned prairie and woodland environments. Idaho fescue (*Festuca idahoensis*) was steadily recovering to about two-thirds of its preburn cover. Persistent aliens included the annuals six-weeks fescue (*Festuca bromoides*), soft brome (*Bromus mollis*), and sheep sorrel (*Rumex acetosella*), and perennials velvetgrass and spotted cats-ear (*Hypochaeris radicata*). It seems unlikely that fire could be frequently used without encouraging alien species to persist on these sites.

Management Implications

Oak woodlands have been significantly altered by human activity for millennia in the Pacific Northwest. The oak woodlands of 1800 have seen additional change: increasing density of oaks in savannas, invasion of conifers in woodland types (Fig. 12.8A,B), increasing shrub cover, and a conversion from primarily perennial to primarily annual herbs due to heavy livestock grazing. Questions have been raised about oak woodlands: Is the "ecosystem in jeopardy or is it already too late?" (Reed and Sugihara 1987). Their answer was that prescribed fire could be used to reduce the conifer invasions documented by Kertis (1986), Barnhardt et al. (1987), and others. However, action had to be initiated soon, as the Douglas-fir and other conifers would overtop the oaks within 20–30 years.

The form of the Oregon white oak woodlands can be restored through a combination of manual and prescribed fire techniques (Fig. 12.8C). The Washington Department of Natural Resources has used a combination of these techniques at Oak Patch Natural Area Preserve. Conifers have been removed or killed by prescribed fire, and the understory has been opened considerably with topkill of shrubs. The understory is likely drier, since winds can freely move through the woodland now. Initial monitoring reveals an understory dominated by alien plants with few of the perennial grasses that probably were dominants before 1800. Successive fires, if native herbs recover, may not encourage the annuals/aliens as much, but the annuals/aliens might also further increase. At present, we cannot predict the outcome. More intensive treatment—for example, seeding native perennials, such as Idaho fescue—may help restore presettlement conditions, but this treatment is as yet unproven. The fescue, in the presence of alien annuals, will not be favored by annual burning, so 5–10 years is the most likely fire-return interval.

The Oregon white oak woodlands to the south appear to be more stable than those to the north. In California they are often on rocky outcrops and have not changed as much as ones that appear to be more anthropogenically maintained, such as those in the coastal Bald Hills of California, in the Willamette Valley, and in Washing-

A

FIG. 12.8. A: *Early invasion of oak woodlands by Douglas-fir, appearing relatively benign (San Juan Island, Washington).*

B

B: *Same site, 20 years later. Douglas-fir is now a major threat to the survival of oak.*

C

C: *Oak woodland that has maintained its form thanks to repeated prescribed burning (Fort Lewis, Washington).*

ton. Without prescriptive treatment, up to 50 percent of the threatened oak woodlands could be beyond help by the year 2010. The window of opportunity narrows each year.

JUNIPERUS OCCIDENTALIS WOODLANDS

Western juniper (*Juniperus occidentalis*) woodlands are found in the intermountain region of the Pacific Northwest, primarily in Oregon but also in Washington, Idaho, Nevada, and California (Fig. 12.1). There are other juniper species found in these states: shrubby common juniper (*J. communis*) in all five states, and Rocky Mountain juniper (*J. scopulorum*) in all but California (Little 1971). This section focuses on woodlands in which western juniper is a dominant. In Oregon *Juniperus occidentalis* savanna or woodland occupies about 1.15 million ha (Dealy et al. 1978). It is the most xeric of all the tree-dominated zones in the Pacific Northwest (Franklin and Dyrness 1973), with annual precipitation averaging 20–25 cm.

Young trees have needlelike leaves, while more mature trees have scalelike leaves (Vasek 1966). The crown form is conical in seedling and sapling-sized trees (<30 years), becoming rounded in mature trees (>50 years) and almost spherical in old-growth trees (>150 years) (Fig. 12.9A,B). Typical heights are 12–14 m for mature trees.

Western juniper is generally the climax species on sites where other conifers are absent. A confusing set of climax/seral classifications has been presented for these communities. Driscoll (1964a) was the first to classify the suite of plant communities dominated by western juniper, identifying nine communities. Franklin and Dyrness (1973) selected five of the most common of Driscoll's communities: juniper/sage/wheatgrass (*Juniperus/Artemisia/Agropyron*), juniper/fescue (*Juniperus/Festuca*), juniper/wheatgrass (*Juniperus/Agropyron*), juniper/sage-bitterbrush (*Juniperus/Artemisia-Purshia*), and juniper/sage/wheatgrass–milk-vetch (*Juniperus/Artemisia/Agropyron-Astragalus*). Variants of the five community types exist across the range of these woodlands (Dealy et al. 1978). Judging from the common members of these communities, all are fairly closely

FIG. 12.9. A: *Conical crown form of a typical expansion juniper.* B: *Rounded crowns of older trees, common in rimrock areas.* C: *Fire-proofing of a large western juniper, which had purged its herbaceous understory through moisture competition and thereby avoided a wildfire. The dashed line at the arrow shows the margin of burned and unburned area, with patchy spread in the burned area. This tree has a fire scar with an apparently authentic carved inscription dated 1906.*

related to one another and differ by the relative dominance of a shrub or herb component. Big sagebrush (*Artemisia tridentata*) is the dominant shrub in most communities, with rabbitbrush (*Chrysothamnus* spp.) on some sites. Bitterbrush occurs most commonly on mesic east to southeast aspects, with Idaho fescue (*Festuca idahoensis*) on north aspects and bluebunch wheatgrass (*Agropyron spicatum*) on south aspects (Driscoll 1964b). In Washington western juniper has a more disjunct distribution and in some areas appears to be associated with actively moving sand dunes (Long et al. 1979).

Franklin and Dyrness (1973) described the juniper/sage/wheatgrass community as a climatic climax and the others as topoedaphic climaxes. In Idaho, Burkhardt and Tisdale (1969) described climax stands as those with all-aged juniper trees and seral stands as those invaded by juniper since the late 1800s. However, the true seral component of these stands is not the juniper, which is the potential vegetation on all these sites; juniper is seral only in stands where ponderosa pine can establish, although it has been identified as coclimax with ponderosa pine at times (Hall 1978). Juniper appears to be expressing its climax potential as a result of land use history over the past century (Fig. 12.10).

The Fire Regime

The fire regimes of the Pacific Northwest were described by Martin (1982) as having a U-shaped trend, with the trough of the U in ponderosa pine forest. Increasingly long fire-return intervals are found in wetter communities because the weather is less conducive to burning and in drier communities because low productivity limits fuel availability. *Juniperus occidentalis* woodlands are in the latter category.

Conclusions about fire regimes in juniper woodlands are based on limited data. Burning by Native Americans in eastern Oregon was documented by Shinn (1980, p. 418), who quotes a traveler in the valleys of the Blue Mountains of the 1830s: "The Indians have fired the prairie, and the whole country for miles around is most brilliantly illuminated. The very heavens themselves appear ignited." East of the Deschutes River in central Oregon, the Ogden party in September 1826 noted recently burned land. Near present-day Pau-

A

B

FIG. 12.10. *Ecological change in the vicinity of the Keystone Ranch east of Prineville, Oregon, between 1910 and 1990.* A: *Juniper invasion under way in 1910.* B: *Full development of juniper woodland in 1990. Note the riparian willows in the 1910 photo and rechannelization and lack of riparian trees in 1990 photo.*
(Photo [A] courtesy of Bowan Museum, Crook County Historical Society, Prineville, Oregon. Photo [B] courtesy Dr. Stuart Garrett, Bend, Oregon.)

lina, Oregon, in the upper Crooked River they noted (Shinn 1980, p. 418): "The Indians crossed the river . . . in the night and set fire to the plain within ten yards of our camp . . . it was blowing a gale at the time." Shinn lists three other instances of Native American burning recorded by Ogden during the 1827–29 period. Fires were used for signaling, hunting, collecting insects, and probably for many other reasons. Other early travelers left records that can be interpreted to mean that fires were locally variable across the region. For example, Vale (1975) quotes travelers around Walla Walla, Washington, in the 1850s who commented that sagebrush stems 15 cm thick were large enough to be used as fuelwood, implying a long fire-return interval in that area. In the area of the confluence of the Deschutes and Columbia Rivers, grass instead of sagebrush was abundant in the 1840s, perhaps reinforcing the accounts of frequent burning recorded by Shinn. In 1844 James Clyman, a traveler on the Oregon Trail near Baker City, Oregon, mentioned what might be interpreted as a recently burned sagebrush slope: "entire country covered with sage which from some cause or other is nearly all dead" (Evans 1991, p. 119).

Although information documenting at least some Native American burning has been available for many decades, Daubenmire (1970, p. 7) claimed there was "little incentive to burn steppe" for native peoples. He concluded that although fire had clear effects on vegetation, the distribution of vegetation types and species was related to soils and climate more than to presence of fire. Clearly the dynamics of juniper woodlands involve factors other than fire, but fires apparently were more common than assumed by some investigators. How "frequent" such fires were is yet to be established.

Fires in *Juniperus occidentalis* woodlands may leave little evidence that remains decades or centuries later to reconstruct historical fire patterns. The grasses and shrubs resprout and leave at most charred stubs. Western juniper trees are thin-barked and easily killed by fire—unlike the sprouting alligator juniper (*Juniperus deppeana*)—so that only those western junipers in rather permanently fuel-limited sites are capable of recording multiple fire scars. Junipers may be killed, or if large enough to create a herb-free zone around the drip line of the crown, they may avoid the fire and any scarring altogether. Junipers transpire significant moisture during the winter, using much stored soil moisture before herbaceous vegetation has an opportunity to exploit it (Jeppesen 1978). This "fireproofing" of large junipers was observed in a wildfire at John Day Fossil Beds National Monument in Oregon, where most of the younger junipers were killed (J. K. Agee, personal observation, 1987; Fig. 12.9C). Fires may leave no evidence of their presence, making fire history in juniper woodlands inherently imprecise (Young and Evans 1981).

The fire history of western juniper was evaluated in a Nevada landscape where juniper grew with big sagebrush on the hills and with low sagebrush (*Artemisia arbuscula*) on broad flats (Young and Evans 1981). The oldest trees with fire scars were in the low sagebrush community. Some trees were scarred in 1650, and others were scarred in 6 of the 11 decades between 1750 and 1860. Wright

(1985) interpreted these data to mean that fire intervals "may be as long as 90 years." Alternatively, if we assume that the record before 1750 may not be as representative as the 1750–1860 interval, the fire-return interval would shorten to 15–20 years. In the northern Yellowstone area (outside the range of western juniper but a sagebrush area), Houston (1973) estimated a 20–25 year fire-return interval for sage-dominated lands, assuming that many of the fire-scarred trees he found did not record each fire that passed through the area. In *Juniperus occidentalis* woodlands of Lava Beds National Monument, California, fire-return intervals of 7–17 years were recorded in a juniper woodland transitional to ponderosa pine (Martin and Johnson 1979).

Ridges and rimrock, two edaphic fuel-limited juniper sites, were studied by Burkhardt and Tisdale (1976) in southwestern Idaho. Each of their study areas was about 260 ha on the 200,000 ha Owyhee Plateau. They found an average of 12 fires per site over the period 1650–1950 recorded on fire-scarred junipers, for an average fire-return interval of 25 years per 260 ha site. The nineteenth century appeared to have about a 15 year fire-return interval per site. Because of the large size of each site, these estimates should not be compared to point estimates of fire frequency in other forest types (e.g., ponderosa pine). These studies suggest that fire did burn these woodland sites, but perhaps only every 15–25 years.

A life-history approach to estimating fire frequency was developed by Wright (1985). He assumes that the absence of plants that respond vigorously to fire for several decades, such as horsebrush (*Tetradymia canescens*), indicates that historical fire-return intervals must therefore have been longer than several decades in sagebrush-grassland communities. However, in landscapes unburned for many decades, the absence of such species may represent recent history rather than prehistoric fire regimes. Sagebrush is killed by fire and will reestablish slowly beginning the first year after burning (Daubenmire 1970), while the remaining shrubs sprout. The oldest stems, although a reliable indicator of the minimum time passed since the last fire, may be useless in reconstructing the return interval of multiple fires.

The intensity of these fires is a function of fuel availability. The longer these communities are fire-free, total fuel increases, but the nature of fuel shifts from fine grassy fuel to juniper or shrub fuel with coarser leaves, thicker stems, and more vertical than horizontal continuity. Fire severity is usually classified as moderate, in that most plant components of these ecosystems sprout after fire (Daubenmire 1970). However, two of the climax dominants, big sagebrush and western juniper, do not sprout, and fire can therefore have a significant effect on the structure of woodland communities that have remained unburned for decades. Intensities of prescribed fire will vary considerably in these fuels (Sapsis and Kauffman 1991). Spring fires may have flame lengths of 1.74 m (883 kW m^{-1}), while autumn fires, with more cured grass fuel, may have flame lengths >4 m (>6,000 kW m^{-1}).

Fire spread is more likely to be extensive if grasses rather than shrubs are the dominant fuel. Threshold responses for spread in grassy fuels were defined by Clark et al. (1985). As fine-fuel load (e.g., grasses) decreased, similar fire spread rates required drier fuels (Table 12.1). As shrub and tree cover increased, higher windspeeds were likely required to maintain fire spread.

Fires can burn from spring through fall in juniper woodlands. Generally, there is lower consumption of fuel in spring than fall burns, even with similar dead fuel moisture, because of greener grass/herb biomass and higher (>175 percent in spring, <100 percent in fall) sage foliar moisture (Sapsis and Kauffman 1991).

Stand Development Patterns

Western juniper appears to have undergone major expansion within its range over the last century (Fig. 12.10). In Idaho, Burkhardt and Tisdale (1969) suggested that it occupies twice the area it did in 1860. Four hypotheses have been advanced to explain the phenomenon: (1) absence of fire, which allows fire-sensitive juniper to survive; (2) overgrazing by domestic livestock, which opens up competition-free microsites for juniper; (3) reestablishment of juniper in places from which it was logged; and (4) climatic change (Young and Evans 1981, Quinsey 1984). Each of these may play a role at a particular place or time.

The expansion appears to be greatest in areas where juniper was either absent or present in low numbers before the twentieth century. In old-growth stands (often called climax stands in the litera-

TABLE 12.1. FUEL AND MOISTURE CONDITIONS RESULTING IN FIRES THAT CARRY ACROSS FINE FUELS (<0.3 CM) ABOUT 80 PERCENT OF THE TIME

Fine Fuel Load (*kg ha*$^{-1}$)	*Windspeed at 2 m Height* (*km hr*$^{-1}$)	*Fuel Moisture Content* (*percent*)
2,000	10	20
1,000	10	10
1,000	10	10
500	10	10
250	5	10
200	10	10

SOURCE: Clark et al. 1985.

NOTE: These values assume uniform fuel and terrain.

ture), tree densities are variable, ranging from 28 to 237 trees ha^{-1} (Burkhardt and Tisdale 1969, Young and Evans 1981). Sites where major juniper expansion has occurred appear to have higher tree densities at present, ranging from 150 to 424 trees ha^{-1} (Young and Evans 1981, Quinsey 1984, Doescher et al. 1987). The age-class distributions of these stands also differ significantly (Fig. 12.11). The size structure of junipers in areas where they have expanded is dominated by small trees; Quinsey (1984) found that almost 75 percent of the juniper density in Crooked River National Grassland comprised seedling- or sapling-sized trees.

Juniper seeds will disperse downslope by gravity (Burkhardt and Tisdale 1969). Young and Evans (1981) suggested that birds can also be a major dispersal vector, as they found little difference in establishment dates of juniper at increasing distance from seed source (Table 12.2). They concluded that expansion occurred from seed dispersal by robins (*Turdus migratorius*) eating juniper seeds from older established trees in the adjacent community, instead of by a few trees establishing and then producing seed for further expansion, since junipers less than 50 years old rarely produce seed.

Existing shrub or tree cover is helpful to juniper establishment, as young seedlings are vulnerable to heat damage in unprotected microsites. Soil surface temperatures in bare ground and under perennial grass cover can reach 60°C, while under sagebrush and juniper maximum summer temperatures were 33°C and 26°C (Burkhardt and Tisdale 1976). Once established, juniper seedlings grow slowly,

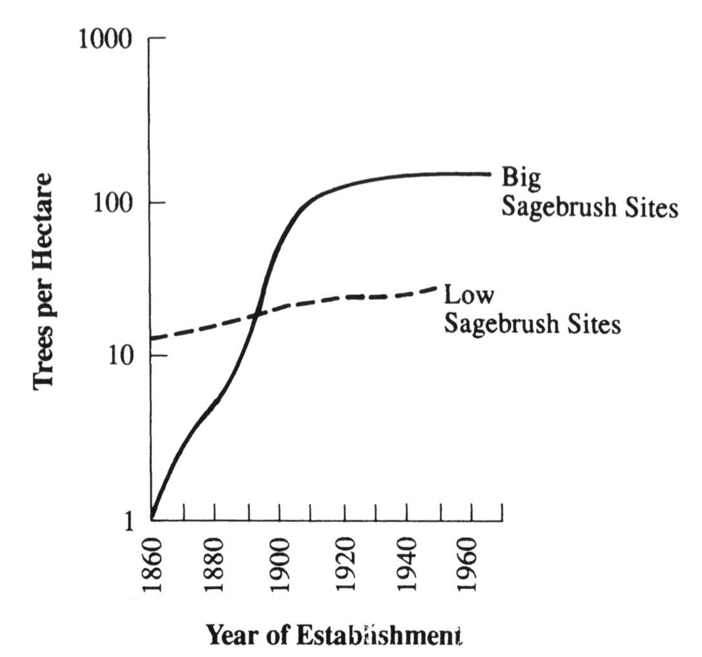

FIG. 12.11. *Age-class distributions of old-growth and expansion stands, showing nearly continuous, slow recruitment in the old-growth stand (low sagebrush* [Artemisia arbuscula] *sites) and more recent, pulsed regeneration in the expansion stand (big sagebrush* [Artemisia tridentata] *sites). Steeper slopes of curves represent higher recruitment rates during that period.*
(From Young and Evans 1981)

only several centimeters per year at first, and they are very susceptible to fire through the sapling stage, which may last through age 30–40. Lifetime height growth averaged 6–9 cm yr^{-1} for trees established between 1880 and 1930, and 2–4 cm yr^{-1} for younger trees (Eddleman 1987), suggesting that intraspecific competition for moisture was becoming important.

The fire hypothesis (irregular fires at 15–25 years) appears to explain much of the juniper expansion into the sagebrush-steppe ecosystem. Prescribed fires and wildfires will kill trees by crown scorch or basal heating, and mortality of trees below about 4 m height and/or <50 years old is substantial under most conditions (Burkhardt and Tisdale 1976, Martin 1978); hot summer wildfires may result in 100 percent mortality of all size classes of western

TABLE 12.2. PERCENTAGE OF STAND ESTABLISHMENT OF WESTERN JUNIPER PER DECADE IN PLOTS LOCATED AT VARIOUS DISTANCES FROM LOW SAGEBRUSH SITES

DECADE	ESTABLISHMENT IN INDICATED SITE (PERCENT OF TOTAL)					
	1	*2*	*3*	*4*	*5*	*6*
1880–90	2	1	3	2	2	3
1890–1900	28	30	30	26	31	30
1900–1910	36	29	34	38	36	36
1910–20	20	21	18	22	19	22

SOURCE: Young and Evans 1981.

NOTE: Site 1 is closest to seed source, site 6 is farthest away.

juniper. First-year survival of larger trees may be significant under moderate conditions (Fig. 12.12, Table 12.3). In central Oregon Quinsey (1984) found 100 percent mortality of juniper on xeric sites, but only 25 percent mortality on mesic sites that burned in a patchy manner. At least one tree survived with 75 percent crown scorch by volume, and on other trees epicormic sprouting was observed on some scorched branches.

Juniper reestablishes after fire, but regeneration may not be immediate. Prolonged herbaceous and shrub stages may occur (Everett 1987; Fig. 12.13). Junipers killed by fire act as "safe sites" (in the sense used by Harper 1977) for juniper regeneration. Most regeneration occurs under the crowns of the fallen trees, probably less because of residual seed (Burkhardt and Tisdale 1976) than as a result of the shading impact of the crown (Quinsey 1984), reducing surface soil temperatures. In central Oregon, Eddleman found over 80 percent of juniper regeneration occurring under either sagebrush or other juniper plants. In Idaho, Burkhardt and Tisdale (1976) found no seedlings six years after fire, while Quinsey (1984) found a minimum four-year delay. These delays may be associated with the time necessary for trees to begin to topple and produce "safe sites"; sagebrush recovery would require much longer before live shrubs would provide much microclimatic amelioration. Once juniper is established on a site, it changes mineral cycling patterns and reduces total nitrogen in surface soils (Doescher et al. 1987). The effect on subsequent succession is as yet unknown.

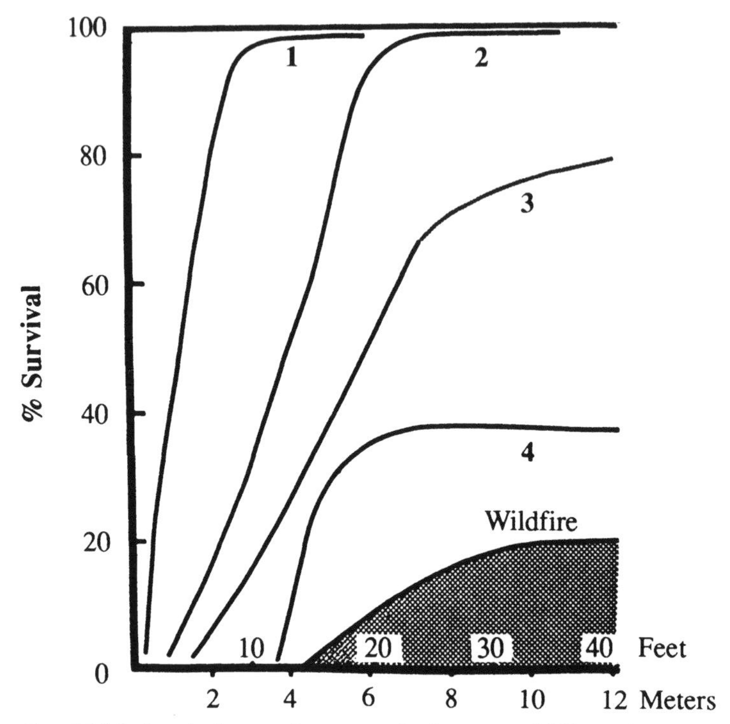

FIG. 12.12. *Survival curves for western juniper trees of different heights. Numbers are associated with prescription conditions defined in Table 12.3.* (From Martin 1978)

Big sagebrush is also killed by fire (Blaisdell 1953), but it will usually regenerate from seed. Initial recovery may be slow; it is normally complete in 30 years (Harniss and Murray 1973). Bitterbrush is a weak sprouter, which may sprout after moist spring burns (Driver 1982); summer to autumn burning usually results in complete mortality (Clark et al. 1982, Britton and Clark 1985). Rabbitbrush, as a strong sprouter, usually increases immediately after burning (Chadwick and Dalke 1965) and may persist on sites for over 50 years (Bunting 1985).

Grasses can be slightly to significantly damaged by fire. Perennial bunchgrasses show the most deleterious effects. Idaho fescue has dense, fine-textured culms around its base that can smolder and damage the meristematic tissue beneath. Most research has focused on single fires burning plants protected from fire for a long time, so

TABLE 12.3. PRESCRIBED BURNING CONDITIONS ASSOCIATED WITH CURVES IN FIGURE 12.12

Prescription Variable	BACKING FIRES		HEADING FIRES	
	Curve 1	*Curve 2*	*Curve 3*	*Curve 4*
Temperature (°C)	21	21–24	24–29	24
Windspeed (km hr^{-1})	8–16	8–16	8–16	8–19
Wind gusts	24	21	24	24
Relative humidity (%)	25–30	25–30	18–20	10

SOURCE: Martin 1978.

that a large amount of basal fuel is available. Summer and fall burns have been shown to kill 0–25 percent of plants and reduce basal area by 35–80 percent (Conrad and Poulton 1966, Countryman and Cornelius 1957, Britton et al. 1983). Early spring burns may be conducted with little first-year impact on production (Blaisdell et al. 1982), which may be related to less smoldering time under moist conditions. In Idaho return to preburn conditions took about 30 years (Harniss and Murray 1973). On drier sites, Idaho fescue is less resilient than on more mesic sites (Clark and Starkey 1990). Bluebunch wheatgrass is more tolerant of fire (McShane and Sauer 1985), likely because of its larger, coarser culms, which allow less smoldering to occur around the base of the plant. Bluebunch wheatgrass production often returns to normal within three years after burning (Wright 1985, Uresk et al. 1976). Little work has been done on repeated burning, which has been hypothesized to have historically kept the sagebrush-steppe free of western juniper.

One of the major reasons why frequent burning is not purposefully practiced is the increase in alien annual grasses that accompanies such disturbance. Many of these grasses, and cheatgrass in particular, increased in the Intermountain West as severe overgrazing decimated the perennial bunchgrass ranges in the late nineteenth century (Galbraith and Anderson 1991, Harris 1991). Cheatgrass germinates in autumn, except in very arid portions of the Great Basin, and uses soil moisture in rosette form as it develops an extensive root system (Young et al. 1987). It produces a fine-textured fuel that is cured by early summer, making these ranges more flammable than they were historically. Increased burning is associated with increasing dominance of annual grasses (Fig. 12.14;

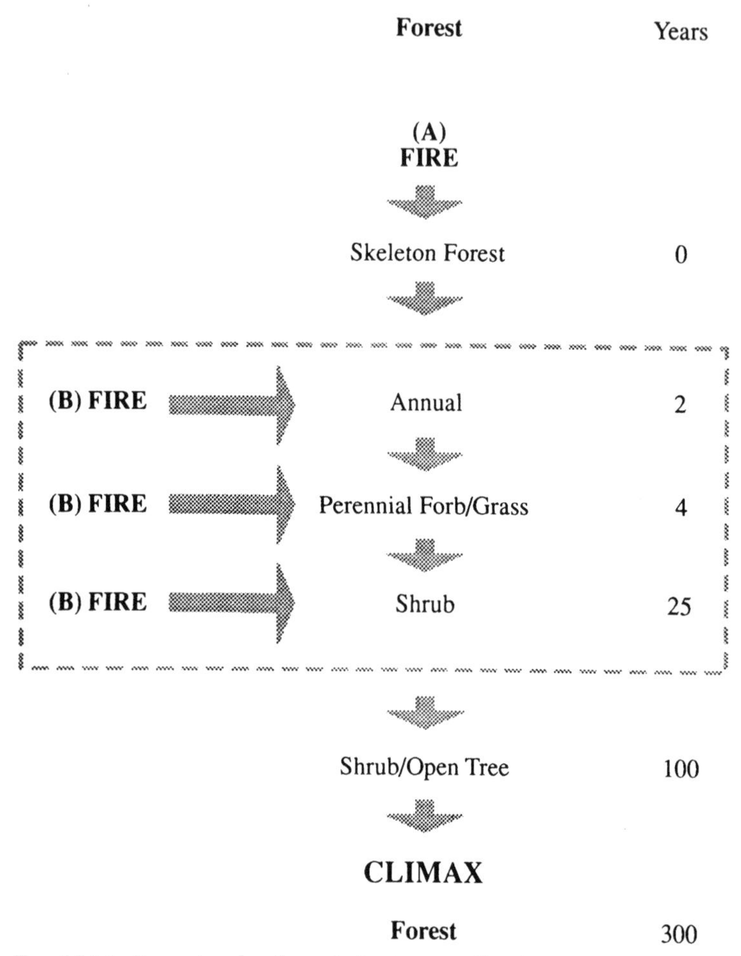

FIG. 12.13. *Successional patterns in juniper woodlands.*
(From Everett 1987)

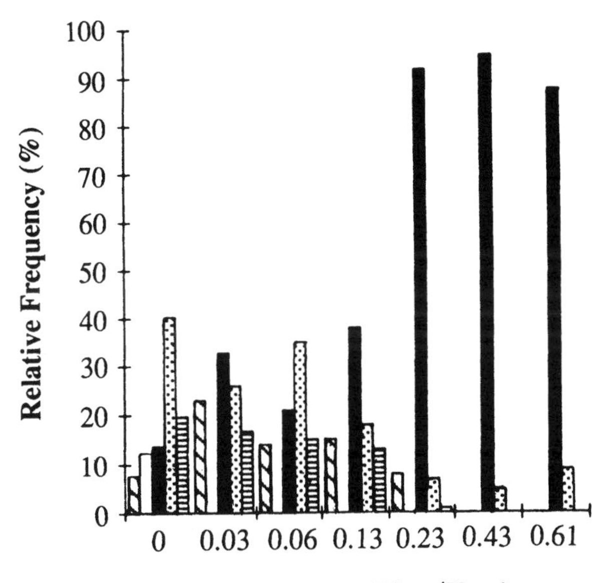

FIG. 12.14. *Relative dominance of different plant life-forms under varying fire frequencies. More frequent fire is associated with increasing dominance by annual grasses.*
(From Whisenant 1990)

Whisenant 1990). Burning can reduce the total seed density available from annuals, but a single seed can produce an open-grown plant that will produce up to 5,000 seeds one year later (Young et al. 1987).

The interactions between grazing history and fire have produced *Juniperus occidentalis* woodlands in which the hypotheses related to juniper expansion may now be impossible to test. The fire hypothesis, by itself, is clearly conceivable but perhaps not provable. The grazing hypothesis, by itself, is also conceivable, but juniper invasion has occurred, though to a lesser extent, even in areas that are thought to have been ungrazed (Quinsey 1984). The hypothesis related to logging of juniper is valid in limited areas. A deforested area of 32 km radius was created around ore smelters in the Eureka Mining District in Nevada (Young and Budy 1979). Young and Evans (1981) noted stumps in their study area that dated to the 1880s, but these were in the old-growth section of their area, not the expansion area. Stumps will normally be present to validate this hypothesis where it is applicable. Climate change encouraging juniper expansion has not been clearly demonstrated (Young and Evans 1981, Quinsey 1984).

Management Implications

Juniper woodlands have changed considerably over the past century, creating much of what we classify as juniper woodland today. Fire was probably once more common in the areas into which juniper has expanded, but reintroduction of fire is no panacea for management problems.

First, there are clearly areas in which fire effects were always minimal; that is why the old-growth juniper communities exist. Some areas show no evidence that a herbaceous component sufficient to allow fire spread ever existed (Clark and Starkey 1990). Second, the invasion of alien species over the last century suggests that the successional outcomes of burning may not be the same as they were historically. Most of these aliens are there because they are adaptable to disturbance, and fire may be no less favorable to them than overgrazing has been.

Intensive management strategies, such as weed control with herbicides, chaining juniper, and drilling seed, can be used to favor perennial grasses where cheatgrass invasion is present (Young et al. 1987). The objective in most of these cases is to increase grazing capacity for domestic livestock, rather than to restore or maintain some semblance of a natural community. Alien species that are spreading rapidly, such as the knapweeds (*Centaurea* spp.), may cause additional problems in treated areas.

There are both similarities and differences in the future management of *Quercus garryana* woodlands and *Juniperus occidentalis* woodlands. Reintroduction of fire will likely create a closer mimic of form rather than function in both ecosystems, due to the competitiveness of alien annual grasses in the understory. Unlike oaks, which are fire *resisters* in oak woodlands, juniper is a fire *avoider* in woodlands where it is present. Reintroducing fire will result in less juniper on the landscape, yet possibly with little semblance of a natural successional pattern in the remaining herbs and shrubs. The chances of restoration are greatest where severe overgrazing has not occurred and where the perennial bunchgrass component remains.

CHAPTER 13

FIRE IN OUR FUTURE

THE BODY OF THIS BOOK has scientifically documented the historic role of fire in Pacific Northwest forests. This concluding chapter attempts to place that information in the real world, a world of conflicting values, a playing field on which the scientist has no inherent advantage. Forest fires will occur in the wildlands of our West, from human and natural origin. Fire origin or presence is less relevant than whether that presence helps achieve ecosystem goals. The progression of knowledge about forests and fire has provided some advantages that were not available in the past. More than 50 years ago, the rudimentary science of fire allowed eminent ecologist Fredric Clements (1935, p. 344) to state: "Under primitive conditions, the great [vegetation] climaxes of the globe must have remained essentially intact, since fires from natural causes must have been both relatively infrequent and localized."

Our knowledge has expanded, but so has the complexity of society and the decisions society attempts to make. Knowing more about fire does not directly lead us to a strategy for management, and in fact it makes management strategies more complex. We know that a variety of fire regimes existed on the pre-European landscapes of the Pacific Northwest, and we know in general terms how we have changed species composition and structure by our management practices. Consensus on fire management, however, requires the incorporation of human values, and most past fire management policies have been derived from the view of fire only as a threat, rather than from a broader perspective of values evident in today's society.

We also know that institutions tend to focus on short-term operational objectives, because such actions are easily visualized and achievable. The best example is our attempt in the twentieth century to minimize the area burned by wildfire. Because of the threat that fire posed to forest management in the early twentieth century, all land management institutions sought to erase it from the landscape (chapter 3). Ecological roles of fire were ignored, and criteria for economic efficiency were rarely applied. The short-term result was successful: the enemy was fire, and we conquered it . . . we thought. Our institutions became so effective at presenting fire as the enemy that they exacerbated the problem they were attempting to solve.

The area burned by wildfire in the West since the 1920s continuously declined into the 1960s (Fig. 13.1). Yet a disturbing U-shaped trend is evident in these data and similar data from Canada (Flannigan and Van Wagner 1991). A simple projection into the future from these data suggests an increasing wildfire problem. Such projections may also be too simplistic without a clear understanding of the possible reasons for recent increases and what might be done to avert a continued problem.

One possible reason for recent increases in area burned in the western United States is that natural fire programs since about 1970 have allowed some fires to grow as "management fires," which under old policies would have been suppressed when small. As weather conditions change, these management fires occasionally become too large or too intense to remain in "prescription." When they are reclassified as wildfires, the total area burned is added to the wildfire statistics (for example, about half the Yellowstone fire area, close to 250,000 ha, or about one-quarter the length of the 1988 bar in Fig. 13.1). Without the natural fire policy, the area increase in recent years might not be so significant, but the increases would still be there (and they are present in the Canadian statistics,

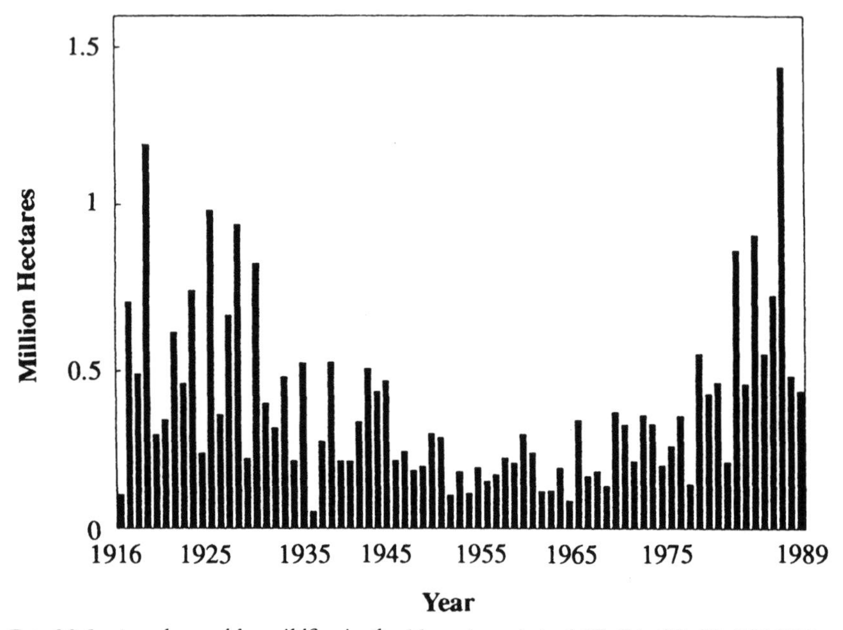

FIG. 13.1. *Area burned by wildfire in the 11 western states (AZ, CA, CO, ID, MT, NM, NV, OR, UT, WA, WY) between 1915 and 1990. Data <1931 do not include national parks or Native American reservations; data <1926 are forested lands only.* (Compiled by B. Mitchell, Bureau of Land Management, BIFC, Boise, ID, and provided by S. F. Arno and J. K. Brown, USDA Forest Service, Forestry Sciences Laboratory, Missoula, MT)

too). Only the peaks of the area burned in recent years would be trimmed.

A second possible explanation for the shape of the curve might be long-term climate shifts: drought in the 1920s and 1930s associated with big fire years, cooler weather in the 1940s and 1950s, and some unusual strings of droughty years in recent decades associated again with big fires. Climate has not been established as the cause of recent increases in burned area, although it is likely the primary cause of the second-order, year-to-year blips in Figure 13.1. A third reason, linked with fire severity as well as fire size, is the buildup of fuel hazards we have fostered through fire exclusion policies. Without fire acting as an agent of decomposition, litter has built up, tree density has increased, and fuel continuity, both vertically and horizontally, is greater than historically. Fires that occur in such fuels are more intense and more difficult to control, even as fire control technology improves. This effect is most pronounced in fire regimes of low to moderate severity, where increased fire area is now burned primarily by high severity fires. The more successful we are at fire control, the worse the problem becomes: a seemingly insoluble problem if we do not look beyond the short-term objectives.

The lesson learned by renewable natural resources managers is to repeatedly check short-term objectives against long-term goals, continually framing and reframing the objectives in terms of the goals. For example, minimizing area burned is best interpreted as a short-term objective leading to a goal, rather than as a goal in itself, as it was in the past. Wildland management strategies are constrained by the fact that natural resources management is an experiment: Natural systems cannot be tuned like an automobile. A natural resources manager dealing with fire must contend with additional uncertainty about the characteristics of fire (will it stay within certain bounds? will its intensity change with shifts in weather?) as well as uncertainty about how the ecosystem will react. This uncertainty is higher than with any other silvicultural option available to the manager. This is a major reason why in the past fire was not more widely applied as a management tool.

A MANAGEMENT FRAMEWORK

The *adaptive management* and *ecosystem management* approaches have considerable value in dealing with an uncertain biological world in a rapidly changing social context. Too often, however, they are viewed as goal-oriented systems rather than process-oriented means of achieving goals. Adaptive management is a system by which experimental techniques can be applied to approach elusive and often changing goals (Walters 1986). Ecosystem management is a framework within which adaptive management can function (Agee and Johnson 1988). Both are value-neutral, so they deal not with setting goals but with achieving them. In contrast, the USDA Forest Service's recent embrace of "ecosystem management" is primarily goal-oriented, and although perhaps quite appropriate, is used in a different context than it is in this chapter.

The ecosystem, simply put, is any part of the universe around which a line is drawn, with the line being a "permeable membrane" across which many things pass: people, animals, plant seeds, and so forth. Ecosystems are temporally variable, suggesting that the form of a problem and its solution may be different in the future than today. Ecosystems are spatially variable, suggesting that the nature of the problem and its solution will vary across the landscape at any point in time. Ecosystems include people, so an ecosystem management approach incorporates socioeconomic values as a component (but these values do not necessarily have priority over the biocentric values). Different ecosystem components have different boundaries, implying a set of overlapping and interacting systems to challenge the land manager. The permeable membrane concept implies that few problems or solutions will be self-contained within a political boundary.

Implications for resources management flow from these basic properties (Holling 1978, Agee and Johnson 1988). The first is that ecosystems are constantly changing. Even apparently stable fire regimes, such as those in ponderosa pine forests, were maintained by a cyclic repetition of low severity fire. The second is that cause and effect may be spatially heterogeneous: Smoke from a forest fire may not be a problem in a stand directly upwind from the fire, but it may be a significant nuisance in a town 10 km downwind. A third implication is that ecosystems may exhibit several levels of stable behavior. An ecological progression over time may not be capable of reversal; although this has been documented mainly for alien plant invasions, it may also be true of past ecosystem states that we define as desirable. For example, once fuels have built up around shallow-rooted, old-growth ponderosa pine, can even low intensity fire be used to restore natural fuel loads and still keep the pines alive? A fourth implication is that although there are organized connections between parts, not everything is connected to everything else. We should not be paralyzed by complexity, but we must not ignore it, either.

Ecosystem management involves regulating internal ecosystem structure and function, plus inputs and outputs, to achieve socially desirable conditions (Agee and Johnson 1988). It is a *process* of understanding ecosystem components (including people) and interactions. It neither defines the desirable conditions nor constrains actions to achieve those conditions. Similarly, adaptive management is an iterative approach to managing uncertainty in both biological and social components of the ecosystem (Walters 1986). A successful approach to resources management requires a four-stage strategy (Agee and Johnson 1988):

1. Have well-defined goals.
2. Define the ecosystem boundaries for primary components.
3. Adopt a management strategy to achieve these goals.
4. Monitor the effectiveness of the management strategies.

The iterative nature of this approach implies that either management strategies or goals may change over time. Both flexibility and accountability must be maintained in the process, so that continually shifting goals do not result in a lack of responsibility or accountability for actions.

In the remainder of this chapter I apply these concepts to several land management challenges in which fire will play an important role. These include park and wilderness management; New Forestry; preservation of the northern spotted owl; forest health in the eastern Cascades; and global warming.

PARK AND WILDERNESS MANAGEMENT

National park and wilderness managers face dual mandates: preservation of nature and the enjoyment/experience of these resources by present and future generations of people. I assume that present and future generations of people will be served well by efforts to "preserve nature." Much of this book describes the interaction of fire in natural ecosystems; if we know how disturbance once interacted with these systems, is the maintenance of that process of disturbance sufficient? In the case of many other types of disturbances, the answer is simple because we are powerless. We did not have to suspend the "let it blow" policy after the Everglades hurricane of 1992. We do not suspend the "let it grind" policy of glaciers, or the "let it erupt" policy for Kilauea or Mount St. Helens. But the human relationship with fire is unique: we can and do control, within limits, a "let it burn" policy. Furthermore, we have an ethical responsibility to do so when in society's interest, and this is rarely true for most other natural disturbances that have catastrophic potential.

Our nature preserves, as thrilling and wonderful as they are, are not truly natural systems anymore, unconstrained by the bounds of society. Few of these systems are large enough that fire can run free in the forest; the 1988 Yellowstone fires, in the largest nature preserve in the lower 48 states, provide ample evidence of this. Both the flames and their smoke may affect adjacent or distant communities. Further, the "natural" system may have had a significant component of Native American ignitions: are these considered natural? Without such ignitions, the structure of the system may be different than with them. This is a social decision, but its resolution has a significant impact on nature. Moreover, there are multiple objectives beyond the scale of the individual park or wilderness, for example, air quality (particularly regional haze) and endangered species legislation. These may place constraints either on ecosystem outputs or on ecosystem states (such as old growth) in the presence of fire.

Fire is in some respects like a spirit in a bottle, but we do not always know whether it is a genie or a demon. It always promises excitement and may result in great fascination and satisfaction; it may also cause death and horror, to humans as well as to other animals and plants (Maclean 1992). We can choose, at least to some degree, the role fire plays as a natural disturbance factor in wildlands, yet how can we choose to control a factor of nature? Stephen Pyne's (1989b) article, "The Summer We Let Wildfire Loose," captures this paradox: We cannot evade the human imprint in parks and wilderness.

Under what conditions may we let the spirit out of the bottle? We must first define our ecosystem goals, outlined by legislation for parks and wilderness, for air quality, for endangered species, and for historic preservation, among other goals. Some of these conflict with one another: the Wilderness Act demands that natural forces be allowed to act "untrammeled" by man, yet that may be in direct conflict with other types of legislation. Many of the conflicts can be resolved by applying spatial and temporal limits to fire: avoiding historic structures, or not burning during severe atmospheric inversions, conditions that do not universally apply across space and time. Such constraints may, however, result in severe limits to fire as a natural process.

Even with nature as a sole constraint, we cannot agree on appropriate fire management goals (Agee and Huff 1986). Significant controversy has arisen over whether fire should be reintroduced as a *process* or to achieve objectives defined in terms of *structure* for park and wilderness ecosystems (Bonnicksen 1985, Bancroft et al. 1985). Are we mainly interested in recreating the natural fire regime, or do we want the vegetation and other structural ecosystem components that a natural fire regime should have produced?

Process-related goals are often most applicable to high severity fire regimes, particularly those that have not been significantly affected by fire exclusion policies. In large reserves with little edge effect, reintroduction of most naturally occurring fires may be sufficient to achieve the objective. Where large edge effects occur—for example, the absence of natural fires moving into the reserve from outside as they did historically—prescribed fires might be substituted in a process-oriented framework.

Structure can be defined in terms of individual stand architecture or at the large scale of the patches that make up a landscape. Structural goals have been recommended most often in low severity fire regimes with small patch sizes, where individual trees or groups of trees may be selectively affected by burning. For example, a prescribed fire might be applied to remove all the white fir trees below 10 m height in a mixed-conifer forest while preserving the larger sugar pine and ponderosa pine trees. Even in high severity fire regimes, structural goals may be formulated, such as maintaining a certain mix of patches of varying sizes and ages throughout a watershed by limiting natural fire spread or igniting prescribed fire.

National policy for natural fires has become more conservative since 1988. Managers must now certify on a daily basis that they have the ability to control any fire in their reserve, or suppression action must be initiated. The result of this policy will be longer fire-return intervals with a trend toward later successional stages. Our choice is to accept this eventually "unnatural" landscape mosaic or to produce one that might be closer to natural, either by introducing fire ourselves or by creating fuel-limited zones around areas in which natural fires will be allowed to burn under prescription. This should increase the window of "ability to control" and increase the confidence of managers to allow more natural ignitions to burn. In park and wilderness fire management, we are faced with Hobson's choice, since none of the alternatives are truly natural.

This realization in fact frees us from the dilemma of defining *natural*. We know that today's landscapes are different from those of distant millennia, and we know that a variety of future natural landscape configurations is possible, although some may be more desirable than others (Christensen et al. 1989, Sprugel 1991). Human intervention and manipulation will be necessary to preserve the natural processes, and knowledge of natural fire regimes will help in the definition of acceptable envelopes of natural ecological conditions in the park and wilderness areas of the West.

NEW FORESTRY

New Forestry is an attempt to incorporate ecological values with commodity (timber) production. Such an approach offers an alternative to what has become a traditional emphasis on timber production on public lands, particularly those managed by the USDA Forest Service. The principles and practices (Hopwood 1991) and the scientific basis (Franklin 1992) of New Forestry suggest that much of the rationale has evolved from experiences in the western Cascades area of Washington and Oregon. These forests have high severity fire regimes, grading to moderate severity on drier landscapes of this subregion. Application of New Forestry principles to other areas should be done by adapting these principles as necessary rather than by adopting them wholesale.

The scientific basis for New Forestry rests on the examination of natural forests and streams as ecosystems (Franklin 1992). An underlying assumption is that the suite of natural forests we inherited had many of the biodiversity characteristics we wish to maintain in the future: viable populations of animal populations, or the structural diversity of forests. Maintaining biological diversity may depend on maintaining a mosaic of patches of different successional stages and making the transition between them less abrupt.

Older forest as described in much New Forestry literature is usually the most structurally diverse, with large live trees, large snags, large downed logs, and a multilayered understory. These features are characteristic of the natural old-growth forests of the *Tsuga heterophylla* zone and perhaps some of the *Abies amabilis* zone, but certainly not of ponderosa pine or many mixed-conifer forests, because those natural old-growth forests developed with much more frequent and low severity fire. These drier forests had more open understories, fewer down logs, and possibly fewer snags.

The concept of biological legacies—living and dead organic materials that are carried over into new stands after disturbance—may be relevant to biodiversity. Sometimes, however, this paradigm conflicts with what we know about natural forests. For example, Franklin (1992) notes that most natural disturbances, whether wind or fire, leave behind a significant component of green trees as well as a legacy of snags and logs (Fig. 13.2). Franklin's intent is to suggest that by maintaining structural complexity during a harvest operation, desirable postlogging structural complexity may be restored in a much shorter period of time (80 years) than if the forest is left to recover after clearcutting. Yet natural fire has not always left a significant component of green trees on the landscape in the *Tsuga heterophylla* or *Abies amabilis* zones (chapters 7 and 8). Where it has left significant overstory in drier forests (chapter 10), logs on the forest floor may have been consumed on a regular basis. Nature, therefore, may not be the best guide. Rather, the identification or maintenance of desirable structures to achieve ecosystem objectives is preferable. This may mean, in a New Forestry concept, that significant green trees should be left in a natural fire regime where postfire green trees were historically rare or absent. It may mean that downed logs should be preserved where once they were less abundant.

The "natural forest" paradigm around which much of New Forestry is implicitly organized is a less meaningful guide than a more definitive identification of desirable ecosystem character, whether or not it is natural. Understanding the natural system is essential to managing it. But using the patterns of the past, even when correctly interpreted, as a narrow guide to the future is far too constraining. We must *adapt* mimics of natural pattern if we are to be successful in the future. New Forestry has much to offer if it is more than natural forestry.

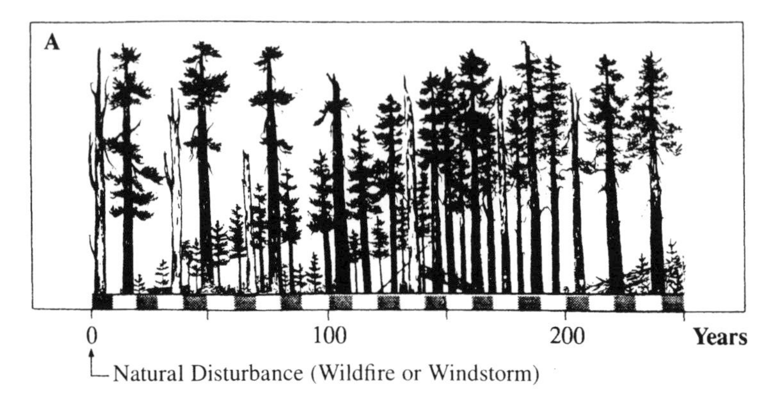

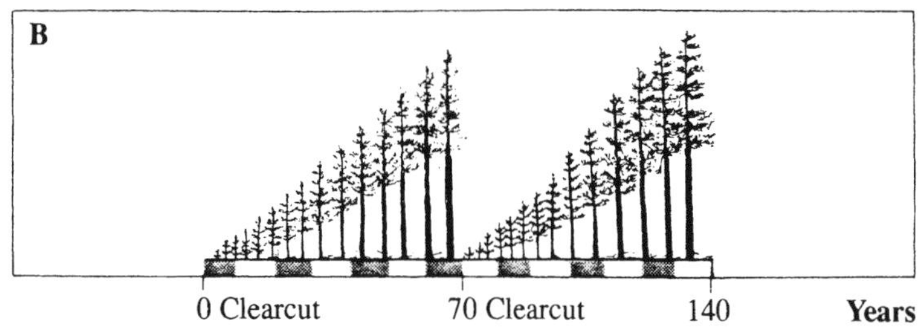

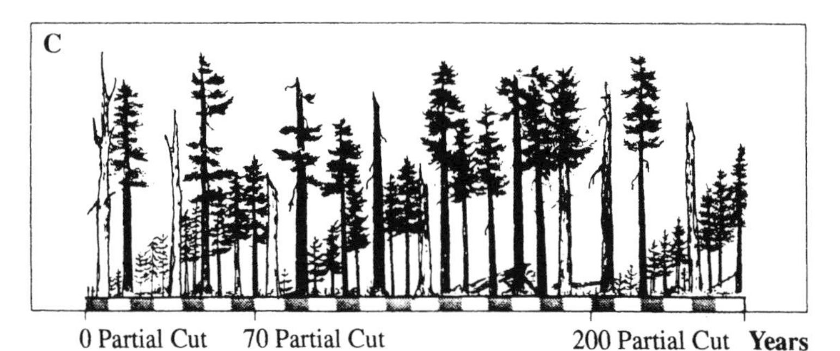

FIG. 13.2. *Temporal development of forest stands using a New Forestry paradigm.* A: *Natural recovery following a moderate intensity disturbance, such as wildfire or windstorm. This may not be a dominant intensity level in much of the western Cascades area.* B: *Intensive management, in which early seral stages are shortened and later seral stages are absent.* C: *Stand-level application of New Forestry designed to emulate structural diversity found in unmanaged, natural forests. Practices include snag and green tree retention, mixed-species regeneration, long rotations, and creation of snags by topping green trees.*

(From Hopwood 1991, by permission of the Information and Extension Resources Branch, Ministry of Forests, Province of British Columbia)

PRESERVATION OF THE NORTHERN SPOTTED OWL

The northern spotted owl (*Strix occidentalis caurina*), one of the best-known owls in the world, was listed as a "threatened" species under the Endangered Species Act in June 1990. The primary threat to the owl is the fragmentation of its habitat in forests of Washington, Oregon, and northern California. It typically uses old-growth forests and other forests with similar characteristics for nesting, breeding, and rearing young. As of late 1992, a draft recovery plan for the northern spotted owl was in circulation, proposing 196 designated conservation areas (DCAs) over 3 million ha of federal forestland in the region. The DCAs include currently suitable owl habitat plus younger forest that may mature into suitable habitat. Each DCA is intended to provide habitat for at least 20 owl pairs. Federal lands between DCAs (called matrix lands) will be managed under a 50-11-40 rule (50 percent of the landscape must be covered by forest averaging 11 in [28 cm] dbh with 40 percent average cover).

The plan is designed to provide suitable habitat so that viable populations of owls will persist into the future. Good habitat appears to be associated with older forest that has large live trees, standing snags, downed logs, and multilayered canopies. But the act of breaking up the landscape into a series of adjacent DCA polygons from Washington into northern California makes the implicit assumption that this habitat will remain stable over time. That is, it assumes that owl habitat will be created from young stands within DCAs and maintained in old stands. This requires effective forest protection strategies, which begin with a clear understanding of natural disturbance regimes.

Forest protection strategies for protection of spotted owl habitat were discussed by Agee and Edmonds (1992), and much of the following discussion is summarized from that report, an appendix to the spotted owl recovery plan. They divided the range of the owl into three large subregions: West Cascades, Klamath, and East Cascades (Fig. 13.3), based primarily on characteristics of historical disturbance regimes. Based on these regimes, they rated the risk of common disturbance agents in each subregion (Table 13.1).

The West Cascades subregion has the highest probability of a

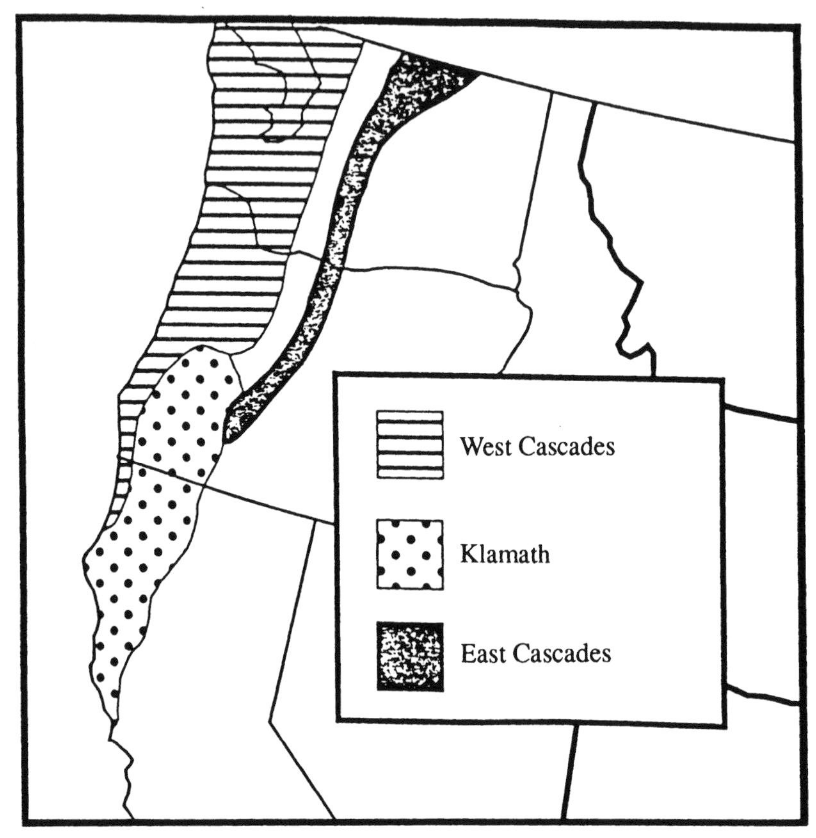

FIG. 13.3. *The subregions defined for forest protection strategies for northern spotted owls.*
(From Agee and Edmonds 1992)

successful fire suppression strategy for DCAs. This subregion receives more precipitation than the other two, and mature to old-growth forests have relatively low potential for surface fire behavior. There is a chance of a large fire complex due to unusual lightning storm/east wind events, but the probability of such an event is low and it is likely to be beyond management control if it occurs. Global climate change, if it creates more lightning or increases fire behavior potential, may alter predicted fire disturbance patterns. An aggressive fire suppression policy is recommended for DCAs of the West Cascades subregion. This may conflict with park and wilderness management objectives in the subregion, which

TABLE 13.1. LEVELS OF CATASTROPHIC RISK FROM FOUR NATURAL DISTURBANCE AGENTS IN THE THREE FOREST SUBREGIONS IDENTIFIED FOR NORTHERN SPOTTED OWL HABITAT

Disturbance Agent	*West Cascades Subregion*	*Klamath Subregion*	*East Cascades Subregion*
Fire	low	high	high
Wind	moderate-low[a]	low[b]	low
Insects	low	moderate	high
Diseases	low	moderate	high

SOURCE: Agee and Edmonds 1992.

[a] May be high in areas close to the coast.
[b] May be moderate in areas close to the coast.

otherwise might allow some natural fires to burn in old-growth forest.

In the East Cascades and Klamath subregions, fire exclusion has helped create a broader landscape pattern of multiple-canopied stands with thick understories, thought to be suitable for northern spotted owl habitat. It appears that at the same time that clearcutting and fragmentation have reduced owl habitat in these areas, a policy of fire exclusion has helped to increase owl habitat in protected areas. Whether on balance owl habitat has increased or decreased in any local area due to these offsetting factors has not been determined and is irrelevant to future management; what is important is how the existing or potential habitat can be protected into the future. These same two subregions have the highest potential for habitat loss through wildfire. The primary forest protection strategy employed in the recovery plan is to increase the number of DCAs in those areas and provide for the protection of territorial single owls outside DCAs. Given the uniform nature of the problem, this is as effective as adding more dominoes to a closely packed line; it is likely that all will eventually fall unless some are managed under a different strategy. A longer term adaptive management strategy for fire potential will be needed to conserve owl habitat.

Recent experience with large, uncontrolled fire events in the Klamath subregion is likely to continue if a total fire protection strategy is attempted across the whole of this subregion. Such a strategy might be successful in the coastal or high elevation portions of the subregion—the *Sequoia sempervirens* coastal belt, and the *Abies concolor* and *Abies magnifica* zones—but not across the widespread mixed-evergreen and mixed-conifer forests. Through effective fire exclusion, the moderate severity fire regimes of the past have been replaced with high severity fire regimes. When fire historically occurred, owl habitat may have been damaged for a decade or two and destroyed for longer times only in limited areas. The wildfires of today, burning in higher fuel loads and more uniform multilayered canopies, have resulted in an increased proportion of stand replacement fire, which will destroy owl habitat. Some sort of fuel management program is recommended to increase landscape diversity relative to fire, so that the potential for catastrophic fire is reduced. Some of the fuel reduction could be integrated with timber removal, but such removal must be planned to achieve owl habitat objectives, with commodity production as a result rather than an objective.

The East Cascades subregion is at most risk from catastrophic fire effects because of its lower species diversity as compared to the Klamath subregion and the generally frequent fire-return intervals documented in this area. Additionally, forest health problems verge on an epidemic, which increases wildfire potential while at the same time reducing owl habitat. As in the Klamath subregion, fire regimes have been shifted from low-moderate severity to moderate-high severity as a result of effective fire exclusion.

Two types of fuel management strategies are proposed for the Klamath and East Cascades subregions: underburning and fuelbreaks. Each addresses different objectives. Understory burning reduces dead fuel loads and vertical fuel continuity within a treatment area. Although this reduces catastrophic fire potential for some time, the elimination of a multilayered understory may result in suboptimum owl habitat at that site, so it should not be done over wide areas of any DCA in the same decade. Where evergreen hardwoods make up the understory, as in the Klamath subregion, regeneration of understory canopy will occur more rapidly than where the understory is composed of conifers (more typical of the East Cascades subregion). Fuelbreaks are designed to compartmentalize units by creating a zone of reduced fuel between them, which allows safe access for fire suppression forces during wildfires. Fuelbreaks may also allow better containment possibilities for wildfires within given compartments. They are normally installed on ridges and can be visually pleasing if well designed.

Paradoxically, owl habitat may have to be reduced to save it: A cyclic fuel maintenance program will help to protect large areas from stand replacement fire events. The strategies employed will have to be developed for individual DCAs. The efficacy of these treatments, however, has not been well established. The mix of prescribed burning and fuelbreaks will have to be experimentally applied to landscapes and monitored over time, with the more successful treatments being applied more widely over time. This is an example of adaptive management at its best. Replicated treatments will be necessary to account for variability in landscapes and the sensitivity of surrounding communities to forest fire smoke.

FOREST HEALTH IN THE EASTERN CASCADES

The choices to be made in fire management associated with spotted owl protection are in one sense easier than those made in fire management for forest health. There is a single primary objective in owl management: maintaining viable populations of owls through maintenance of habitat. The forest health issue is complicated by the fuzziness of the term "forest health" as well as the number of competing interests for the outputs of the forests east of the Cascades. In its simplest form, a healthy forest is one that is capable of maintaining desirable character or condition (visually attractive, and relatively resistant to insects, disease, and fire) while sustaining desirable outputs (recreation, timber, wildlife). Forest health is a global concern, but regionally the focus has been on the drier forests of eastern Oregon and Washington because of the obvious forest decline seen there. Insect populations seem to remain at high levels; disease pockets, once localized, are expanding in scope; and fires that once spread lightly across forest floors are now stand replacement events.

A recent analysis of eastside forest health (Gast et al. 1991) concluded that fire played an important role in the evolution of natural ecosystems, and that it is an important option in managing for long-term productivity of the forest. Two major constraints to implementing this option are intermingled private ownerships and smoke. Larger blocks of uniform public ownership will be easier to manage, as fire will be safer and easier to apply in situations where public/private property boundaries do not fragment the landscape.

The most significant constraint is smoke. Although frequent maintenance burns may produce little smoke, initial restoration burns will consume biomass that has accumulated for many decades. Restoration burns will kill smaller live trees, which will add dead fuels to the ground over time. These sites will have to be underburned when moist, and combustion will be relatively inefficient, producing more smoke per ton of fuel consumed. A rough idea of the magnitude of potential consumption, at least for the first round of restoration burns, can be estimated by evaluating the ponderosa pine and mixed-conifer types in eastern Washington (Table 13.2). Prescribed fire-return intervals determine the proportion of the landscape to be treated each year over the first cycle. Tonnage is estimated from restoration burns in similar forest types, and the addition of "activity fuels" from ongoing harvest is added at the 1987–91 level (DNR 1992), a time of decline in tonnage due primarily to air quality concerns. The total consumption is about three and a half times the 1977–91 average, or eight and a half times the 1987–91 average.

After the first 15 years, when a second round of burning in ponderosa pine sites would start, tonnages would only slightly decline, because additions to the dead fuel category from the first burn would be consumed by the second. Measuring the total impact of such a program requires adding effects on wildfire area and fuel consumption. During the first cycle of burning, little immediate positive impact on area burned by wildfire would result, but results might begin to show after the first 10–15 years. These figures are very rough but are presented as a means of gaining perspective on the magnitude of the problem. They suggest that if air quality constraints remain high, forest health concerns will have to be prioritized.

Forest health and public health are to some extent mutually exclusive objectives under current fuel and regulatory conditions. We will have to pick our highest priority sites on which to reintroduce fire and expand treatment on lower priority sites only to the limits of social acceptance. Using the example already given, if we were to treat only 25 percent of the landscape in these forest types,

TABLE 13.2. PROJECTED BIOMASS CONSUMPTION FROM A PRESCRIBED BURNING PROGRAM IN PONDEROSA PINE AND MIXED-CONIFER FORESTS (EASTERN CASCADES, WASHINGTON)

Forest Type	*Area*[a] *(thousand ha)*	*Fire Cycle (yrs)*	*Consumption*[b] *($t\ ha^{-1}$)*	*Annual Consumption (million metric tons)*
Ponderosa pine	1,438	15	50	4.8
Mixed-conifer	504	30	100	1.7
1987–91 activity fuels				0.85
Total				7.35

NOTE: Annual average of 1977–91 activity fuel was 1.036 million metric tons (DNR 1992).

[a] Forest type area taken from Table 1.2.

[b] Consumption from restoration burns interpreted from Sackett (1980), Covington and Sackett (1984), and Thomas and Agee (1986).

we would be close to the tonnage of the 1977–91 period. Recognizing that the natural role of fire cannot be reapplied across the landscape, even if biologically sound, is a biological adjustment to the forest health solution. Putting up with more prescribed fire smoke, in part possibly offset in the long run by less wildfire smoke, is one of the social adjustments to be made.

Timber salvage operations may be an important part of the process of restoring forest health. Partial cuts that remove the shade-tolerant trees will produce wood, though of smaller diameter, and will remove coarse fuel from the sites. Medium and fine fuels (branches and leaves) will be left onsite and will cause increases in those categories of fuel. A key to success is to leave the more fire-tolerant, and generally larger, species—such as larch and ponderosa pine—whenever possible. Adapting or constructing lumber mills to use the smaller true fir and to some extent Douglas-fir will be another social adjustment to forest health solutions, but it will require a social contract to supply such materials in predictable quantities.

The transition to ecosystem management in eastside forests requires complex adjustments in traditional institutions: the Forest Service, the forest industry, conservation organizations, and the communities of the region. Everybody wants a solution, but the solutions are not cost-free. Current problems of forest decline are not future scenarios; they are here, at least in some areas like the Blue Mountains. Business as usual is costly to everyone, so there is a good chance that adaptive management approaches can be adopted. Fire will be a central theme in any such approach.

GLOBAL CLIMATE CHANGE

The forests of the Pacific Northwest will be subject to significant environmental change if current predictions about global warming are accurate (Franklin et al. 1991). Greenhouse gases, including carbon dioxide and methane, which trap outgoing radiation from the earth, have been increasing because of agricultural activities and fossil fuel consumption. Mature forests are better buffered against such change than regenerating forests, because the environment is more critical for seedlings than for mature trees. Mature trees will alter growth and morphology patterns as a first response to environmental change; seedlings will often die.

Significant forest change will occur after a forest has been disturbed by logging or by a natural disturbance factor, such as fire, when new seedlings must cope with a changed climate. An interesting feedback mechanism in this process of global change is that fires will also add carbon dioxide to the atmosphere, so that if global warming accelerates fire activity, fires could also accelerate the warming process. Most ecological predictions assume that global warming will result in an increase in fire as a disturbance factor. This may be a simplified assumption. My intent here is not to disprove this assumption but to show the complexity of making such an assumption.

For both Pacific Northwest and Rocky Mountain forests, changes in future climate have been projected, although climate and ecological reconstructions at a regional level appear speculative: If climate scenario A occurs, the forests are projected to change in one way, while if scenario B occurs, the forest should change in a different way. For the Rocky Mountains, one of the scenarios is a warmer, drier climate, either with or without greater efficiency of plant water use (due to higher CO_2 levels; Leverenz and Lev 1987). Under this scenario, fires are projected to increase in frequency and severity (Romme and Turner 1991), but no evidence to support this conjecture is provided. In the Pacific Northwest, temperature changes up to 5°C are projected to result in significant alterations in the distribution of forest types of the region (Fig. 13.4). Plant communities will probably not move as intact units across the landscape, as the authors recognize. They note that some species will be capable of faster migration than others, but the large magnitude of potential change is illustrated well by Figure 13.4. Increased frequency of fire is claimed to be certain, and increases in intensity are stated as probable (Franklin et al. 1991). The apparent rationale for these statements is that as environments now found to the south "move" north with global warming, so will at least portions of the associated fire regimes. The increase in intensity is assumed to be due to increased fuel loads of stressed forests with higher dead tree components. The major assumptions in common between these projections are that the controlling factor for fire activity is moisture and that this is fairly represented by potential evapotranspiration. In fact, there are several controlling factors, and not all are correlated with moisture.

The seasonal distribution of moisture is important in such predictions, but existing models cannot project seasonal precipitation with any certainty. Lower annual precipitation but an increased proportion in the dry season may result in no net change or even less fire activity. The fire cycle model (see chapter 7) generates fire activity in the western Cascades based on seasonal moisture, lightning activity, and presence of east wind. Offsetting changes in the latter two factors could negate any increase in fire activity due to drier summers. Two new runs of the fire cycle model were made, both assuming a 25 percent decrease in summer precipitation, with the first assuming no change in thunderstorm or east wind activity and the second assuming a 25 percent decrease in both factors (Table 13.3). Other factors remaining constant, a decrease in summer precipitation significantly increases the occurrence of fires across the region from 40 percent to 90 percent. However, if other factors associated with fires decrease, the net change ranges from a 10 percent decrease to a 20 percent increase. Until we have a better mechanism to project changes in synoptic weather associated with global warming, reasonably justified projections about fire activity are not possible.

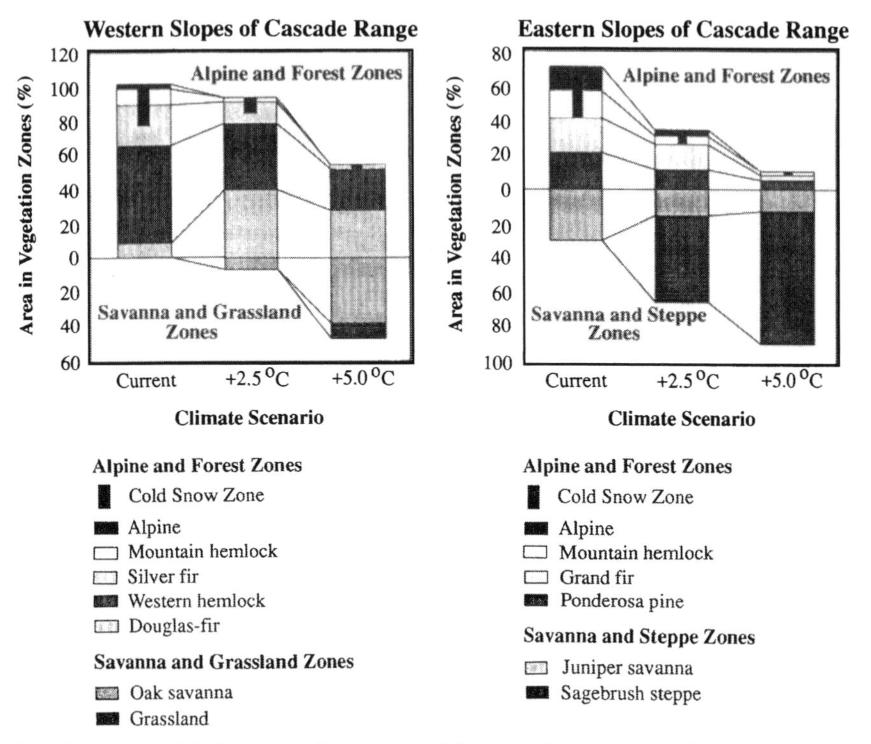

FIG. 13.4. *Potential changes in forest zones of the central Oregon Cascades under two altered climate scenarios.*
(From Franklin et al. 1991)

Recent projections of the effects of global climate change on Canadian forest fire activity have concluded that increased temperature due to a doubling of carbon dioxide is most likely to drive changes in fire activity (Flannigan and Van Wagner 1991). Precipitation is less important in projections of increased fire activity in Canada because it was projected to increase for many Canadian stations, and due to lack of data, sequences of dry periods were assumed to mimic those of the past. Increasing temperature affects the drying phase of the "moisture codes" of the Canadian Forest Fire Weather Index in two ways. First, assuming a constant dew point, relative humidity and equilibrium moisture content decline and, in the Canadian system, the rate at which fuel dries after a rain increases. Second, assuming little change in relative humidity (that is,

TABLE 13.3. CHANGES IN OCCURRENCE OF FOREST FIRES AT FOUR WESTERN CASCADE LOCATIONS WITH GLOBAL WARMING.

Location	*Base Case*[a] *(no. of fires)*	*Scenario 1*[b] *(percent change from base case)*	*Scenario 2*[c] *(percent change from base case)*
Western Olympics	1,236	+52	+2
Wind River	2,998	+89	+23
McKenzie	4,946	+67	+3
Southwest Oregon	10,342	+38	−10

NOTE: See Figure 7.5 for location of sites; southwest Oregon site is Kerby-Jacksonville area.

[a] Base case is essentially an index of number of fires per unit area, based on twentieth-century weather patterns (see chapter 7 for more detail).

[b] Scenario 1 is base case with 25 percent less summer precipitation.

[c] Scenario 2 is base case with 25 percent less summer precipitation, thunderstorm activity, and occurrence of east winds.

dew point increases along with temperature), a fire danger index called "seasonal severity rating" is projected to increase 46 percent across Canada with a possible similar increase in area burned. Cranbrook, British Columbia, the most westerly station, has a 30 percent projected increase.

We have only a rudimentary understanding of the interactions between climate change and fire activity. Although it is tempting to envision a global warming scenario with larger, more intense wildfires across the region, there is little evidence to suggest that the inferno is inevitable. We need to focus research on more precise climate scenarios, including those affecting fire ignitions, before we can draw realistic inferences about fire.

A tantalizing precursor to conclusions about global changes and fire activity in the Pacific Northwest is the seeming historical relation between large, intense fires and global cooling episodes. In the past millennium there have been three major sunspot minima: the Wolf, Sporer, and Maunder (Fig. 13.5). During such times, the energy output of the sun is slightly less. Mid-latitude glaciers have advanced during such times, and these are considered periods of global cooling. Superimposition of large fire events of the past results in a surprising correspondence between sunspot minima and large, intense fires. Whether this is a cause-effect association or not is unknown, but it may represent a change in the frequency of long- and short-term drought, thunderstorm occurrence, or the frequency of east winds across the Pacific Northwest. This adds another cautionary note to simple assumptions about global warming and fire.

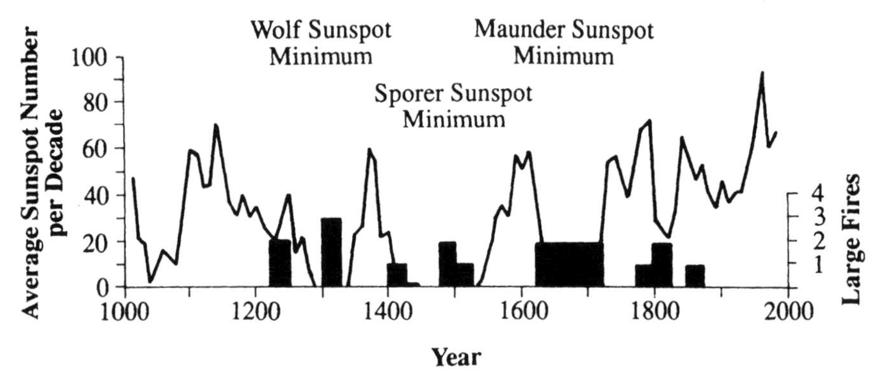

FIG. 13.5. *Occurrence of large fires in the Pacific Northwest and sunspot minima.* (Pre-1850 fires drawn by the author from published and unpublished sources in western Oregon and western Washington. Sunspot graph from Henderson et al. 1989)

APPENDIX A

COMMON CONVERSION FACTORS

LENGTH
1 cm = 0.39 in
1 m = 3.3 ft
1 km = 1,000 m = 3,280 ft = 0.6213 mi

AREA
1 m^2 = 10.763 ft^2
1 ha = 10,000 m^2 = 2.471 ac
1 km^2 = 0.386 mi^2 = 247.1 ac

MASS
1 kg = 2.205 lb

VOLUME
1 m^3 ha^{-1} = 14.28 ft^3 ac^{-1}

RATE OF SPREAD
1 m sec^{-1} = 3.6 km hr^{-1} = 2.23 mi hr^{-1} = 196.85 ft min^{-1}

HEAT AND HEAT CONTENT

1 cal = 4.18 Joules = 0.0039 BTU

1 cal g^{-1} = 1.8 BTU lb^{-1}

FUEL LOADING

1 tonne ha^{-1} = 1,000 kg ha^{-1} = 0.1 kg m^{-2} = 0.45 tons ac^{-1}

1 g m^2 = 8.91 lbs ac^{-1}

POWER/IRRADIANCE

1 watt (W) = 1 Joule sec^{-1} = 0.239 cal sec^{-1}

1 kW m^{-2} = 5.3 BTU ft^{-2} min^{-1}

FIRELINE INTENSITY

1 kW m^{-1} = 0.29 BTU ft^{-1} sec^{-1} = 0.238 kcal m^{-1} sec^{-1}

APPENDIX B

NAMES OF PLANTS MENTIONED IN TEXT

TREES

Abies amabilis
 Pacific silver fir
Abies concolor
 white fir
Abies grandis
 grand fir
Abies lasiocarpa
 subalpine fir
Abies magnifica
 California red fir
Abies magnifica var. *shastensis*
 Shasta red fir
Abies procera
 noble fir
Acer circinatum
 vine maple
Acer macrophyllum
 bigleaf maple
Alnus rubra
 red alder
Arbutus menziesii
 Pacific madrone
Betula spp.
 birch
Calocedrus decurrens
 incense-cedar
Castanopsis chrysophylla
 chinquapin
Chamaecyparis lawsoniana
 Port Orford–cedar
Chamaecyparis nootkatensis
 Alaska yellow-cedar

Cornus nuttallii
Pacific dogwood
Cupressus bakeri
Baker's cypress
Eucalyptus spp.
eucalyptus
Fraxinus latifolia
Oregon ash
Juniperus deppeana
alligator juniper
Juniperus occidentalis
western juniper
Juniperus scopulorum
Rocky Mountain juniper
Larix lyallii
subalpine larch
Larix occidentalis
western larch
Lithocarpus densiflorus
tanoak
Picea engelmannii
Engelmann spruce
Picea mariana
black spruce
Picea sitchensis
Sitka spruce
Pinus albicaulis
whitebark pine
Pinus attentuata
knobcone pine
Pinus banksiana
jack pine
Pinus contorta
lodgepole pine
Pinus echinata
shortleaf pine
Pinus flexilis
limber pine
Pinus jeffreyi
Jeffrey pine
Pinus michoacana
Michoacan pine
Pinus montezumae
Montezuma pine
Pinus monticola
western white pine
Pinus muricata
bishop pine
Pinus palustris
longleaf pine
Pinus ponderosa
ponderosa pine
Pinus radiata
Monterey pine
Pinus resinosa
red pine
Pinus rigida
pitch pine
Pinus sabiniana
digger pine
Pinus taeda
loblolly pine
Populus spp.
cottonwood, aspen
Populus tremuloides
quaking aspen
Populus trichocarpa
black cottonwood
Prunus spp.
cherry
Pseudotsuga macrocarpa
bigcone spruce
Pseudotsuga menziesii
Douglas-fir
Quercus agrifolia
coast live oak
Quercus chrysolepis
canyon live oak
Quercus douglasii
blue oak
Quercus engelmannii
Engelmann oak
Quercus garryana
Oregon white oak

Quercus kelloggii
black oak
Quercus lobata
valley oak
Quercus wislizenii
interior live oak
Salix scouleriana
Scouler's willow
Sequoia sempervirens
coast redwood
Sequoidendron giganteum
giant sequoia
Taxus brevifolia
Pacific yew
Thuja plicata
western redcedar
Tsuga heterophylla
western hemlock
Tsuga mertensiana
mountain hemlock
Umbellularia californica
California bay, laurel

SHRUBS AND HERBS

Adenostoma fasciculatum
chamise
Agoseris spp.
false-dandelion
Agropyron spicatum
bluebunch wheatgrass
Amelanchier alnifolia
serviceberry
Anaphalis margaritacea
pearly everlasting
Andropogon spp.
broomsedge
Arctostaphylos spp.
manzanita
Arctostaphylos uva-ursi
bearberry, kinnikinnick
Arnica cordifolia
arnica
Arrenatherum elatius
oatgrass
Artemisia arbuscula
low sagebrush
Artemisia tridentata
big sagebrush
Astragalus spp.
milkvetch
Atriplex spp.
saltbush
Berberis repens
Oregon grape
Bromus carinatus
California brome
Bromus mollis
soft chess
Bromus tectorum
cheatgrass
Calamagrostis rubescens
pinegrass
Camassia leichtlenii
Leichtlin's camas
Camassia quamash
camas
Carex geyeri
Geyer's sedge
Carex pensylvanica
long-stolon sedge
Carex spectabilis
showy sedge
Cassiope mertensiana
white heather
Ceanothus cordulatus
mountain whitethorn
Ceanothus integerrimus
deerbrush
Ceanothus prostatus
squaw carpet
Ceanothus sanguineus
redstem ceanothus
Ceanothus velutinus
snowbrush

Centaurea spp.
knapweed
Cercocarpus betuloides
mountain mahogany
Chimaphila umbellata
prince's pine
Chrysothamnus spp.
rabbitbrush
Cirsium spp.
thistle
Collomia heterophylla
varied-leaf collomia
Conzya canadensis
horseweed
Coptis laciniata
Oregon goldthread
Corylus cornuta
hazelnut
Cynosurus echinatus
hedgehog dogtail
Cytisus scoparius
Scotch broom
Dactylis glomerata
orchardgrass
Delphinium spp.
larkspur
Elymus cinereus
Great Basin wildrye
Elymus glaucus
blue wildrye
Epilobium angustifolium
fireweed
Epilobium paniculatum
autumn willowherb
Festuca bromoides
six-weeks fescue
Festuca idahoensis
Idaho fescue
Festuca ovina
hard fescue
Festuca viridula
green fescue
Gaultheria shallon
salal
Gnaphalium microcephalum
slender cudweed
Goodyera oblongifolia
rattlesnake plantain
Hieracium albiflorum
trail plant
Holcus lanatus
velvetgrass
Holodiscus discolor
ocean spray
Hypericum perforatum
St. John's wort
Hypochaeris radicata
spotted cats-ear
Ilex glabra
gallberry
Juniperus communis
mountain juniper
Lactuca serriola
prickly lettuce
Linnaea borealis
twinflower
Lolium multiflorum
annual ryegrass
Lolium perenne
perennial ryegrass
Lotus spp.
deervetch
Lupinus spp.
lupine
Madia spp.
tarweed
Melilotus officinalis
yellow sweetclover
Menziesia ferruginea
rusty menziesia
Muhlenbergia montana
mountain muhly
Oplopanax horridum
devil's club

Oxalis oregana
oxalis
Pachistima myrsinites
Oregon boxwood
Panicum spp.
panicgrass
Phyllodoce empetriformis
red heather
Physocarpus malvaceus
ninebark
Poa sandbergii
Sandberg's bluegrass
Polystichum munitum
sword fern
Pteridium aquilinium
bracken fern
Purshia tridentata
bitterbrush
Pyrola spp.
wintergreen
Rhododendron macrophyllum
western rhododendron
Rhus spp.
sumac
Rhus diversiloba
poison-oak
Ribes spp.
gooseberry, currant
Rosa gymnocarpa
baldhip rose
Rubus leucodermis
black raspberry
Rubus nivalis
snow bramble
Rubus parviflorus
thimbleberry
Rubus spectabilis
salmonberry
Rubus ursinus
trailing blackberry
Rumex acetosella
sheep sorrel
Senecio jacobea
tansy ragwort
Senecio sylvaticus
woodland groundsel
Serenoa repens
palmetto
Spartina patens
wiregrass
Spiraea betulifolia
birchleaf spirea
Spraguea umbellata
pussypaws
Stipa comata
needle-and-thread
Stipa occidentalis
western needlegrass
Stipa thurberiana
Thurber's needlegrass
Symphoricarpos albus
common snowberry
Symphoricarpos mollis
creeping snowberry
Synthyris reniformis
snowqueen
Tetradymia canescens
horsebrush
Tiarella unifoliata
foamflower
Trientalis latifolia
starflower
Vaccinium spp.
huckleberry
Vaccinium alaskense
Alaska huckleberry
Vaccinium deliciosum
Cascade huckleberry
Vaccinium globulare
blue huckleberry
Vaccinium membranaceum
thinleaf huckleberry
Vaccinium ovatum
evergreen huckleberry

Vaccinium parvifolium
red huckleberry
Valeriana sitchensis
Sitka valerian
Vicia americana
American vetch
Viola sempervirens
evergreen violet
Whipplea modesta
whipple vine
Xerophyllum tenax
beargrass

GLOSSARY

Adiabatic Of or denoting change in volume or pressure, with no heat gain or loss with the surrounding air.

Aspect The direction in which a slope faces.

Avoider A life-history strategy of plants that have little adaptation to fire.

Climax Species or communities representing the final (or an indefinitely prolonged) stage of a *sere*.

Conduction The movement of heat from one molecule to another.

Convection The movement of heat by currents in liquids or gases.

Crown fire A fire burning into the crowns of the vegetation, generally associated with an intense *understory fire*.

Endurer A life-history strategy of plants in which the plants resprout following a fire or endure the effects of fire.

Evader A life-history strategy of plants in which long-lived propagules are stored in soil or canopy and germinate after fire.

Fire cycle A fire-return interval calculated using a negative exponential distribution, applied using current age-class structure on the landscape.

Fire frequency The return interval of fire.

Fire predictability A measure of variation in *fire frequency*, expressed as a range, standard deviation, or standard error.

Fire regime The combination of fire frequency, predictability, intensity, seasonality, and extent characteristic of fire in an ecosystem.

Fire severity The effect of fire on plants. For trees, severity is often measured as percentage of basal area removed.

Fireline intensity The rate of heat release along a unit length of fireline, measured in kW m^{-1}.

Foehn wind A dry wind associated with windflow down the lee side of a plateau or mountain range and with adiabatic warming.

Habitat type The land area capable of supporting a single plant association.

Initial floristics A process of succession where seeds or plants of later successional stages are present from the outset but are subordinate to other species. See *Relay floristics.*

Invader A life-history strategy of plants in which plants, through highly dispersive propagules, invade a site after fire.

Mass transfer The movement of heat by burning firebrands, as used in fire literature.

Natural fire rotation A fire-return interval calculated as the quotient of a time period and the proportion of a study area burned in that time period.

Plant association The basic abstract unit in the classification of potential vegetation, described by overstory/ understory indicator species.

Plant community An assemblage of plant species that occur widely enough across the landscape to be recognized as a unit. This assemblage can be a pioneer group of species, a late successional group, or a combination of both.

Plant series Aggregations of plant associations that have the same overstory dominant.

Prescribed fire A fire ignited under known conditions of fuel, weather, and topography to achieve specified objectives.

Prescribed natural fire A fire ignited by natural processes (usually lightning) and allowed to burn within specified parameters of fuels, weather, and topography to achieve specified objectives.

Radiation The movement of heat by electromagnetic wave motion.

Rate of spread The rate at which a fire moves across the landscape, usually measured in m sec^{-1}.

Relay floristics A process of succession where one set of species prepares the site and is replaced by a new set (like passing the baton in a relay race). See *Initial floristics.*

Resister A life-history strategy of plants in which the plants, through adaptations such as thick bark, survive low intensity fire relatively unscathed.

Seral A plant species or community that will be replaced by another plant community if protected from disturbance.

Sere The product of succession: a sequence of plant communities that successively occupy and replace one another in a particular environment over time.

Spotting Mass transfer of firebrands ahead of a fire front.

Succession The process of change in plant communities.

Surface fire A fire burning along the surface without significant movement into the understory or overstory, with flame length usually below 1 m.

Timelag class A method of categorizing fuels by the rate at which they are capable of moisture gain or loss, indexed by size class of fuel.

Understory fire A fire burning in the understory, more intense than a surface fire and with flame lengths of 1–3 m.

Vegetation zone A land area with a single overstory dominant as the primary climax dominant. Occasionally zones are named after major seral species. Other climax types may exist in the zone.

Water repellency The resistance to soil wettability, which can be increased by intense fires.

Wildfire A fire, naturally caused or caused by humans, that is not meeting land management objectives.

REFERENCES

Agee, J. K. 1973. Prescribed fire effects on physical and hydrologic properties of mixed-conifer forest floor and soil. University of California Water Resources Center Contr. 143.

Agee, J. K. 1980. Issues and impacts of Redwood National Park expansion. Environ. Manage. 4:407–23.

Agee, J. K. 1981a. Fire effects on Pacific Northwest forests: Flora, fuels, and fauna. In Northwest Forest Fire Council proceedings: pp. 54–66. Northwest Forest Fire Council, Portland, OR.

Agee, J. K. 1981b. Initial effects of prescribed fire in a climax *Pinus contorta* forest: Crater Lake National Park. Nat. Park Serv. Coop. Park Studies Unit Rep. CPSU/UW 81-4. Coll. of For. Resour., University of Washington, Seattle.

Agee, J. K. 1982. Fuel weights of understory-grown conifers in southern Oregon. Can. J. For. Res. 13:648–56.

Agee, J. K. 1987. The forests of San Juan Island National Historical Park. Nat. Park Serv. Coop. Park Studies Unit Rep. CPSU/UW 88-1. University of Washington, Seattle.

Agee, J. K. 1988. Successional dynamics in forest riparian zones. In Raedeke, K. (ed.), Streamside management: Riparian wildlife and forestry interactions: pp. 31–45. Inst. For. Resour. Contr. 59. University of Washington, Seattle.

Agee, J. K. 1989a. A conceptual plan for the forest landscape of Fort Clatsop National Memorial. Nat. Park Serv. Rep. CPSU/UW 89-2. College of For. Resour., University of Washington, Seattle.

Agee, J. K. 1989b. Wildfire in the Pacific West: A brief history and its implications for the future. In Proceedings of the symposium on fire and watershed management: pp. 11–16. USDA For. Serv. Gen. Tech. Rep. PSW-109.

Agee, J. K. 1990. The historical role of fire in Pacific Northwest forests. In Walstad, J., et al. (eds.), Natural and prescribed fire in Pacific Northwest forests: pp. 25–38. Corvallis: Oregon State University Press.

Agee, J. K. 1991a. Fire history along an elevational gradient in the Siskiyou Mountains, Oregon. Northwest Sci. 65:188–99.

Agee, J. K. 1991b. Fire history of Douglas-fir forests in the Pacific Northwest. In Ruggiero, L. F., et al. (tech. coords.), Wildlife and vegetation of unmanaged Douglas-fir forests: pp. 25–33. USDA For. Serv. Gen. Tech. Rep. PNW-GTR-285.

Agee, J. K., and H. H. Biswell. 1970. Debris accumulation in a ponderosa pine forest. Calif. Agr. 24(6):6–7.

Agee, J. K., and P. Dunwiddie. 1984. Recent forest development on Yellow Island, Washington. Can. J. Bot. 62:2074–80.

Agee, J. K., and R. E. Edmonds. 1992. Forest protection guidelines for the northern spotted owl. In Recovery plan for the northern spotted owl: app. F. Washington, DC: USDI Fish and Wildlife Service.

Agee, J. K., M. Finney, and R. deGouvenain. 1986. The fire history of Desolation Peak, Washington. USDI Nat. Park Serv. Final Rep. CA-9000-3-0004. Nat. Park Serv. Coop. Park Studies Unit, College of For. Resour., University of Washington, Seattle.

Agee, J. K., M. Finney, and R. deGouvenain. 1990. Forest fire history of Desolation Peak, Washington. Can. J. Forest Res. 20:350–56.

Agee, J. K., and R. Flewelling. 1983. A fire cycle model based on climate for the Olympic Mountains, Washington. Fire and Forest Meteorology Conf. 7:32–37.

Agee, J. K., and M. H. Huff. 1980. First-year ecological effects of the Hoh Fire, Olympic Mountains, Washington. Fire and Forest Meteorology Conf. 6:175–81.

Agee, J. K., and M. H. Huff. 1986a. Structure and process goals for vegetation in wilderness areas. In Proceedings: National wilderness research conference: pp. 17–25. USDA For. Serv. Gen. Tech. Rep. INT-212.

Agee, J. K., and M. H. Huff. 1986b. The care and feeding of increment borers. Nat. Park Serv. Coop. Park Studies Unit, University of Washington Rep. CPSU/UW 86-3.

Agee, J. K., and M. H. Huff. 1987. Fuel succession in a western hemlock/ Douglas-fir forest. Can. J. For. Res. 17:697–704.

Agee, J. K., and D. R. Johnson. 1988. Ecosystem management for parks and wilderness. Seattle: University of Washington Press.

Agee, J. K., and J. Kertis. 1987. Forest types of the North Cascades National Park Service Complex. Can. J. Bot. 65:152–53.

Agee, J. K., and L. Smith. 1984. Subalpine tree establishment after fire in the Olympic Mountains, Washington. Ecology 65:810–19.

Agee, J. K., R. H. Wakimoto, and H. H. Biswell. 1978. Fire and fuel dynamics of Sierra Nevada conifers. For. Ecol. and Manage. 1:255–65.

Ahlgren, I. F., and C. E. Ahlgren. 1960. Ecological effects of forest fires. Bot. Rev. 26:483–533.

Albini, F. 1976. Estimating wildfire behavior and effects. USDA For. Serv. Gen. Tech. Rep. INT-30.

Albini, F. 1979. Spot fire distances from burning trees: A predictive model. USDA For. Serv. Gen. Tech. Rep. INT-156.

Alexander, G. W. 1927. Lightning storms and forest fires in the state of Washington. Monthly Weather Rev. 55:122–29.

Alexander, J. L. 1929. Reforestation influences and yield. In Western Forestry and Conservation Assn., Cooperative forest study of the Grays Harbor area (Washington): pp. 25–37. Portland, OR: Western Forestry and Conservation Assn.

Alexander, M. E. 1979. Bibliography and a resume of current studies on fire history. Canadian Forestry Service, Great Lakes Forest Res. Centre Rep. 0-X-304.

Alexander, M. E. 1980. Bibliography on fire history: A supplement. In Stokes, M. A., and J. H. Dieterich (eds.), Proceedings of the fire history workshop: pp. 132–43. USDA For. Serv. Gen. Tech. Rep. RM-81.

Alexander, W. H. 1924. The distribution of thunderstorms in the United States. Monthly Weather Rev. 52:337–43.

Allen, E. T. 1911a. Report of forester. In Western Forestry and Conservation Assn. 1911 proceedings: pp. 1–2. Portland, OR: Western Forestry and Conservation Assn.

Allen, E. T. 1911b. Resume of forest fire legislation governing the North Pacific States. In Third annual session, Pacific Logging Congress: pp. 52–53. Portland, OR.

Allen, E. T. 1912. Burning slash is a question of increasing importance to loggers. In Fourth annual session, Pacific Logging Congress: pp. 39–40. Portland, OR.

Allen, E. T. 1925. A discussion of fires and logged off lands. In Sixteenth annual session, Pacific Logging Congress: pp. 26–27. Portland, OR.

Allen, M. H., R. W. Berry, D. Gill, G. L. Hayes, P. S. Truesdell, M. Zwolinski, and J. M. Pierovich. 1968. Guide to prescribed fire in the Southwest. Tucson, AZ: Southwest Interagency Fire Council.

Ames, F. E. 1912. Fire prevention and reforestation promoted by the burning of slash. In Fourth annual session, Pacific Logging Congress: p. 40. Portland, OR.

Anderson, H. E. 1982. Aids to determining fuel models for estimating fire behavior. USDA For. Serv. Gen. Tech. Rep. INT-122.

Anderson, H. W. 1954. Suspended sediment discharge as related to streamflow, topography, soil, and land use. Trans. Amer. Geophys. Union 35:268–81.

Anderson, H. W., M. D. Hoover, and K. G. Reinhart. 1976. Forest and water: Effects of forest management on floods, sedimentation, and water supply. USDA For. Serv. Gen. Tech. Rep. PSW-118.

Anderson, L., C. E. Carlson, and R. H. Wakimoto. 1987. Forest fire frequency and western spruce budworm outbreaks in western Montana. For. Ecol. and Manage. 22:251–60.

Andrews, H. J., and R. W. Cowlin. 1940. Forest resources of the Douglas-fir region. USDA Misc. Pub. 389.

Andrews, P. L. 1986. BEHAVE: Fire behavior prediction and fuel modeling system: BURN subsystem, part 1. USDA For. Serv. Gen. Tech. Rep. INT-194.

Andrews, P. L., and R. C. Rothermel. 1982. Charts for interpreting wildland fire behavior characteristics. USDA For. Serv. Gen. Tech. Rep. INT-131.

Anonymous. 1912. Supreme Court of Washington upholds amendment to forest fire act. Timberman 13(9):23.

Anonymous. 1983. The great fire of 500 years ago. Quinault Natural Resources 4(5):23.

Antos, J. A. 1977. Grand fir (*Abies grandis* [Dougl.] Forbes) forests of the Swan Valley, Montana. Thesis, University of Montana, Missoula.

Antos, J. A., and J. R. Habeck. 1981. Successional development in *Abies grandis* (Dougl.) Forbes forests in the Swan Valley, western Montana. Northwest Sci. 55:26–39.

Arend, J. S. 1941. Infiltration rates of forest soils in the Missouri Ozarks as affected by woods burning and litter removal. J. For. 39:726–28.

Arno, S. F. 1976. The historical role of fire on the Bitterroot National Forest. USDA For. Serv. Res. Pap. INT-187.

Arno, S. F. 1980. Forest fire history of the northern Rockies. J. For. 78:460–65.

Arno, S. F. 1986. Whitebark pine cone crops: A diminishing source of wildlife food. West. J. Appl. For. 1:92–94.

Arno, S. F., and D. H. Davis. 1980. Fire history of western redcedar/hemlock forests in northern Idaho. In Stokes, M. A., and J. H. Dieterich (eds.), Proceedings of the fire history workshop: pp. 21–26. USDA For. Serv. Gen. Tech. Rep. RM-81.

Arno, S. F., and J. R. Habeck. 1972. Ecology of alpine larch (*Larix lyallii* Parl.) in the Pacific Northwest. Ecol. Monogr. 42:417–50.

Arno, S. F., and R. P. Hammerly. 1984. Timberline: Mountain and Arctic forest frontiers. Seattle: The Mountaineers.

Arno, S. F., and R. J. Hoff. 1989. Silvics of whitebark pine *(Pinus albicaulis)*. USDA For. Serv. Gen. Tech. Rep. INT-253.

Arno, S. F., and T. D. Peterson. 1983. Variation in estimates of fire intervals: A closer look at fire history on the Bitterroot National Forest. USDA For. Serv. Res. Pap. INT-301.

Arno, S. F., D. G. Simmerman, and R. E. Keane. 1985. Forest succession on four habitat types in western Montana. USDA For. Serv. Gen. Tech. Rep. INT-177.

Arno, S. F., and K. M. Sneck. 1977. A method for determining fire history in coniferous forests of the Mountain West. USDA For. Serv. Gen. Tech. Rep. INT-42.

Atzet, T., and D. L. Wheeler. 1982. Historical and ecological perspectives on fire activity in the Klamath Geological Province of the Rogue River and Siskiyou National Forests. USDA For. Serv., Pacific Northwest Reg., pub. R-6-Range-102. Portland, OR.

Atzet, T., D. Wheeler, and R. Gripp. 1988. Fire and forestry in southwest Oregon. FIR Report 9(4):4–7.

Aufderheide, R., and W. G. Morris. 1948. Broadcast slash burning after a rain. West Coast Lumberman 75 (9):79–80, 82–83.

Austin, R. C., and D. H. Baisinger. 1955. Some effects of burning on forest soils of western Oregon and Washington. J. For. 53:275–80.

Axelrod, D. I. 1967. Evolution of the Californian closed-cone pine forest. In Philbrick, R. N. (ed.), Proceedings of the symposium on the biology of the California islands: pp. 93–149. Santa Barbara, CA: Santa Barbara Botanic Garden.

Baker, W. L. 1989. Effect of scale and spatial heterogeneity on fire-interval distributions. Can. J. For. Res. 19:700–706.

Bancroft, L., T. Nichols, D. Parsons, D. Graber, B. Evison, and J. W. van Wagtendonk. 1985. Evolution of the natural fire management program at Sequoia and Kings Canyon National Parks. In Lotan, J. E., et al. (tech. coords.), Proceedings: Symposium and workshop on wilderness fire: pp. 174–80. USDA For. Serv. Gen. Tech. Rep. INT-182.

Barbour, M. G., B. M. Pavlik, and J. A. Antos. 1990. Seedling growth and survival of red and white fir in a Sierra Nevada ecotone. Amer. J. Bot. 77:927–38.

Barbour, M. G., and R. A. Woodward. 1985. The Shasta red fir forest of California. Can. J. For. Res. 15:570–76.

Barnhardt, S. J., J. R. McBride, C. Cicero, P. da Silva, and P. Warner. 1987. Vegetation dynamics of the northern oak woodland. In Plumb, T. R., and N. H. Pillsbury (tech. coords.), Proceedings of the symposium on multiple use management of California's hardwood resources: pp. 53–58. USDA For. Serv. Gen. Tech. Rep. PSW-100.

Barnosky, C. W., P. M. Anderson, and P. J. Bartlein. 1989. The northwestern U.S. during deglaciation: Vegetational history and paleoclimatic implications. In Ruddiman, W. F., and H. E. Wright, Jr. (eds.), North American and adjacent oceans during the last deglaciation: pp. 289–321. Geology of North America, vol. K-3. Boulder: Geological Society of America.

Barrett, J. W. 1966. A record of ponderosa pine seed flight. USDA For. Serv. Res. Note PNW-38.

Barrett, J. W. 1973. Latest results from the Pringle Falls ponderosa pine spacing study. USDA For. Serv. Res. Note PNW-209.

Barrett, J. W. 1979. Silviculture of ponderosa pine in the Pacific Northwest: The state of our knowledge. USDA For. Serv. Gen. Tech. Rep. PNW-97.

Barrett, S., and S. Arno. 1982. Indian fires as an ecological influence in the northern Rockies. J. For. 80:647–51.

Barrett, S. W., and S. F. Arno. 1988. Increment-borer methods for determining fire history in coniferous forests. USDA For. Serv. Gen. Tech. Rep. INT-244.

Barrett, S. W., S. F. Arno, and C. H. Key. 1991. Fire regimes of western larch–lodgepole pine forests in Glacier National Park, Montana. Can. J. For. Res. 21:1711–20.

Baver, L. D. 1940. Soil physics. New York: John Wiley and Sons.

Beadle, N.C.W. 1940. Soil temperatures during forest fires and their effect on the survival of vegetation. J. Ecol. 28:180–92.

Beaton, J. D. 1959. The influence of burning on the soil in the timber range area of Lac Le Jeune, British Columbia, I: Physical properties. Can. J. Soil Sci. 39:1–5.

Beck, C. B., K. Coy, and R. Schmid. 1982. Observations on the fine structure of *Callixylon* wood. Amer. J. Bot. 69:54–76.

Bega, R. V. 1978. Diseases of Pacific Coast conifers. USDA Agr. Handbook 521.

Bendell, J. F. 1974. Effects of fire on birds and mammals. In Kozlowski, T. T., and C. E. Ahlgren (eds.), Fire and ecosystems: pp. 73–138. New York: Academic Press.

Berry, C. W. 1970. Enumeration and identification of the microbial populations from burned and unburned pine forest soil. M.S. thesis, Louisiana Technical University, Ruston.

Beschta, R. L. 1990. Effect of fire on water quantity and quality. In Walstad, J., et al. (eds.), Natural and prescribed fire in Pacific Northwest forests: pp. 219–32. Corvallis: Oregon State University Press.

Bethlalmy, N. 1974. Water supply as affected by micro- and macro-watershed management decisions on forest lands. Northwest Sci. 48: 1–8.

Bevins, C. D., and R. J. Barney. 1980. Lightning fire densities and their management implications on Northern Region national forests. Fire and Forest Meteorology Conf. 6:127–31.

Billings, W. D. 1969. Vegetational pattern near alpine timberline as affected by fire-snowdrift interactions. Vegetatio 19:191–207.

Bingham, B. B., and J. O. Sawyer. 1988. Volume and mass of decaying logs in an upland old-growth redwood forest. Can. J. For. Res. 18:1649–51.

Biswell, H. H. 1960. Danger of wildfire reduced by prescribed burning in ponderosa pine. Calif. Agr. 14(10):5–6.

Biswell, H. H. 1973. Fire ecology in ponderosa pine–grassland. Tall Timbers Fire Ecol. Conf. 12:69–96.

Biswell, H. H. 1974. The effects of fire on chaparral. In Kozlowski, T. T., and C. E. Ahlgren (eds.), Fire and ecosystems: pp. 321–64. New York: Academic Press.

Biswell, H. H. 1989. Prescribed burning in California wildlands vegetation management. Berkeley: University of California Press.

Biswell, H. H., H. R. Kallander, R. Komarek, R. J. Vogl, and H. Weaver. 1973. Ponderosa fire management. Tall Timbers Research Station Misc. Pub. 2. Tallahassee, FL.

Biswell, H. H., and P. C. Lemon. 1943. Effect of fire on seedstalk production of range grasses. J. For. 41:844.

Biswell, H. H., A. M. Schultz, and J. L. Launchbaugh. 1955. Brush control in ponderosa pine. Calif. Agr. 9(1):3–14.

Biswell, H. H., R. D. Taber, D. W. Hedrick, and A. M. Schultz. 1952. Management of chamise brushlands for game in the north coast region of California. Calif. Fish and Game 38:453–84.

Blaisdell, J. P. 1953. Ecological effects of planned burning on sagebrush-grass range on the upper Snake River plains. USDA Tech. Bull. 1075.

Blaisdell, J. P., R. B. Murray, and E. D. McArthur. 1982. Managing Intermountain rangelands—sagebrush-grass ranges. USDA For. Serv. Gen. Tech. Rep. INT-134.

Bockheim, J. G. 1972. Effects of alpine and subalpine vegetation on soil development, Mount Baker, Washington. Ph.D. diss., University of Washington, Seattle.

Boe, K. N. 1965. Windfall after experimental cuttings in old-growth redwood. Soc. Amer. For. Proc. 1965:59–63.

Boerker, R. H. 1912. Light burning versus forest management in northern California. For. Quarterly 10(2):184–94.

Bonnicksen, T. M. 1985. Ecological information base for park and wilderness fire management planning. In Lotan, J. E., et al. (tech. coords.), Proceedings: Symposium and workshop on wilderness fire: pp. 168–73. USDA For. Serv. Gen. Tech. Rep. INT-182.

Bonnicksen, T. M. 1989. Fire gods and federal policy. Amer. For. (July–August): 14–16, 66–68.

Bonnicksen, T. M., and E. C. Stone. 1981. The giant sequoia–mixed conifer forest community characterized through pattern analysis as a mosaic of aggregations. For. Ecol. and Manage. 3:307–28.

Bonnicksen, T. M., and E. C. Stone. 1982. Managing vegetation within U.S. national parks: A policy analysis. Environ. Manage. 6:101–2, 109–22.

Borchers, J. G., and D. A. Perry. 1990. Effects of prescribed fire on soil organisms. In Walstad, J., et al. (eds.), Natural and prescribed fire in the Pacific Northwest: pp. 143–57. Corvallis: Oregon State University Press.

Bork, J. 1985. Fire history in three vegetation types on the east side of the Oregon Cascades. Ph.D. diss., Oregon State University, Corvallis.

Bossard, C. C. 1990. Secrets of an ecological interloper: Ecological studies on *Cytisus scoparius* (scotch broom) in California. Ph.D. diss., University of California, Davis.

Botkin, D. B., J. F. Janak, and J. R. Wallis. 1972. Some ecological consequences of a computer model of forest growth. J. Ecol. 60:849–72.

Boyce, J. S. 1929. Deterioration of wind-thrown timber on the Olympic Peninsula, Washington. USDA Tech. Bull. 104.

Boyd, R. 1986. Strategies of Indian burning in the Willamette Valley. Can. J. Anthropology 5:65–86.

Braun-Blanquet, J., and E. Furrer. 1913. Sur l'étude des associations. Bull. de la Société Languedocienne de Géographie 36:1–22.

Bridge, J. L. 1912. Washington Forest Fire Association advocates fall burning of slash. In Fourth annual session, Pacific Logging Congress: p. 40. Portland, OR.

Britton, C. M., and R. G. Clark. 1985. Effects of fire on sagebrush and bitterbrush. In Sanders, K., et al. (eds.), Rangeland fire effects: pp. 22–26. Boise, ID: USDI Bureau of Land Management.

Britton, C. M., R. G. Clark, and F. A. Sneva. 1983. Effects of soil moisture on burned and clipped Idaho fescue. J. Range Manage. 36:708–10.

Brockley, R. P., and E. Elmes. 1987. Barking damage by red squirrels in juvenile-spaced lodgepole pine stands in south-central British Columbia. For. Chron. 63:28–31.

Brown, A. I., and S. M. Marco. 1958. Introduction to heat transfer. New York: McGraw-Hill.

Brown, J. K., and S. F. Arno. 1991. The paradox of wildland fire. Western Wildlands (Spring): 40–46.

Brubaker, L. B. 1986. Responses of tree populations to climatic change. Vegetatio 67:119–30.

Brubaker, L. B. 1991. Climate change and the origin of old-growth Douglas-fir forests in the Puget Sound lowland. In Ruggiero, L., et al. (eds.), Wildlife and vegetation of unmanaged Douglas-fir forests: pp. 17–24. USDA For. Serv. Gen. Tech. Rep. PNW-GTR-285.

Bruce, D. 1923. Light burning: Report of the California Forestry Committee. J. For. 21:129–30.

Buckley, R. 1983. The role of fire in maintaining a small scale vegetation gradient: A test of Mutch's hypothesis. Oikos 41:291–92.

Bunting, S. C. 1985. Fire in sagebrush-grass ecosystems: Successional changes. In Sanders, K., et al. (eds.), Rangeland fire effects: pp. 7–11. Boise: USDI Bureau of Land Management.

Burgan R. E. 1979. Fire danger/fire behavior computations with the Texas Instruments TI-59 calculator: A user's manual. USDA For. Serv. Gen. Tech. Rep. INT-61.

Burgan, R. E., and R. C. Rothermel. 1984. BEHAVE: Fire behavior prediction and fuel modeling subsystem: Fuel subsystem. USDA For. Serv. Gen. Tech. Rep. INT-167.

Burke, C. J. 1979. Historic fires in the central western Cascades, Oregon. M.S. thesis, College of Forestry, Oregon State University, Corvallis.

Burkhardt, J. W., and E. W. Tisdale. 1969. Nature and successional status of western juniper vegetation in Idaho. J. Range Manage. 22:264–70.

Burkhardt, J. W., and E. W. Tisdale. 1976. Causes of juniper invasion in southwestern Idaho. Ecology 57:472–84.

Burns, P. Y. 1952. Effects of fire on forest soils in the Pine Barren region of New Jersey. Yale University School of Forestry Bull. 57.

Byram, G. M. 1959. Combustion of forest fuels. In Davis, K. P., Forest fire: Control and use: pp. 61–89. New York: McGraw-Hill.

Cain, M. D. 1984. Height of stem-bark char underestimates flame length in prescribed burns. Fire Manage. Notes 45(1):17–21.

Chadwick, H. W., and P. D. Dalke. 1965. Plant succession on dune stands in Fremont County, Idaho. Ecology 46:765–80.

Chandler, C., P. Cheney, P. Thomas, L. Trabaud, and D. Williams. 1983. Fire in forestry, vols. 1 and 2. New York: John Wiley and Sons.

Chappell, C. B. 1991. Fire ecology and seedling establishment in Shasta red fir (*Abies magnifica* var. *shastensis*) forests of Crater Lake National Park, Oregon. M.S. thesis, University of Washington, Seattle.

Chase, C. H. 1981. Spot fire distance equations for pocket calculators. USDA For. Serv. Res. Note INT-310.

Chen, J. 1991. Edge effects: Microclimatic pattern and biological response in old-growth Douglas-fir forests. Ph.D. diss., University of Washington, Seattle.

Chew, R. M., B. B. Butterworth, and R. Grechman. 1958. The effects of fire on the small mammal populations of chaparral. J. Mammalogy 40:253.

Christensen, N. L. 1985. Shrubland fire regimes and their evolutionary consequences. In S.T.A. Pickett and P. White (eds.), The ecology of natural disturbance and patch dynamics: pp. 85–100. New York: Academic Press.

Christensen, N. L. 1988. Succession and natural disturbance: Paradigms, problems, and preservation of natural ecosystems. In Agee, J. K., and D. R. Johnson (eds.), Ecosystem management for parks and wilderness: pp. 62–86. Seattle: University of Washington Press.

Christensen, N. L., J. K. Agee, P. F. Brussard, J. Hughes, D. H. Knight, G. W. Minshall, J. M. Peek, S. J. Pyne, F. J. Swanson, J. W. Thomas, S. Wells, S. E. Williams, and H. A. Wright. 1989. Interpreting the Yellowstone fires. Bioscience 39:678–85.

Christensen, N. L., and C. H. Muller. 1975. Effects of fire on factors controlling plant growth in *Adenostoma* chaparral. Ecol. Monogr. 45:29–55.

Clagg, H. B. 1975. Fire ecology in high-elevation forests in Colorado. M.S. thesis, Colorado State University, Fort Collins.

Clar, C. R. 1959. California government and forestry. Sacramento: Division of Forestry, Dept. of Natural Resources.

Clar, C. R. 1969a. Evolution of California's wildland fire protection system. Sacramento, CA: Division of Forestry, Department of Conservation.

Clar, C. R. 1969b. California government and forestry II: During the Young and Rolph administrations. Sacramento, CA: Division of Forestry, Department of Conservation.

Clark, J. S. 1988. Effect of climate change on fire regimes in northwestern Minnesota. Nature 334:233–35.

Clark, J. S. 1990. Fire and climate change during the last 750 years in northwestern Minnesota. Ecol. Monogr. 60:135–59.

Clark, P. J., and F. C. Evans. 1954. Distance to nearest neighbor as a measure of spatial relationships in populations. Ecology 35:445–53.

Clark, R. G., C. M. Britton, and F. A. Sneva. 1982. Mortality of bitterbrush after burning and clipping in eastern Oregon. J. Range Manage. 35:711–14.

Clark, R. G., and E. E. Starkey. 1990. Use of prescribed fire in rangeland ecosystems. In Walstad, J., et al. (eds.), Natural and prescribed fire in Pacific Northwest forests: pp. 81–91. Corvallis: Oregon State University Press.

Clark, R. G., H. A. Wright, and F. H. Roberts. 1985. Threshold requirements

for fire spread in grassland fuels. In Sanders, K., et al. (eds.), Rangeland fire effects: pp. 27–32. Boise: USDI Bureau of Land Management.
Clements, F. E. 1910. The life history of lodgepole burn forests. USDA For. Serv. Bull. 79.
Clements, F. E. 1916. Plant succession: An analysis of the development of vegetation. Carnegie Inst. Pub. 242. Washington, DC.
Clements, F. E. 1935. Experimental ecology in the public service. Ecology 16:342–63.
Clements, F. E. 1936. Nature and structure of the climax. J. Ecol. 24:252–84.
Clevinger, W. R. 1951. S.W. Washington's legendary "Big Fire." Seattle Times Magazine (October 28): 2.
Cobb, D. F. 1988. Development of mixed western larch, lodgepole pine, Douglas-fir, and grand fir stands in eastern Washington. M.S. thesis, University of Washington, Seattle.
Coblentz, B. E. 1980. Production of Oregon white oak acorns in the Willamette Valley. Wildl. Soc. Bull. 8(4):348–50.
Cochran, P. H. 1970. Seeding ponderosa pine. In R. K. Hermann (ed.), Regeneration of ponderosa pine: pp. 28–35. Oregon State University School of Forestry Pap. 681. Corvallis, OR.
Cochran, P. H., and C. M. Bernsten. 1973. Tolerance of lodgepole and ponderosa pine seedlings to low night temperatures. For. Sci. 19:272–80.
Cochran, P. H., L. Boersma, and C. T. Youngberg. 1967. Thermal properties of a pumice soil. Soil Sci. Soc. Amer. Proc. 31:454–59.
Cochran, P. H., and W. E. Hopkins. 1991. Does fire exclusion increase productivity of ponderosa pine? In Harvey, A. E., and L. F. Neuenschwander (comps.), Proceedings: Management and productivity of western-montane forest soils: pp. 224–28. USDA For. Serv. Gen. Tech. Rep. INT-280.
Cole, W. E., and G. D. Amman. 1980. Mountain pine beetle dynamics in lodgepole pine forests, part I: Course of an infestation. USDA For. Serv. Gen. Tech. Rep. INT-89.
Connell, J. H., and R. O. Slayter. 1977. Mechanisms of succession in natural communities and their role in community stability and organization. Amer. Naturalist 111:1119–44.
Connor, M. A., and K. P. Cannon. 1989. The mountains burnt: Forest fires and site formation processes. N. Amer. Archeologist 10:293–310.
Conrad, C. E., and C. E. Poulton. 1966. Effects of a wildfire on Idaho fescue and bluebunch wheatgrass. J. Range Manage. 19:138–41.
Cooper, C. F. 1960. Changes in vegetation, structure, and growth of southwestern pine forests since white settlement. Ecol. Monogr. 30:129–64.

Cooper, C. F. 1961a. Pattern in ponderosa pine forests. Ecology 42:493–99.
Cooper, C. F. 1961b. The ecology of fire. Sci. Amer. 204(4):150–60.
Cooper, D. W. 1965. The coast redwood (*Sequoia sempervirens*) and its ecology. Calif. Agr. Ext. Serv., Humboldt Co.
Cooper, S. V., K. E. Neiman, and D. W. Roberts. 1991. Forest habitat types of northern Idaho: A second approximation. USDA For. Serv. Gen. Tech. Rep. INT-236.
Cooper, W. S. 1926. The nature of vegetation change. Ecology 7:391–413.
Cope, M. J., and W. G. Chaloner. 1985. Wildfire: An interaction of biological and physical processes. In Tiffney, B. H. (ed.), Geological factors and the evolution of plants: pp. 257–77. New Haven: Yale University Press.
Countryman, C. M., and D. R. Cornelius. 1957. Some effects of fire on a perennial range type. J. Range Manage. 10:39–41.
Covington, W. W., and S. S. Sackett. 1984. The effect of a prescribed burn in southwestern ponderosa pine on organic matter and nutrients in woody debris and forest floor. For. Sci. 30:183–92.
Cowan, C. S. 1961. The enemy is fire. Seattle: Superior Publishing Co.
Cowlin, R. W., P. A. Briegleb, and F. L. Moravets. 1942. Forest resources of the ponderosa pine region. USDA Misc. Pub. 490.
Cramer, O. P. (ed.). 1974. Environmental effects of forest residues management in the Pacific Northwest: A state-of-knowledge compendium. USDA For. Serv. Gen. Tech. Rep. PNW-24.
Cramer, O. P., and H. E. Graham. 1971. Cooperative management of smoke from forest fires. J. For. 69:327–31.
Crandell, D. R. 1971. Postglacial lahars from Mount Rainier volcano, Washington. U.S. Geol. Survey Prof. Pap. 677.
Crane, M. F., J. R. Habeck, and W. C. Fischer. 1983. Early postfire revegetation in a western Montana Douglas-fir forest. USDA For. Serv. Res. Pap. INT-319.
Crane, W.J.B. 1972. Urea-nitrogen transformations, soil reactions, and elemental movement via leaching and volatilization, in a coniferous forest ecosystem following fertilization. Ph.D. diss., University of Washington, Seattle.
Cremer, K. W., and A. B. Mount. 1965. Early stages of plant succession following the complete felling and burning of *Eucalyptus regnans* forest in the Florentine Valley, Tasmania. Austral. J. Bot. 13:303–22.
Crocker, R. L., and J. Major. 1955. Soil development in relation to vegetation and surface age of Glacier Bay, Alaska. J. Ecol. 43:427–48.
Cwynar, L. C. 1987. Fire and the forest history of the North Cascade range. Ecology 68:791–802.
Dale, V., and M. Hemstrom. 1984. CLIMACS: A computer model of forest

stand development for western Oregon and Washington. USDA For. Serv. Res. Pap. PNW-327.

Dale, V., M. Hemstrom, and J. F. Franklin. 1986. Modeling the long-term effects of disturbances on forest succession, Olympic Peninsula, Washington. Can. J. For. Res. 16:56–67.

Daniel, T. W. 1942. The comparative transpiration rates of several western conifers under controlled conditions. Ph.D. diss., University of California, Berkeley.

Daniel, T. W., and J. Schmidt. 1972. Lethal and non-lethal effects of the organic horizons of forest soils on germination of seeds from several associated conifer species of the Rocky Mountains. Can. J. For. Res. 2:179–84.

Daubenmire, R. 1947 and 1959. Plants and environment. 1st and 2nd eds. New York: John Wiley and Sons.

Daubenmire, R. 1952. Forest vegetation of northern Idaho and adjacent Washington, and its bearing on the concepts of vegetation classification. Ecol. Monogr. 22:301–30.

Daubenmire, R. 1968a. Ecology of fire in grasslands. Adv. Ecol. Res. 5:209–66.

Daubenmire, R. 1968b. Plant communities. New York: Harper and Row.

Daubenmire, R. 1970. Steppe vegetation of Washington. Washington State University Agr. Exp. Sta. Tech. Bull. 62. Pullman, WA.

Daubenmire, R., and J. Daubenmire. 1968. Forest vegetation of eastern Washington and northern Idaho. Washington State University Agr. Exp. Sta. Tech. Bull. 60. Pullman, WA.

Davidson, J.G.N. 1971. Pathological problems in redwood regeneration from seed. Ph.D. diss., University of California, Berkeley.

Davies, J. 1980. Douglas of the forests. Seattle: University of Washington Press. Davis, J. B. 1984. Burning another empire. Fire Manage. Notes 45(4):12–17.

Davis, K. M., B. D. Clayton, and W. C. Fischer. 1980. Fire ecology of Lolo National Forest habitat types. USDA For. Serv. Gen. Tech. Rep. INT-79.

Davis, K. P. 1959. Forest fire: Control and use. New York: McGraw-Hill.

Davis, L., S. Aaberg, and A. Johnson. 1992. Archeological fieldwork at Yellowstone's Obsidian Cliff. Park Sci. 12(2):26–27.

Davis, M. B. 1981. Quaternary history and the stability of forest communities. In West, D. C., et al. (eds.), Forest succession: Concepts and application: pp. 132–53. New York: Springer-Verlag.

Day, R. J. 1963. Spruce seedling mortality caused by adverse summer microclimate in the Rocky Mountains. Can. Dept. For. Resour. Branch Pub. 1003. Ottawa, Ont.

Day, R. J. 1964. The microenvironments occupied by spruce and fir regeneration in the Rocky Mountains. Can. Dept. For. Resour. Branch Pub. 1037. Ottawa, Ont.

Dealy, J. E., J. M. Geist, and R. S. Driscoll. 1978. Communities of western juniper in the Intermountain Northwest. In Martin, R. E., et al. (eds.), Proceedings of the western juniper ecology and management workshop: pp. 11–29. USDA For. Serv. Gen. Tech. Rep. PNW-74.

DeBano, L. F., and C. E. Conrad. 1978. The effect of fire on nutrients in a chaparral ecosystem. Ecology 59:489–97.

DeBano, L. F., and J. S. Krammes. 1966. Water repellent soils and their relation to wildfire temperatures. Bull. IASH 11(2):14–19.

DeBano, L. F., L. D. Mann, and D. A. Hamilton. 1970. Translocation of hydrophobic substances into soil by burning organic matter. Soil Sci. Soc. Amer. Proc. 34:130–33.

DeBano, L. F., R. M. Rice, and C. E. Conrad. 1979. Soil heating in chaparral fires: Effects on soil properties, plant nutrients, erosion, and runoff. USDA For. Serv. Res. Pap. PSW-145.

DeBano, L. F., S. M. Savage, and D. A. Hamilton. 1976. The transfer of heat and hydrophobic substances during burning. Soil Sci. Soc. Amer. Proc. 40:779–82.

DeBell, D. S., and C. W. Ralston. 1970. Release of nitrogen by burning light forest fuels. Soil Sci. Soc. Amer. Proc. 34:936–38.

DeBenedetti, S. H., and D. J. Parsons. 1979. Natural fire in subalpine meadows: A case description from the Sierra Nevada. J. For. 77:477–79.

DeByle, N. V. 1973. Broadcast burning of logging residues and the water repellency of soils. Northwest Sci. 47:77–87.

DeByle, N. V., and P. E. Packer. 1972. Plant nutrient and soil losses in overland flow from burned soil clearcuts. In Proceedings: Symposium on watersheds in transition: pp. 296–307. Fort Collins, CO: Amer. Water Resour. Assn. and Colorado State University.

Dell, J. D. 1969. Lengthening the slash burning season in the Douglas-fir region. Northwest Forest Fire Council 1969: 52–58.

Dell, J. D. 1976. Fuels and fire management. . . . Fire use on the national forests in the Pacific Northwest Region. Tall Timbers Fire Ecology Conf. 15:119–28.

Department of Natural Resources. 1992. Washington smoke management program annual report: 1991. Division of Fire Control, Washington Department of Natural Resources, Olympia.

Despain, D. G. 1983. Nonpyrogenous climax lodgepole pine communities in Yellowstone National Park. Ecology 64:231–34.

Despain, D. G., and R. E. Sellers. 1977. Natural fire in Yellowstone National Park. West. Wildlands 4:20–24.

Dickman, A., and S. Cook. 1989. Fire and fungus in a mountain hemlock forest. Can. J. Bot. 67:2005–16.

Dieterich, J. H. 1979. Recovery potential of fire-damaged southwestern ponderosa pine. USDA For. Serv. Res. Note RM-379.

Dieterich, J. H. 1980a. Chimney Springs forest fire history. USDA Gen. Tech. Rep. RM-220.

Dieterich, J. H. 1980b. The composite fire interval: A tool for more accurate interpretation of fire history. In Stokes, M. A., and J. H. Dieterich (tech. coords.), Proceedings of the fire history workshop: pp. 8–14. USDA For. Serv. Gen. Tech. Rep. RM-81.

Dieterich, J. H., and T. W. Swetnam. 1984. Dendrochronology of a fire-scarred ponderosa pine. For. Sci. 30:238–47.

Doescher, P. S., L. E. Eddleman, and M. R. Vaitkus. 1987. Evaluation of soil nutrients, pH, and organic matter in rangelands dominated by western juniper. Northwest Sci. 61:97–102.

Donaghey, J. 1969. The properties of heated soils and their relation to giant sequoia (*Sequoiadendron giganteum*) germination and growth. M.A. thesis, San Jose State University, San Jose, CA.

Douglas, G. W., and T. M. Ballard. 1971. Effect of fire on alpine plant communities in the North Cascades, Washington. Ecology 52:1058–64.

Driscoll, R. S. 1964a. A relict area in central Oregon juniper zone. Ecology 45:345–53.

Driscoll, R. S. 1964b. Vegetation-soil units in the central Oregon juniper zone. USDA For. Serv. Res. Pap. PNW-19.

Driver, C. H. 1982. Potentials for the management of bitterbrush habitats by the use of prescribed fire. In Proceedings: Research and management of bitterbrush and cliffrose in western North America: pp. 137–41. USDA For. Serv. Gen. Tech. Rep. INT-152.

Drucker, P. 1937. The Tolowa and their southwest Oregon kin. University of California Pub. Amer. Archeol. and Ethnol. 36:221–300.

Drury, W. H., and I.C.T. Nisbet. 1973. Succession. J. Arnold Arboretum 54:331–68.

Dunn, P. H., and L. F. DeBano. 1977. Fire's effect on the biological properties of chaparral soils. In Mooney, H., et al. (tech coords.), International symposium on the environmental consequences of fire and fuel management in Mediterranean-climate ecosystems: pp. 75–85. USDA For. Serv. Gen. Tech. Rep. WO-3.

Dunwiddie, P. W. 1986. A 6,000-year record of forest history on Mount Rainier, Washington. Ecology 67:58–68.

Dutton, C. E. 1881. Physical geology of the Grand Cañon district. In USDI Geol. Survey, Second annual report: pp. 49–166. Washington, DC: Government Printing Office.

Dyrness, C. T. 1976. Effects of fire on soil wettability in the High Cascades of Oregon. USDA For. Serv. Res. Pap. PNW-202.

Dyrness, C. T., and C. T. Youngberg. 1957. The effect of logging and slash burning on soil structure. Soil Sci. Soc. Amer. Proc. 21:444–47.

Eddleman, L. E. 1987. Establishment and stand development of western juniper in central Oregon. In Everett, R. L. (comp.), Proceedings: Pinyon-juniper conference: pp. 255–59. USDA For. Serv. Gen. Tech. Rep. INT-215.

Eggers, D. E. 1986. Management of whitebark pine as potential grizzly bear habitat. In Contreras, G. P., and K. E. Evans (comp.), Proceedings: Grizzly bear habitat symposium: pp. 170–75. USDA For. Serv. Gen. Tech. Rep. INT-207.

Egler, F. E. 1954. Vegetation science concepts, I: Initial floristic composition—a factor in old field vegetation development. Vegetatio 4:412–17.

Elliott, F. A. 1911. Mr. Elliott's address. In Western Forestry and Conservation Assn. 1911 proceedings: pp. 9–10. Portland, OR: Western Forestry and Conservation Assn.

Emmingham, W., and N. Halverson. 1981. Community types, productivity, and reforestation: Management implications for the Pacific silver fir zone of the Cascade Mountains. In Oliver, C. D., and R. M. Kenady (eds.), Proceedings of the biology and management of true fir in the Pacific Northwest: pp. 291–303. Inst. For. Resour. Contr. 45. University of Washington, Seattle.

Evans, J. W. 1991. Powerful Rockey: The Blue Mountains and the Oregon Trail, 1811–83. LaGrande: Eastern Oregon State College.

Everett, R. L. 1987. Plant response to fire in the pinyon-juniper zone. In Everett, R. L. (comp.), Proceedings: Pinyon-juniper conference: pp. 152–57. USDA For. Serv. Gen. Tech. Rep. INT-215.

Everitt, B. L. 1968. Use of the cottonwood in an investigation of the recent history of a flood plain. Amer. J. Sci. 266:417–39.

Fahnestock, G. R. 1976. Fires, fuel, and flora as factors in wilderness management: The Pasayten case. Tall Timbers Fire Ecology Conf. 15:33–70.

Fahnestock, G. R., and J. K. Agee. 1983. Biomass consumption and smoke production by prehistoric and modern forest fires in western Washington. J. For. 81:653–57.

Fahnestock, G. R., and R. C. Hare. 1964. Heating of tree trunks in surface fires. J. For. 62:799–805.

Faurot, J. L. 1977. Estimating merchantable volume and stem residue in four timber species: Ponderosa pine, lodgepole pine, western larch, Douglas-fir. USDA For. Serv. Res. Pap. INT-196.

Ferry, G., R. Mutch, F. Abeita, H. Ryan, L. Bancroft, J. W. van Wagtendonk, W. L. McCleese, R. Wilson, H. N. Miller, and F. Zontek. 1985. Prescribed fire smoke management guide. National Wildfire Coord. Group NFES 1279. Boise Interagency Fire Center, Boise, ID.

Filip, G. M., L. D. Bryant, and C. A. Parks. 1989. Mass movement of river ice causes severe tree wounds along the Grande Ronde River in northeastern Oregon. Northwest Sci. 63:211–13.

Finch, R. B. 1984. Fire history of selected sites on the Okanogan National Forest. USDA For. Serv., Pacific Northwest Reg., Okanogan National Forest, Okanogan, WA.

Finch, T. L. 1948. Effect of bark growth in measurement of periodic growth of individual trees. USDA For. Serv. Res. Note NRM-60.

Finney, M. A. 1986. Effects of low intensity fire on the successional development of seral lodgepole pine forests in the North Cascades. M.S. thesis, University of Washington, Seattle.

Finney, M. A. 1991. Ecological effects of prescribed and simulated fire on the coast redwood (*Sequoia sempervirens* [D. Don. Endl.]). Ph.D. diss., University of California, Berkeley.

Finney, M. A., and R. E. Martin. 1989. Fire history in a *Sequoia sempervirens* forest at Salt Point State Park, California. Can. J. For. Res. 19:1451–57.

Finnis, J. M. 1970. "Brown and burn" as a means of site preparation. Western Soc. Weed Sci. Proc. 23:47.

Fischer, W. C. 1980. Prescribed fire and bark beetle attack in ponderosa pine forests. Fire Manage. Notes 41 (Spring): 10–12.

Fischer, W. C., and B. D. Clayton. 1983. Fire ecology of Montana forest habitat types east of the Continental Divide. USDA For. Serv. Gen. Tech. Rep. INT-141.

Flannigan, M. D., and C. E. Van Wagner. 1991. Climate change and wildfire in Canada. Can. J. For. Res. 21:66–72.

Flinn, M., and R. Wein. 1977. Depth of underground plant organs and theoretical survival during fire. Can. J. Bot. 55:2250–54.

Flint, H. R. 1925. Fire resistance of northern Rocky Mountain conifers. Idaho Forester 7(7):7–10, 41–43.

Fonda, R. W. 1974. Forest succession in relation to river terrace development in Olympic National Park, Washington. Ecology 55:927–42.

Fonda, R. W. 1979. Fire-resilient forests of Douglas-fir in Olympic National Park: An hypothesis. In Linn, R. (ed.), Proceedings: First conference on scientific research in national parks: pp. 1239–42. USDI Nat. Park Serv. Trans. Proc. Ser. 5.

Fonda, R. W., and L. C. Bliss. 1969. Forest vegetation of the montane and subalpine zones, Olympic Mountains, Washington. Ecol. Monogr. 39:271–301.

Fosberg, M. A., and J. Deeming. 1971. Derivation of the 1- and 10-hour timelag fuel moisture calculations for fire-danger rating. USDA For. Serv. Res. Note RM-207.

Fowells, H. A. (comp.). 1965. Silvics of forest trees of the United States. USDA Agr. Handbook 271.

Fowells, H. A., and R. E. Stephenson. 1934. Effect of burning on forest soils. Soil Sci. 38:175–81.

Fowler, P. M., and D. O. Asleson. 1984. The location of lightning-caused wildland fires, northern Idaho. Phys. Geog. 5:240–52.

Fowler, W. B., and J. D. Helvey. 1978. Changes in the thermal regime after prescribed burning and select tree removal (Grass Camp, 1975). USDA For. Serv. Res. Pap. PNW-234.

Franklin, J. F. 1992. Scientific basis for new perspectives in forests and streams. In Naiman, R. J. (ed.), Watershed management: Balancing sustainability and environmental change: pp. 25–72. New York: Springer-Verlag.

Franklin, J. F., and C. T. Dyrness. 1973. Natural vegetation of Oregon and Washington. USDA For. Serv. Gen. Tech. Rep. PNW-8.

Franklin, J. F., and M. A. Hemstrom. 1981. Aspects of succession in the coniferous forests of the Pacific Northwest. In West, D. C., et al. (eds.), Forest succession: pp. 222–29. New York: Springer-Verlag.

Franklin, J. F., A. McKee, F. J. Swanson, J. Means, and A. Brown. 1979. Age structure analysis of old-growth Douglas-fir stands: Data versus conventional wisdom. Bull. Ecol. Soc. Amer. 60:102.

Franklin, J. F., W. H. Moir, G. W. Douglas, and C. Wiberg. 1971. Invasion of subalpine meadows by trees in the Cascade Range, Washington and Oregon. Arctic and Alp. Res. 3:215–24.

Franklin, J. F., W. H. Moir, M. A. Hemstrom, S. E. Greene, and B. G. Smith. 1988. The forest communities of Mount Rainier National Park. USDI Nat. Park Serv. Monogr. Ser. 19.

Franklin, J. F., F. C. Sorensen, and R. K. Campbell. 1978. Summarization of the ecology and genetics of the noble and California red fir complex. In Proceedings: IUFRO joint meeting work party, vol. 1: pp. 133–39. Victoria, B.C.: British Columbia Ministry of Forestry, Info. Serv. Branch.

Franklin, J. F., F. J. Swanson, M. E. Harmon, D. A. Perry, T. A. Spies, V. H. Dale, A. McKee, W. K. Ferrell, J. E. Means, S. V. Gregory, J. D. Lattin, T. D. Schowalter, and D. Larsen. 1991. Effects of global climatic change on forests in northwestern North America. Northwest Envir. J. 7:233–54.

Franklin, J. F., and R. Waring. 1979. Evergreen coniferous forests of the Pacific Northwest. Science 204:1380–86.

Fritts, H. C. 1976. Tree rings and climate. New York: Academic Press.

Fritz, E. 1929. Some popular fallacies concerning California redwood. Madrono 1:221–23.

Fritz, E. 1932. The role of fire in the redwood region. University of California Agr. Exp. Sta. Circ. 323.

Fritz, E. 1951. Bear and squirrel damage to young redwood. J. For. 49:651–52.

Froehlich, R. C., C. S. Hodges, Jr., and S. S. Sackett. 1978. Prescribed burning reduces severity of annosus root rot in the South. For. Sci. 24:93–100.

Fujimora, T. 1971. Primary productivity of a young *Tsuga heterophylla* stand and some speculations about biomass of forest commmunities on the Oregon coast. USDA For. Serv. Res. Pap. PNW-123.

Fujimora, T. 1977. Stem biomass and structure of a mature *Sequoia sempervirens* stand of the Pacific coast of northern California. J. Jap. For. Soc. 59:435–41.

Fuquay, D., A. R. Taylor, R. G. Hawk, and C. W. Schmid, Jr. 1972. Lighting discharges that caused forest fires. J. Geophys. Res. 77:2156–58.

Furniss, R. L., and V. M. Carolin. 1977. Western forest insects. USDA Misc. Pub. 1339.

Gabriel, H. W., III. 1976. Wilderness ecology: The Danaher Creek drainage, Bob Marshall Wilderness, Montana. Ph.D. Diss., University of Montana, Missoula, MT.

Galbraith, W. A., and E. W. Anderson. 1991. Grazing history of the Northwest. Rangelands 13:213–18.

Gara, R. I., J. K. Agee, W. R. Littke, and D. R. Geiszler. 1986. Fire wounds and beetle scars: Distinguishing between the two can help reconstruct past disturbances. J. For. 84:47–50.

Gara, R. I., D. R. Geiszler, and W. R. Littke. 1984. Primary attraction of the mountain pine beetle to lodgepole pine in Oregon. Ann. Entomol. Soc. Amer. 77:333–34.

Gara, R. I., W. R. Littke, J. K. Agee, D. R. Geiszler, J. D. Stuart, and C. H. Driver. 1985. Influence of fires, fungi, and mountain pine beetles on development of a lodgepole pine forest in south-central Oregon. In Baumgartner, D. M. et al. (eds.), Lodgepole pine: The species and its management: pp. 153–62. Pullman: Washington State University.

Gast, W. R., Jr., D. W. Scott, C. Schmitt, D. Clemens, S. Howes, C. G. Johnson, R. Mason, F. Mohr, and R. A. Clapp, Jr. 1991. Blue Mountains forest health report: New perspectives in forest health. USDA For. Serv., Pacific Northwest Region, Malheur, Umatilla, and Wallowa-Whitman National Forests.

Geiszler, D. R., and R. I. Gara. 1978. Mountain pine beetle attack dynamics in lodgepole pine. In Baumgartner, D. M. (ed.), Theory and practice of mountain pine beetle management in lodgepole pine forests: pp. 182–87. Pullman: Washington State University Coop. Ext. Serv.

Geiszler, D. R., R. I. Gara, C. H. Driver, V. F. Gallucci, and R. E. Martin. 1980. Fire, fungi, and beetle influences on a lodgepole pine ecosystem of south-central Oregon. Oecologia 46:239–43.

Gill, A. M. 1974. Towards an understanding of fire-scar formation: Field observation and laboratory simulation. For. Sci. 20:198–205.

Gill, A. M. 1980. Adaptive responses of Australian vascular plant species to fires. In Gill, A. M., et al. (eds.), Fire and the Australian biota: pp. 243–71. Canberra: Australian Academy of Science.

Gill, A. M., and D. H. Ashton. 1968. The role of bark type in relative tolerance to fire of three eucalypt species. Austral. J. Bot. 16:491–98.

Gill, J. D., and F. L. Pogge. 1974. *Cytisus scoparius,* scotch broom. In Schopmeyer, C. S. (ed.), Seeds of woody plants in the United States: pp. 370–71. USDA Handbook 450.

Gilmour, D. A., and N. P. Cheney. 1968. Experimental prescribed burn on radiata pine. Austral. For. 32:171–78.

Gleason, H. A. 1917. The structure and development of the plant association. Bull. Torrey Bot. Club 43:463–81.

Glickstein, D. 1987. Twain's great Northwest tour. Washington Mag. 2(2):45–47.

Gordon, D. T. 1970. Natural regeneration of white and red fir . . . influence of several factors. USDA For. Serv. Res. Pap. PSW-258.

Gordon, D. T. 1973. Damage from wind and other causes in mixed white fir–red fir stands adjacent to clearcuttings. USDA For. Serv. Res. Pap. PSW-90.

Gordon, D. T. 1978. White and red fir cone production in northeastern California: Report of a 16-year study. USDA For. Serv. Res. Note PSW-330.

Gorman, M. W. 1899. Eastern part of Washington Forest Reserve. In USDI Geol. Survey, Nineteenth annual report, part V: Forest reserves: pp. 315–50. Washington, DC: Government Printing Office.

Gracz, M. 1989. Reestablishment of *Abies lasiocarpa* after fire in the Pacific Northwest. M.S. thesis, University of Washington, Seattle.

Graham, R. L., and K. Cromack. 1982. Mass, nutrient content, and decay rate of dead boles in rain forests of Olympic National Park. Can. J. For. Res. 12:511–21.

Gratkowski, H. J. 1956. Windthrow around staggered settings in old growth Douglas-fir. For. Sci. 2:60–74.

Graumlich, L. 1985. Long-term records of temperature and precipitation in the Pacific Northwest derived from tree rings. Ph.D. diss., University of Washington, Seattle.

Graves, H. T. 1920a. Graves terms light burning "Piute forestry." Timberman 21(3):35.

Graves, H. T. 1920b. The torch in the timber. Sunset 44(4):37–40, 80–82.

Gray, A. 1990. Forest structure on the Siouxon burn, southern Washington Cascades: Comparison of single and multiple burns. M.S. thesis, University of Washington, Seattle.

Greene, S. E. 1982. Neskowin Crest Research Natural Area. Suppl. 13 to Franklin, J. F., et al. (1972), Federal Research Natural Areas in Oregon and Washington: A guidebook for scientists and educators. USDA For. Serv., Pacific Northwest For. and Range Exp. Sta, Portland, OR.

Greene, S. E., and W. H. Emmingham. 1986. Early lessons from commercial thinning in a 30-year-old Sitka spruce–western hemlock forest. USDA For. Serv. Res. Note PNW-448.

Greenlee, J. M. 1983. Vegetation, fire history, and fire potential of Big Basin Redwoods State Park, California. Ph.D. diss., University of California, Santa Cruz.

Greig-Smith, P. 1952. The use of random and contiguous quadrats in the study of the structure of plant communities. Ann. Bot. 16:293–316.

Grier, C. C. 1975. Wildfire effects on nutrient distribution and leaching in a coniferous ecosystem. Can. J. For. Res. 5:599–608.

Grier, C. C. 1978. A *Tsuga heterophylla/Picea sitchensis* ecosystem of coastal Oregon: Decomposition and nutrient balances of fallen logs. Can. J. For. Res. 2:198–206.

Grier, C. C. 1989. Effects of prescribed springtime underburning on production and nutrient status of a young ponderosa pine stand. In Multiresource management of ponderosa pine forests: pp. 71–76. USDA For. Serv. Gen. Tech. Rep. RM-185.

Grier, C. C., and R. S. Logan. 1977. Old growth *Pseudotsuga menziesii* communities of a western Oregon watershed: Biomass distribution and production budgets. Ecol. Monogr. 47:373–400.

Grier, C. C., K. A. Vogt, M. R. Keyes, and R. L. Edmonds. 1981. Biomass distribution and above- and below-ground production in young and mature *Abies amabilis* zone ecosystems of the Washington Cascades. Can. J. For. Res. 11:155–67.

Griffin, J. R. 1977. Oak woodland. In Barbour, M. G., and J. Major (eds.), Terrestrial vegetation of California: pp. 384–415. Calif. Native Plant Soc. Special Pub. 9. Sacramento, CA.

Griffin, J. R. 1980. Sprouting in fire-damaged valley oaks, Chews Ridge, California. In Plumb, T. R. (ed.), Proceedings of the symposium on the ecology, management, and utilization of California oaks: pp. 216–19. USDA For. Serv. Gen. Tech. Rep. PSW-44.

Gripp, R. A. 1976. An appraisal of critical fire weather in northwestern California. M.S. thesis, Humboldt State University, Arcata, CA.

Gross, E., I. Steinblums, C. Ralston, and H. Jubas. 1989. Emergency watershed treatments on burned lands in southwestern Oregon. In Berg, N. H. (tech. coord.), Proceedings of the symposium on fire and watershed management: pp. 109–14. USDA For. Serv. Gen. Tech. Rep. PSW-109.

Gruell, G. E., W. C. Schmidt, S. F. Arno, and W. J. Reich. 1982. Seventy years of vegetative change in a managed ponderosa pine forest in western Montana: Implications for resource management. USDA For. Serv. Gen. Tech. Rep. INT-130.

Gunther, E. 1973. Ethnobotany of western Washington. Seattle: University of Washington Press.

Habeck, J. R. 1961. The original vegetation of the mid-Willamette Valley, Oregon. Northwest Sci. 35:65–77.

Habeck, J. R. 1962. Forest succession in Monmouth Township, Polk County, Oregon, since 1850. Proceedings of the Montana Academy of Arts and Sciences 21:7–17.

Habeck, J. R., and R. W. Mutch. 1973. Fire-dependent forests in the northern Rocky Mountains. Quat. Res. 3:408–24.

Haefner, H. F. 1917. Chaparral areas on the Siskiyou National Forest. Proc. Soc. of Amer. For. 12:82–95.

Hagenstein, W. 1951. What should be the state responsibility on unburned restocked areas? In Western Forestry and Conservation Assn., Forty-second annual meeting: pp. 44–45. Portland, OR: Western Forestry and Conservation Assn.

Haig, I. T., K. P. Davis, and R. H. Weidman. 1941. Natural regeneration in the western white pine type. USDA Tech. Bull. 767.

Hall, F. C. 1967. Vegetation-soil relations as a basis for resource management on the Ochoco National Forest of central Oregon. Ph.D. diss., Oregon State University, Corvallis.

Hall, F. C. 1976. Fire and vegetation in the Blue Mountains: Implications for land managers. Tall Timbers Fire Ecol. Conf. Proc. 15:155–70.

Hall, F. C. 1978. Western juniper in association with other tree species. In Martin, R. E., et al. (eds.), Proceedings of the western juniper ecology and management workshop: pp. 31–36. USDA For. Serv. Gen. Tech. Rep. PNW-74.

Hall, J. A. 1972. Forest fuels, prescribed fire, and air quality. USDA For. Serv., Pacific Northwest Forest and Range Exp. Sta., Portland, OR.

Hallam, S. A. 1975. Fire and hearth. Australian Inst. Aboriginal Studies No. 58. Canberra: Australian Inst. Aboriginal Studies.

Halpern, C. B. 1989. Early successional patterns of forest species: Interactions of life history traits and disturbance. Ecology 70:704–20.

Halpern, C. B., and J. F. Franklin. 1990. Physiognomic development of *Pseudotsuga* forests in relation to initial structure and disturbance intensity. J. Veg. Sci. 1:475–82.

Hansen, E. M. 1979. Survival of *Phellinus weirii* in Douglas-fir stumps after logging. Can. J. For. Res. 9:484–88.

Harcombe, P. A. 1986. Stand development in a 130-year-old spruce-hemlock forest based on age structure and 50 years of mortality data. For. Ecol. and Manage. 14:41–58.

Hare, R. C. 1965. Bark surface and cambium temperatures in simulated forest fires. J. For. 63:437–40.

Harmon, M. E., and J. F. Franklin. 1989. Tree seedlings on logs in *Picea-Tsuga* forests of Oregon and Washington. Ecology 70:48–59.

Harmon, M. E., J. F. Franklin, F. J. Swanson, P. Sollins, S. V. Gregory, J. D. Lattin, N. H. Anderson, S. P. Cline, N. G. Aumen, J. R. Sedell, G. N. Lienkaemper, K. Cromack, Jr., and K. W. Cummins. 1986. Ecology of coarse woody debris in temperate ecosystems. Adv. Ecol. Res. 15: 133–302.

Harniss, R. O., and R. B. Murray. 1973. Thirty years of vegetal change following burning of sagebrush-grass range. J. Range Manage. 26:322–25.

Harper, J. L. 1961. Approaches to the study of plant competition. In F. L. Milthorpe (ed.), Mechanisms in biological competition: pp. 1–39. Symposium of the Society for Experimental Biology 15. Southampton: University of Southampton.

Harper, J. L. 1977. The population biology of plants. New York: Academic Press.

Harr, D. 1980. Streamflow after patch cutting in small drainages within the Bull Run Municipal Watershed, Oregon. USDA For. Serv. Res. Pap. PNW-268.

Harr, R. D., W. C. Harper, and J. T. Krygier. 1975. Changes in storm hydrographs after road building and clearcutting in the Oregon coast range. Water Resour. Res. 11:436–44.

Harrington, M. G. 1982. Stand, fuel, and potential fire behavior characteristics in an irregular southeastern Arizona ponderosa pine stand. USDA For. Serv. Res. Note RM-418.

Harrington, M. G. 1987. Predicting reduction of natural fuels by prescribed burning under ponderosa pine in southeastern Arizona. USDA For. Serv. Res. Note RM-472.

Harrington, M. G., and F. G. Hawksworth. 1990. Interactions of fire and dwarf mistletoe on mortality of southwestern ponderosa pine. In Krammes, J. S. (tech. coord.), Effects of fire management of southwestern natural resources: pp. 234–40. USDA For. Serv. Gen. Tech. Rep. RM-191.

Harris, G. A. 1991. Grazing lands of Washington state. Rangelands 13:222–27.

Harris, T. M. 1958. Forest fire in the Mesozoic. J. Ecol. 46:447–53.

Harris, T. M. 1981. Burnt ferns from the English Wealden. Proc. Geol. Assn. 92:47–58.

Harvey, A. E., M. F. Jurgensen, and M. J. Larsen. 1981. Organic reserves: Importance to ectomycorrhizae in forest soils of western Montana. For. Sci. 27:442–45.

Harvey, A. E., M. F. Jurgensen, M. J. Larsen, and J. A. Schlieter. 1986. Distribution of active ectomycorrhizae short roots in forest soils of the inland Northwest: Effects of site and disturbance. USDA For. Serv. Res. Pap. INT-374.

Heacox, E. F. 1951. Should limited clearance be granted by the state on unburned slash areas which, because of silvicultural reasons, do not require hazard reduction? In Western Forestry and Conservation Assn., Forty-second annual meeting: pp. 43–44. Portland, OR: Western Forestry and Conservation Assn.

Hebda, R. J., and R. W. Mathewes. 1984. Holocene history of cedar and native Indian cultures of the North American Pacific Coast. Science 225:211–13.

Hedin, A. 1979. Nineteen seventy-eight annual summary of prescribed burning activities conducted under the Washington Smoke Management Plan. Washington Dept. Nat. Resour., Olympia.

Heinrichs, J. 1983. Tillamook. J. For. 81:442–44, 446.

Heinselman, M. L. 1973. Fire in the virgin forests of the Boundary Waters Canoe Area, Minnesota. Quat. Res. 3:329–82.

Heinselman, M. L. 1981. Fire intensity and frequency as factors in the distribution and structure of northern ecosystems. In Mooney, H. A., et al. (eds.), Fire regimes and ecosystem properties: pp. 7–57. USDA For. Serv. Gen. Tech. Rep. WO-26.

Helgerson, O. 1988. Historic fire year for Oregon and California. Oregon State University FIR Rep. 9(4):2–4.

Helvey, J. D., A. R. Tiedemann, and W. B. Fowler. 1976. Some climatic and hydrologic effects of wildfire in Washington state. Tall Timbers Fire Ecol. Conf. 15:210–22.

Hemstrom, M. A. 1979. A recent disturbance history of forest ecosystems at Mount Rainier National Park. Ph.D. diss., Oregon State University, Corvallis.

Hemstrom, M. A. 1982. Fire in the forest of Mount Rainier National Park. In Starkey, E. E., et al. (eds.), Ecological research in the national parks of the Pacific Northwest: pp. 121–25. Corvallis: Oregon State University For. Res. Lab.

Hemstrom, M. A., and J. F. Franklin. 1982. Fire and other disturbances of the forests in Mount Rainier National Park. Quat. Res. 18:32–51.

Henderson, J. A. 1973. Composition, distribution, and succession of subalpine meadows in Mount Rainier National Park, Washington. Ph.D. diss., Oregon State University, Corvallis.

Henderson, J. A. 1981. Ecology of subalpine fir. In Oliver, C. D., and R. M. Kenady (eds.), Proceedings of the biology and management of true fir in the Pacific Northwest: pp. 53–58. Inst. of For. Resour. Contr. 45. University of Washington, Seattle.

Henderson, J. A., and D. Peter. 1981. Preliminary plant associations and habitat types of the Shelton Ranger District, Olympic National Forest. USDA For. Serv., Pacific Northwest Region.

Henderson, J. A., D. H. Peter, R. D. Lesher, and D. C. Shaw. 1989. Forested plant associations of the Olympic National Forest. USDA For. Serv. R6 ECOL Tech. Pap. 001-88.

Hennon, P. E., E. M. Hansen, and C. G. Shaw III. 1990. Causes of basal scars on *Chamaecyparis nootkatensis* in southeast Alaska. Northwest Sci. 64:45–54.

Herman, F. R., C. E. Smith, and J. E. Firth. 1972. Freezing decayed wood to facilitate ring counts and width measurements. USDA For. Serv. Res. Note PNW-187.

Hessburg, P. F., and R. L. Everett. In prep. Forest pathogens and insects as catalysts of change in fire-restricted northern spotted owl landscapes. Ecology.

Heusser, C. J. 1978. Palynology of quaternary deposits of the lower Bogachiel River area, Olympic Peninsula, Washington. Can. J. Earth Sci. 15:1568–78.

Heyward, F., and R. M. Barnette. 1934. Effect of frequent fires on chemical composition of forest soils in the longleaf pine region. University of Florida Agr. Exp. Sta. Tech. Bull. 265.

Heyward, F. D. 1934. Comments on the effect of fire on feeding roots of pine. Naval Stores Rev. 44(19):4.

Heyward, F. D. 1937. The effect of frequent fires on profile development of longleaf pine forest soils. J. For. 35:23–27.

Hibbs, D. E., and B. J. Yoder. 1993. Development of Oregon white oak seedlings. Northwest Sci. 67:30–36.

Hinckley, T. M., G. H. Goldstein, F. Meinzer, and R. O. Teskey. 1985. Environmental constraints at arctic, temperate-maritime, and tropical treelines. In Establishment and tending of subalpine forest: pp. 21–30. Int. Union of For. Res. Organizations Workshop P.1.07.00.

Hines, W. W. 1971. Plant communities in the old growth forests of the north coast of Oregon. M.S. thesis, Oregon State University, Corvallis.

Hofmann, J. V. 1917a. Natural reproduction from seed stored in the forest floor. J. Agr. Res. 11:1–26.
Hofmann, J. V. 1917b. The relation of brush fires to natural reproduction: Applegate Division of the Crater National Forest. USDA For. Serv., Wind River Exp. Sta., WA.
Hofmann, J. V. 1922. Discussion of Mr. Joy's comments. In Thirteenth annual session, Pacific Logging Congress: pp. 31–32. Portland, OR.
Hofmann, J. V. 1924. Natural regeneration of Douglas-fir in the Pacific Northwest. USDA Bull. 1200.
Holling, C. S. 1978. Adaptive environmental assessment and management. London and New York: John Wiley and Sons.
Hopwood, D. 1991. Principles and practices of new forestry. B.C. Ministry of Forests Land Manage. Rep. 71.
Houston, D. B. 1973. Wildfires in northern Yellowstone National Park. Ecology 54:1111–17.
Hoxie, G. L. 1910. How fire helps forestry. Sunset 25(7):145–51.
Huff, M. H. 1984. Post-fire succession in the Olympic Mountains, Washington: Forest vegetation, fuels, and avifauna. Ph.D. diss., University of Washington, Seattle.
Huff, M. H. 1988. Mount Rainier: Fire and ice. Park Sci. 8(3):22–23.
Huff, M. H., and J. K. Agee. 1991. Subalpine forest dynamics after fire in the Pacific Northwest national parks. Nat. Park Serv. Coop. Park Studies Unit, College of For. Resour., University of Washington, Seattle.
Huff, M. H., J. K. Agee, M. Gracz, and M. Finney. 1989. Fuel and fire behavior predictions in subalpine forests of Pacific Northwest national parks. Nat. Park Serv. Coop. Park Studies Rep. CPSU/UW 89-4. College of For. Resour., University of Washington, Seattle.
Huff, M. H., J. K. Agee, and D. A. Manuwal. 1985. Postfire succession of avifauna in the Olympic Mountains, Washington. In Lotan, J. E., and J. K. Brown (comps.), Fire's effects on wildlife habitat: Symposium proceedings: pp. 8–15. USDA For. Serv. Gen. Tech. Rep. INT-186.
Hurley, C., and D. J. Taylor. 1974. Brown and burn site preparation in western Washington. Washington Dept. Nat. Resour. Note 8. Olympia, WA.
Hussain, S. B., C. M. Skau, S. M. Bashir, and R. O. Meeuwig. 1969. Infiltrometer studies of water-repellent soils on the east slope of the Sierra Nevada. In DeBano, L. F., and J. Letey (eds.), Water-repellent soils: pp. 127–31. Riverside: University of California, Riverside.
Isaac, L. A. 1930. Seedling survival on burned and unburned surfaces. J. For. 28(4):569–71.
Isaac, L. A. 1935. Life of Douglas-fir seed in the forest floor. J. For. 33:61–66.

Isaac, L. A. 1940. Vegetation succession following logging in the Douglas-fir region with special reference to fire. J. For. 38:716–21.

Isaac, L. A. 1943. Reproductive habits of Douglas-fir. Washington, DC: Charles Lathrop Pack Forestry Foundation.

Isaac, L. A. 1963. Fire: A tool not a blanket rule in Douglas-fir ecology. Tall Timbers Fire Ecol. Conf. 2:1–17.

Isaac, L. A., and H. G. Hopkins. 1937. The forest soil of the Douglas-fir region, and changes wrought upon it by logging and slash burning. Ecology 18(2):264–79.

Jackson, M. L. 1958. Soil chemical analysis. Englewood Cliffs, NJ: Prentice-Hall.

Jacobs, D. F., D. W. Cole, and J. R. McBride. 1985. Fire history and perpetuation of natural coast redwood ecosystems. J. For. 83:494–97.

James, S. 1984. Lignotubers and burls: Their structure, function, and ecological significance in Mediterranean ecosystems. Bot. Rev. 50:225–45.

Jeppesen, D. J. 1978. Competitive moisture consumption by the western juniper. In Martin, R. E., et al. (eds.), Proceedings of the western juniper ecology and management workshop: pp. 83–90. USDA For. Serv. Gen. Tech. Rep. PNW-74.

Jepson, W. L. 1910. The silva of California. University of California Mem. Vol. 2. Berkeley: University of California.

Jepson, W. L. 1916. Regeneration in manzanita. Madrono 1:3–11.

Johnson, C. G., Jr., and R. R. Clausnitzer. 1992. Plant associations of the Blue and Ochoco Mountains. USDA For. Serv., Pacific Northwest Region Rep. R-6-ERW-TP-036-92. Portland, OR.

Johnson, E. A. 1979. Fire recurrence in the subarctic and its implications for vegetation composition. Can. J. Bot. 57:1374–79.

Johnson, E. A., and C.P.S. Larsen. 1991. Climatically induced change in fire frequency in the southern Canadian Rockies. Ecology 72:194–201.

Johnson, E. A., and C. E. Van Wagner. 1985. The theory and use of two fire history models. Can. J. For. Res. 15:214–20.

Jones, R. 1990. Let it burn. Pacific Northwest Magazine (March 1990):28–29, 94–97.

Joy, G. C. 1913. Comments of WFFA forester. In Western Forestry and Conservation Assn. 1913 proceedings: pp. 3–4. Portland, OR: Western Forestry and Conservation Assn.

Joy, G. C. 1922. Forest fire prevention in the camps. In Thirteenth annual session, Pacific Logging Congress: pp. 30–31. Portland, OR.

Joy, G. C. 1925. State Forester Joy's comments on "To burn or not to burn." In Sixteenth annual session, Pacific Logging Congress: pp. 25–26. Portland, OR.

Kauffman, J. B. 1990. Ecological relationships of vegetation and fire in Pacific Northwest forests. In Walstad, J., et al. (eds.), Natural and prescribed fire in Pacific Northwest forests: pp. 39–52. Corvallis: Oregon State University Press.

Kauffman, J. B., and R. E. Martin. 1984. A preliminary investigation on the feasibility of preharvest prescribed burning for shrub control. In Proceedings: Sixth annual forest vegetation management conference: pp. 89–113. Redding, CA.

Kauffman, J. B., and R. E. Martin. 1985. Shrub and hardwood response to prescribed burning with varying season, weather, and fuel moisture. Fire and Forest Meteorology Conf. 8:279–86.

Kauffman, J. B., and R. E. Martin. 1989. Fire behavior, fuel consumption, and forest-floor changes following prescribed understory fires in Sierra Nevada mixed conifer forests. Can. J. For. Res. 19:455–62.

Kauffman, J. B., and R. E. Martin. 1990. Sprouting shrub response to different seasons and fuel consumption levels of prescribed fire in Sierra Nevada mixed conifer ecosystems. For. Sci. 36:748–64.

Kauffman, J. B., and R. E. Martin. 1991. Factors influencing the scarification and germination of three montane Sierra Nevada shrubs. Northwest Sci. 65:180–87.

Kayll, A. J. 1968. Heat tolerance of tree seedlings. Tall Timbers Fire Ecology Conf. 8:89–105.

Keane, R. E., S. F. Arno, and J. K. Brown. 1989. FIRESUM: An ecological process model for fire succession in western conifer forests. USDA For. Serv. Gen. Tech. Rep. INT-266.

Keane, R. E., S. F. Arno, J. K. Brown, and D. F. Tomback. 1990. Modelling stand dynamics in whitebark pine (*Pinus albicaulis*) forest. Ecol. Modelling 51:73–95.

Keen, F. P. 1937. Climatic cycles in eastern Oregon as indicated by tree rings. Monthly Weather Rev. 65:175–88.

Keen, F. P. 1943. Ponderosa pine tree classes redefined. J. For. 41:249–53.

Kercher, J. R., and M. C. Axelrod. 1984. A process model of fire ecology and succession in a mixed-conifer forest. Ecology 65:1725–42.

Kertis, J. 1986. Vegetation dynamics and disturbance history of Oak Patch Natural Area Preserve, Mason County, Washington. M.S. thesis, University of Washington, Seattle.

Kilgore, B. M. 1971. The role of fire in managing red fir forests. Trans. North Amer. Wildlife and Natural Resour. Conf. 36:405–16.

Kilgore, B. M., and H. H. Biswell. 1971. Seedling germination following fire in a giant sequoia forest. Calif. Agr. 25:8–10.

Kilgore, B. M., and G. S. Briggs. 1972. Restoring fire to high elevation forests in California. J. For. 70:266–71.

Kilgore, B. M., and G. A. Curtis. 1987. Guide to understory burning in ponderosa pine–larch–fir forests in the Intermountain West. USDA For. Serv. Gen. Tech. Rep. INT-233.

Kilgore, B. M., and R. Sando. 1975. Crown-fire potential in a sequoia forest after prescribed burning. For. Sci. 21:83–87.

Kilgore, B. M., and D. Taylor. 1979. Fire history of a sequoia–mixed conifer forest. Ecology 60:129–41.

Kimmins, J. P. 1987. Forest ecology. New York: Macmillan.

Kirby, R. E., S. J. Lewis, and S. J. Sexson. 1988. Fire in North American wetland ecosystems and fire-wildlife relations: An annotated bibliography. USDI Fish and Wildlife Serv. Biol. Rep. 88(1).

Kittredge, J. 1938. Comparative infiltration in the forest and open. J. For. 36:1156–57.

Klock, G. O. 1971. Streamflow nitrogen loss following forest erosion control fertilization. USDA For. Serv. Res. Note PNW-169.

Klock, G. O., and J. D. Helvey. 1976. Soil-water trends following wildfire on the Entiat Experimental Forest. Tall Timbers Fire Ecol. Conf. 15:193–200.

Knapp, A. K., and W. K. Smith. 1982. Factors influencing understory seedling establishment of Engelmann spruce (*Picea engelmannii*) and subalpine fir (*Abies lasiocarpa*) in southeast Wyoming. Can. J. Bot. 60:2753–61.

Knudson, R. 1980. Ancient peoples of the Columbia Plateau. J. For. 78:477–79.

Koenig, J. Q., D. S. Covert, T. V. Larson, et al. 1988. Wood smoke: Health effects and legislation. Northwest Envir. J. 4(1):41–54.

Kohnke, H. 1968. Soil physics. New York: McGraw-Hill.

Komarek, E. V. 1967. The nature of lightning fires. Tall Timbers Fire Ecology Conf. 7:5–41.

Komarek, E. V. 1968. Lightning and lightning fires as ecological forces. Tall Timbers Fire Ecology Conf. 8:169–97.

Komarek, E. V. 1969. Fire and animal behavior. Tall Timbers Fire Ecol. Conf. 9:161–207.

Komarek, E. V. 1971. Principles of fire ecology and fire management in relation to the Alaskan environment. In Slaughter, C. W., et al. (eds.), Fire in the northern environment: pp. 3–25. USDA For. Serv., Pacific Northwest For. and Range Exp. Sta., Portland, OR.

Komarek, E. V. 1973. Ancient fires. Tall Timbers Fire Ecol. Conf. 12:219–40.

Koonce, A., and L. F. Roth. 1980. The effects of prescribed burning on dwarf mistletoe in ponderosa pine. Fire and Forest Meteorology Conf. 6:197–203.

Krajina, V. J. (ed.). 1965. Ecology of western North America, vol. 1. Vancouver: University of British Columbia, Dept. of Botany.

Kuchler, W. 1977. Natural vegetation of California (map). In Barbour, M. G., and J. Major (eds.), Terrestrial vegetation of California. Sacramento: California Native Plant Society.

Kuhn, T. S. 1970. The structure of scientific revolutions. Chicago: University of Chicago Press.

Kuramoto, R. T., and L. C. Bliss. 1970. Ecology of subalpine meadows in the Olympic Mountains, Washington. Ecol. Monogr. 40:317–47.

Lamb, F. H. 1925. To burn, or not to burn. In Sixteenth annual session, Pacific Logging Congress: pp. 23–24. Portland, OR.

Landsberg, J. D., P. H. Cochran, M. M. Frink, and R. E. Martin. 1984. Foliar nitrogen content and tree growth after prescribed fire in ponderosa pine. USDA For. Serv. Res. Note PNW-412.

Larson, B. C. 1982. Development of even-aged and uneven-aged mixed conifer stands in eastern Washington. In Means, J. E. (ed.), Forest succession and stand development research in the Northwest: pp. 113–18. Corvallis: Oregon State University For. Res. Lab.

Larson, B. C., and C. D. Oliver. 1981. Forest resources survey, Lower Stehekin Valley, Washington. USDI Nat. Park. Serv., Pacific Northwest Region, Seattle.

Larson, J. W. 1972. Ecological role of lodgepole pine in the upper Skagit River valley, Washington. M.S. thesis, University of Washington, Seattle.

Larson, M. M., and G. H. Schubert. 1969. Root competition between ponderosa pine seedlings and grass. USDA For. Serv. Res. Pap. RM-54.

Larson, W. H. 1966. Fire hazard vs. air pollution in the state of Washington. Soc. Amer. For. Proc. 1966:43–45.

Laudenslayer, W. F., Jr., H. H. Darr, and S. Smith. 1989. Historical effects of forest management practices on eastside pine communities. In Tecle, A., et al. (tech. coords.), Multiresource management of ponderosa pine forests: pp. 26–34. USDA For. Serv. Gen. Tech. Rep. RM-185.

Laven, R. D., P. N. Omi, J. G. Wyant, and A. S. Pinkerton. 1980. Interpretation of fire scar data from a ponderosa pine ecosystem in the central Rocky Mountains, Colorado. In Stokes, M. A., and J. H. Dieterich (tech. coords.), Proceedings of the fire history workshop: pp. 46–49. USDA For. Serv. Gen. Tech. Rep. RM-81.

Lay, D. W. 1956. Effects of prescribed burning on forage and mast production in southern pine forests. J. For. 54:582–84.

Lee, R. G. 1977. Institutional change and fire management. In Mooney, H. A., and C. E. Conrad (eds.), Proceedings of the symposium on the environmental consequences of fire and fuel management in Mediterranean ecosystems: pp. 202–14. USDA For. Serv. Gen. Tech. Rep. WO-3.

Leopold, A. S., S. A. Cain, C. M. Cottam, I. N. Gabrielson, and T. L. Kimball.

1963. Study of wildlife problems in national parks: Wildlife management in the national parks. Trans. North Amer. Wildlife and Natural Resour. Conf. 28:28–45.
Leopold, E. B., R. Nickmann, J. I. Hedges, and J. R. Ertel. 1982. Pollen and lignin records of late quaternary vegetation, Lake Washington. Science 218:1305–7.
Lertzman, K. P., and C. J. Krebs. 1991. Gap-phase structure of a subalpine old-growth forest. Can. J. For. Res. 21:1730–41.
Leverenz, J., and D. J. Lev. 1987. Effects of carbon dioxide–induced changes on the natural ranges of six major commercial tree species in the western United States. In Shands, W. E., and J. S. Hoffman (eds.), The greenhouse effect, climate change, and U.S. forests. Washington, DC: Conservation Foundation.
Lewis, H. 1973. Patterns of Indian burning in California: Ecology and ethnohistory. Ballena Press Anthropological Papers No. 1. Ramona, CA: Ballena Press.
Lindquist, J. L. 1974. Redwood . . . an American wood. USDA For. Serv. Rep. FS-264.
Linsley, E. G. 1943. Attraction of *Melanophila* beetles by fire and smoke. J. Econ. Entomology 36:341–42.
Lipman, P. W., and D. R. Mullineaux, eds. 1981. The 1980 eruptions of Mount St. Helens, Washington. USDI Geol. Survey Prof. Pap. 1250.
Lissoway, J., and J. Propper. 1990. Effects of fire on cultural resources. In Krammes, J. (tech. coord.), Effects of fire management of southwestern natural resources: pp. 25–30. USDA For. Serv. Gen. Tech. Rep. RM-191.
Littke, W. R., and R. I. Gara. 1986. Decay of fire-damaged lodgepole pine in south-central Oregon. For. Ecol. and Manage. 17:279–87.
Little, E. L., Jr. 1971. Atlas of United States trees, vol. 1: Conifers and important hardwoods. USDA Misc. Pub. 1146.
Little, R. 1992. Subalpine tree regeneration following fire: Effects of climate and other factors. M.S. thesis, University of Washington, Seattle.
Liu, C. J. 1986. Rectifying radii on off-center increment cores. For. Sci. 32:1058–61.
Long, G. S. 1910. The logger of the Pacific Northwest: His opportunities and his duties. In Second annual session, Pacific Logging Congress: p. 22. Portland, OR.
Long, G. S. 1911. Comments of Mr. Long. In Western Forestry and Conservation Assn. 1911 proceedings: p. 12. Portland, OR: Western Forestry and Conservation Assn.
Long, J. N., E. G. Schreiner, and N. J. Manuwal. 1979. The role of actively moving sand dunes in the maintenance of an azonal, juniper-dominated community. Northwest Sci. 53:170–79.

Lorenz, R. W. 1939. High temperature tolerance of forest trees. University of Minn. Agr. Exp. Sta. Tech. Bull. 141.

Lotan, J. E., and J. K. Brown (comps.). 1985. Fire's effects on wildlife habitat: Symposium proceedings. USDA For. Serv. Gen. Tech. Rep. INT-186.

Lowery, R. F. 1972. Ecology of subalpine zone tree clumps in the North Cascade mountains of Washington. Ph.D. diss., University of Washington, Seattle.

Luckman, B. H. 1978. Geomorphic work of snow avalanches in the Canadian Rocky Mountains. Arctic and Alpine Res. 10:261–76.

Lunan, J. S., and J. R. Habeck. 1973. The effects of fire exclusion on ponderosa pine communities in Glacier National Park, Montana. Can. J. For. Res. 3:574–79.

Lyon, L. J., H. S. Crawford, E. Czuhai, R. L. Fredricksen, R. F. Harlow, L. J. Metz, and H. A. Pearson. 1978. Effects of fire on fauna. USDA For. Serv. Gen. Tech. Rep. WO-6.

Maclean, N. 1992. Young men and fire. Chicago: University of Chicago Press.

Madany, M. H., T. W. Swetnam, and N. E. West. 1982. Comparison of two approaches for determining fire dates from tree scars. For. Sci. 28:856–61.

Marsden, M. A. 1983. Modeling the effect of wildfire frequency on forest structure and succession in the northern Rocky Mountains. J. Env. Manage. 16:45–62.

Martin, R. E. 1963. Thermal properties of bark. U.S. For. Prod. J. 13:419–26.

Martin, R. E. 1978. Fire manipulation and effects in western juniper (*Juniperus occidentalis* Hook.). In Martin, R. E., et al. (eds.), Proceedings of the western juniper ecology and management workshop: pp. 121–36. USDA For. Serv. Gen. Tech. Rep. PNW-74.

Martin, R. E. 1982. Fire history and its role in succession. In Means, J. E. (ed.), Forest succession and stand development in the Northwest: pp. 92–99. Corvallis: For. Res. Lab., Oregon State University.

Martin, R. E. 1983. Antelope bitterbrush seedling establishment following prescribed burning in the pumice zone of the southern Cascade Mountains. In Tiedemann, A. R., and K. L. Johnson (comps.), Proceedings: Research and management of bitterbrush and cliffrose in western North America: pp. 82–90. USDA For. Serv. Gen. Tech. Rep. INT-152.

Martin, R. E., and C. H. Driver. 1983. Factors affecting antelope bitterbrush reestablishment following fire. In Tiedemann, A. R., and K. L. Johnson (comps.), Proceedings: Research and management of bitterbrush and cliffrose in western North America: pp. 266–79. USDA For. Serv. Gen. Tech. Rep. INT-152.

Martin, R. E., and A. Johnson. 1979. Fire management of Lava Beds National Monument. In Linn, R. M. (ed.), Proceedings of the first conference on scientific research in the national parks: pp. 1209–17. USDI Nat. Park Serv. Trans. and Proc. Ser. 5.

Mason, D. T. 1915. The life history of lodgepole burn forests. USDA Tech. Bull. 154.

Mastrogiuseppe, R., M. E. Alexander, and W. H. Romme. 1983. Forest and rangeland fire history bibliography. USDA For. Serv., Intermountain For. and Range Exp. Sta., Ogden, UT.

Matson, P. A., and R. D. Boone. 1984. Natural disturbance and nitrogen mineralization: Wave-form dieback of mountain hemlock in the Oregon Cascades. Ecology 65:1511–16.

Maul, T. W. 1968. Oregon waiver law: Impact on operations and a look at the future. Northwest Forest Fire Council 1968: sec. 2.

McArdle, R. E. 1930. Effect of fire on Douglas-fir slash. J. For. 28(4):568–69.

McArdle, R. E., and W. H. Meyer. 1930. The yield of Douglas-fir in the Pacific Northwest. USDA Tech. Bull. 201.

McBride, J. R. 1983. Analysis of tree rings and fire scars to establish fire history. Tree Ring Bull. 43:51–67.

McBride, J. R., and R. D. Laven. 1976. Scars as an indicator of fire frequency in the San Bernadino Mountains, California. J. For. 74:439–42.

McCaughley, K. J., and S. A. Cook. 1980. *Phellinus weirii* infestation of two mountain hemlock forests in the Oregon Cascades. For. Sci. 26:23–29.

McColl, J. G., and D. W. Cole. 1968. A mechanism of cation transport in a forest soil. Northwest Sci. 42:134–40.

McConnell, B. R., and J. G. Smith. 1970. Response of understory vegetation to ponderosa pine thinning in eastern Washington. J. Range Manage. 23:208–12.

McDaniels, E. H. 1939. The Yacolt fire. Portland, OR: USDA Forest Service.

McIntosh, R. P. 1981. Succession and ecological theory. In West, D. C., et al. (eds.), Forest succession: Concepts and applications: pp. 10–23. New York: Springer-Verlag.

McKee, A., G. La Roi, and J. F. Franklin. 1982. Structure, composition, and reproductive behavior of terrace forests, South Fork Hoh River, Olympic National Park. In Starkey, E. E., et al. (eds.), Ecological research in national parks of the Pacific Northwest: pp. 22–29. Corvallis: Oregon State University For. Res. Lab.

McLean, A. 1969. Fire resistance of forest species as influenced by root systems. J. Range Manage. 22:120–22.

McMahon, T. E., and D. S. deCalesta. 1990. Effects of fire on fish and wildlife. In Walstad, J., et al. (eds.), Natural and prescribed fire in Pacific Northwest forests: pp. 233–50. Corvallis: Oregon State University Press.

McNab, W. H. 1977. An overcrowded loblolly pine stand thinned with fire. South. J. Applied For. 1:24–26.

McNabb, D. H., and K. Cromack, Jr. 1983. Dinitrogen fixation by a mature *Ceanothus velutinus* stand in the western Oregon Cascades. Can. J. Microbiol. 29:1014–21.

McNabb, D. H., and F. J. Swanson. 1990. Effects of fire on soil erosion. In Walstad, J., et al. (eds.), Natural and prescribed fire in Pacific Northwest forests: pp. 159–76. Corvallis: Oregon State University Press.

McNeil, R. C., and D. B. Zobel. 1980. Vegetation and fire history of a ponderosa pine–white fir forest in Crater Lake National Park. Northwest Sci. 54:30–46.

McShane, M. C., and R. H. Sauer. 1985. Comparison of experimental fall burning and clipping on bluebunch wheatgrass. Northwest Sci. 59:313–18.

Means, J. E. 1982. Developmental history of dry coniferous forests in the central western Cascade Range of Oregon. In Means, J. E. (ed.), Forest succession and stand development research in the Pacific Northwest: pp. 142–58. Corvallis: Oregon State University, For. Res. Lab.

Means, J. E. 1989. Estimating the date of a single bole scar by counting tree rings in increment cores. Can. J. For. Res. 19:1491–96.

Means, J. E., W. A. McKee, W. H. Moir, and J. F. Franklin. 1982. Natural revegetation of the northeastern portion of the devastated area. In Keller, S.A.C. (ed.), Mount St. Helens: One year later: pp. 93–103. Cheney: Eastern Washington University Press.

Merrill, E. H., H. F. Mayland, and J. M. Peek. 1980. Effects of a fall wildfire on herbaceous vegetation on xeric sites in the Selway-Bitteroot Wilderness. J. Range Manage. 33:363–66.

Mersereau, R. C., and C. T. Dyrness. 1972. Accelerated mass wasting after logging and slash burning in western Oregon. J. Soil and Water Cons. 27:112–14.

Methven, I. R. 1971. Prescribed fire, crown scorch and mortality: Field and laboratory studies on red and white pine. Can For. Serv., Petawawa For. Exp. Sta. Inf. Rep. PS-X-31.

Metz, L. J., T. Lotti, and R. A. Klawitter. 1961. Some effects of prescribed burning on Coastal Plain forest soil. USDA For. Serv. Southeastern For. Exp. Sta. Res. Pap. 133.

Miles, S. R., D. M. Haskins, and D. W. Ranken. 1989. Emergency burn rehabilitation: Cost, risk, and effectiveness. In Berg, N. H. (tech. coord.),

Proceedings of the symposium on fire and watershed management: pp. 97–102. USDA For. Serv. Gen. Tech. Rep. PSW-109.
Miller, J. M., and F. P. Keen. 1960. Biology and control of the western pine beetle. USDA Misc. Pub. 800.
Miller, M. 1977. Response of blue huckleberry to prescribed fires in western Montana larch-fir forest. USDA For. Serv. Res. Pap. INT-188.
Miller, M. 1978. Effect of growing season on sprouting of blue huckleberry. USDA For. Serv. Res. Note INT-240.
Miller, P. C. 1970. Age distributions of spruce and fir in beetle-killed forests on the White River Plateau, Colorado. Amer. Midl. Naturalist 86:206–12.
Minore, D. 1972. The wild huckleberries of Oregon and Washington: A dwindling resource. USDA For. Serv. Res. Pap. 143.
Minore, D. 1979. Comparative autecological characteristics of northwestern tree species: A literature review. USDA For. Serv. Gen. Tech. Rep. PNW-87.
Minore, D., A. W. Smart, and M. E. Dubrasich. 1979. Huckleberry ecology and management research in the Pacific Northwest. USDA For. Serv. Gen. Tech. Rep. PNW-193.
Moir, W. H. 1969. The lodgepole pine zone in Colorado. Amer. Midl. Naturalist 81:87–98.
Molnar, A. C., and R. G. McMinn. 1960. The origin of basal scars in the British Columbia interior white pine type. For. Chron. 36:50–61.
Monserud, R. A. 1979. Relations between inside and outside bark diameter at breast height for Douglas-fir in northern Idaho. USDA For. Serv. Res. Note INT-266.
Moore, K., and B. Macdonald. 1974. Final report of the fringe-strip study team. British Columbia Fish and Wildlife Branch, Nanaimo, B.C., unpub. rep.
Morgan, P., and S. C. Bunting. 1990. Fire effects in whitebark pine forests. In Schmidt, W. C., and K. J. McDonald (comps.), Proceedings: Symposium on whitebark pine ecosystems: Ecology and management of a high-mountain resource: pp. 166–70. USDA For. Serv. Gen. Tech. Rep. INT-270.
Morris, W. G. 1932. A preliminary report, giving some of the results obtained in a study of lightning storm occurrence and behavior on the national forests of Oregon and Washington. USDA For. Serv. For. Res. Note 10. Pacific Northwest For. Exp. Sta.
Morris, W. G. 1934a. Lightning storms and fires on the national forests of Oregon and Washington. USDA For. Serv., Pacific Northwest For. and Range Exp. Sta., Portland, OR.

Morris, W. G. 1934b. Forest fires in Oregon and Washington. Oregon Historical Quarterly 35:313–39.

Morris, W. G. 1958. Influence of slash burning on regeneration, other plant cover, and fire hazard in the Douglas-fir region. U.S. For. Serv. Pac. Northwest For. and Range Expt. Sta. Res. Pap. 29. Portland, Or.

Morris, W. G. 1966. Guidelines offered for slash burning. For. Ind. 93 (10):62–63.

Morris, W. G. 1970. Effects of slash burning in overmature stands of the Douglas-fir region. For. Sci. 16:258–70.

Morris, W. G., and E. L. Mowat. 1958. Some effects of thinning a ponderosa pine thicket with a prescribed fire. J. For. 56:203–9.

Morrison, P., and F. J. Swanson. 1990. Fire history and pattern in a Cascade Range landscape. USDA For. Serv. Gen. Tech. Rep. PNW-GTR-254.

Morrow, R. J. 1985. Age structure and spatial pattern of old-growth ponderosa pine in Pringle Falls Experimental Forest, Central Oregon. M.S. thesis, Oregon State University, Corvallis.

Mowat, E. L. 1960. No serotinous cones on central Oregon lodgepole pine. J. For. 58:118–19.

Muller, C. H. 1951. The significance of vegetative reproduction in *Quercus*. Madrono 11:129–37.

Munger, T. T. 1917. Western yellow pine in Oregon. USDA Bull. 418.

Munger, T. T. 1925. Rainfall probability during the fire season in western Washington and Oregon. Monthly Weather Rev. 53:394–97.

Munger, T. T. 1940. The cycle from Douglas fir to hemlock. Ecology 21:451–59.

Munger, T. T. 1944. Out of the ashes of the Nestucca. Amer. For. 50:342–47.

Munger, T. T. 1951. Is slash burning silviculturally desirable? In Western Forestry and Conservation Assn., Forty-second annual meeting: pp. 34–36. Portland, OR: Western Forestry and Conservation Assn.

Munger, T. T., and D. N. Matthews. 1939. Flashes from "Slash disposal and forest management after clear cutting in the Douglas fir region." USDA For. Serv. Pacific Northwest For. and Range Exp. Sta. Forest Res. Notes 27.

Murray, K. 1968. The pig war. Tacoma: Washington State Historical Society.

Mutch, R. 1970. Wildland fires and ecosystems: A hypothesis. Ecology 51:1046–51.

Neal, J. L., E. Wright, and W. B. Bollen. 1965. Burning Douglas-fir slash: Physical, chemical, and microbial effects in the soil. Corvallis: Oregon State University, For. Res. Lab.

Newton, M., B. A. El Hassan, and J. Zavitkowski. 1968. Role of red alder in western Oregon forest succession. In Trappe, J., et al. (eds.), Biology of

alder: pp. 73–84. Northwest Sci. Ann. Mtg. Symp. Proc. 1967. USDA For. Serv., Pacific Northwest For. and Range Exp. Sta., Portland.
NOAA. 1982. Monthly normals of temperature, precipitation, and heating and cooling degree days 1951–80. NOAA Env. Info. Serv. Climatography of the U.S. 81. Asheville, NC: Nat. Climate Center.
Noble, I. R., and R. O. Slayter. 1980. The use of vital attributes to predict successional changes in plant communities subject to recurrent disturbance. Vegetatio 43:5–21.
Nord, E. A. 1965. Autecology of bitterbrush in California. Ecol. Monogr. 35:307–34.
Nord, E. C., and L. R. Green. 1977. Low-volume and slow-burning vegetation for planting on clearings in California chaparral. USDA For. Serv. Res. Pap. PNW-124.
Nordin, V. J. 1958. Basal fire scars and the occurrence of decay in lodgepole pine. For. Chron. 34:257–65.
Norton, H. 1979. The association between anthropogenic prairies and important food plants in western Washington. Northwest Anthropological Res. Notes 13:175–200.
Norum, R. 1977. Preliminary guidleines for prescribed burning under standing timber in western larch/Douglas-fir forests. USDA For. Serv. Res. Note INT-229.
Noste, N. V., and C. L. Bushey. 1987. Fire response of shrubs of dry forest habitat types in Montana and Idaho. USDA For. Serv. Gen. Tech. Rep. INT-239.
Odum, E. P. 1969. The strategy of ecosystem development. Science 164:262–70.
Oliver, C. D. 1981. Forest development in North America following major disturbances. For. Ecol. and Manage. 3:153–68.
Oliver, C. D., and B. C. Larson. 1990. Forest stand dynamics. New York: McGraw-Hill.
Oliver, W. W. 1986. Growth of California red fir advance regeneration after overstory removal and thinning. USDA For. Serv. Res. Pap. PSW-180.
Olson, D. F., Jr. 1974. *Quercus*. In Schopmeyer, C. S. (tech. coord.), Seeds of woody plants in the United States: pp. 692–703. USDA Agr. Handbook 450.
Oosting, H. J., and W. D. Billings. 1943. The red fir forest of the Sierra Nevada: *Abietum magnificae*. Ecol. Monogr. 13:259–74.
Orr, T. J., Jr. 1977. The importance of prescribed fire in the Pacific Northwest. Northwest Forest Fire Council 1977:56–61.
Oswald, D. D. 1968. The timber resources of Humboldt County, California. USDA For. Serv. Resour. Bull. PNW-26.

Owens, T. E. 1982. Postburn regrowth of shrubs related to canopy mortality. Northwest Sci. 56:34–40.

Packee, E. C., C. D. Oliver, and P. D. Crawford. 1981. Ecology of Pacific silver fir. In Oliver, C. D., and R. M. Kenady, Proceedings of the biology and management of true fir in the Pacific Northwest: pp. 19–34. Inst. of For. Resour. Contr. 45. University of Washington, Seattle.

Parker, V. T. 1987. Can native flora survive prescribed burns? Fremontia 15(2):3–6.

Parmeter, J. R., and B. Uhrenholdt. 1976. Effects of smoke on pathogens and other fungi. Tall Timbers Fire Ecol. Conf. 14:299–304.

Parsons, D. J., and S. H. DeBenedetti. 1979. Impact of fire suppression on a mixed-conifer forest. For. Ecol. and Manage. 2:21–33.

Parsons, D. J., D. M. Graber, J. K. Agee, and J. W. van Wagtendonk. 1986. Natural fire management in national parks. Environ. Manage. 10:21–24.

Pase, C. P. 1958. Herbage production and competition under immature ponderosa pine stands in the Black Hills. J. Range Manage. 11:238–43.

Peek, J. M., F. D. Johnson, and N. N. Pence. 1978. Successional trends in a ponderosa pine/bitterbrush community related to grazing by livestock, wildlife, and to fire. J. Range Manage. 31:49–53.

Perry, D. A., and J. E. Lotan. 1977. Opening temperatures in serotinous cones of lodgepole pine. USDA For. Serv. Res. Note INT-228.

Perry, D. A., and S. L. Rose. 1983. Soil biology and soil productivity: Opportunities and constraints. In Ballard, R., and S. Gessel (eds.), IUFRO symposium on forest site and continuous productivity. USDA For. Serv. Gen. Tech. Rep. PNW-163.

Perry, D. A., S. L. Rose, D. Pilz, and M. M. Schoenberger. 1984. Reduction of natural ferric iron chelators in disturbed forest soils. Soil Sci. Soc. Amer. Proc. 48:379–82.

Perry, J. P., Jr. 1991. The pines of Mexico and Central America. Portland, OR: Timber Press.

Peterson, D. L. 1985. Crown scorch volume and scorch height: Estimates of postfire tree condition. Can. J. For. Res. 15:596–98.

Peterson, D. L., and M. J. Arbaugh. 1986. Postfire survival of Douglas-fir and lodgepole pine: Comparing the effects of crown and bole damage. Can. J. For. Res. 16:1175–79.

Peterson, D. L., M. J. Arbaugh, G. H. Pollock, and L. J. Robinson. 1991. Postfire growth of *Pseudotsuga menziesii* and *Pinus contorta* in the northern Rocky Mountains, USA. Intl. J. Wildland Fire 1:63–71.

Peterson, D. L., and K. C. Ryan. 1986. Modeling post-fire conifer mortality for long-range planning. Environ. Manage. 10:797–808.

Peterson, J. L., and R. D. Ottmar. 1991. Computer applications for pre-

scribed fire and air quality management in the Pacific Northwest. Fire and Forest Meteorology Conf. 11:455–59.
Philpot, C. W. 1969. Seasonal changes in heat content and ether extractive content of chamise. USDA For. Serv. Res. Pap. INT-61.
Philpot, C. W. 1970. Influence of mineral content on the pyrolysis of plant materials. For. Sci. 16:461–71.
Philpot, C. W. 1977. Vegetative features as determinants of fire frequency and intensity. In Mooney, H. A., and C. E. Conrad (eds.), Proceedings of the symposium on the environmental consequences of fire and fuel management in Mediterranean-climate ecosystems: pp. 12–16. USDA For. Serv. Gen. Tech. Rep. WO-3.
Philpot, C. W., B. Leonard, et al. 1988. Report on fire management policy. U.S. Depts. of Interior and Agriculture. Washington, DC: Government Printing Office.
Pickford, S. G. 1981. West-side true fir and prescribed fire. In Oliver, C. D., and R. M. Kenady, Proceedings of the biology and management of true fir in the Pacific Northwest: pp. 315–18. Inst. of For. Resour. Contr. 45. University of Washingon, Seattle.
Pickford, S. G., G. R. Fahnestock, and R. Ottmar. 1980. Weather, fuels, and lightning fires in Olympic National Park. Northwest Sci. 54:92–105.
Picton, H. D., D. M. Mattson, B. M. Blanchard, and R. R. Knight. 1986. Climate, carrying capacity, and the Yellowstone grizzly bear. In Contreras, G. P., and K. E. Evans (comps.), Proceedings: Grizzly bear habitat symposium: pp. 129–35. USDA For. Serv. Gen. Tech. Rep. INT-207.
Pillers, M. D., and J. D. Stuart. In review. Leaf litter accretion and decomposition in interior and coastal old-growth redwood stands.
Pitcher, D. C. 1987. Fire history and age structure in red fir forests of Sequoia National Park, California. Can. J. For. Res. 17:582–87.
Potash, L. L. 1989. Sprouting of red heather (*Phyllodoce empetriformis*) in response to fire. M.S. thesis, University of Washington, Seattle.
Prater, J. D. 1962. Some views on present slash disposal methods. Northwest Forest Fire Council 1962:31–32.
Pratt, M. B. 1911. Results of "light burning" near Nevada City, California. For. Quarterly 9(3):420–22.
Pyne, S. J. 1982. Fire in America: A cultural history of wildland and rural fire. Princeton, NJ: Princeton University Press.
Pyne, S. J. 1989a. Burning questions and false alarms about wildfires in Yellowstone. Forum 4(2):31–40.
Pyne, S. J. 1989b. The summer we let wildfire loose. Natural History (August): 45–49.
Quaye, E. 1982. The structure and dynamics of old-growth Sitka spruce

(*Picea sitchensis*) forest of the Oregon Coast Range. Ph.D. diss., Oregon State University, Corvallis.

Quick, C. 1977. Ceanothus seeds and seedlings on burns. Madrono 15:79–81.

Quinsey, S. 1984. Fire and grazing effects in western juniper woodlands of central Oregon. M.S. thesis, University of Washington, Seattle.

Rakestraw, L. 1979. A history of forest conservation in the Pacific Northwest 1891–1913. New York: Arno Press.

Randall, C. E. 1937. Oregon forest fires. Nat. Fire Protection Assn. Quarterly (January): 226–28.

Rawson, R. P., P. R. Billing, and S. F. Duncan. 1983. The 1982–83 forest fires in Victoria. Austral. For. 46:163–72.

Ream, C. H. (comp.). 1981. The effects of fire and other disturbances on small mammals and their predators: An annotated bibliography. USDA For. Serv. Gen. Tech. Rep. INT-106.

Reaves, J. L., C. G. Shaw III, R. E. Martin, and J. E. Mayfield. 1984. Effects of ash leachates on growth and development of *Armillaria mellea* in culture. USDA For. Serv. Res. Note PNW-418.

Redington, P. 1920. What is the truth? The Forest Service and Stewart Edward White agree to study forest fire danger. Sunset Magazine 44(6):56–57.

Reed, L. J., and N. G. Sugihara. 1987. Northern oak woodlands: Ecosystem in jeopardy or is it already too late? In Plumb, T. R., and N. H. Pillsbury (tech. coords.), Proceedings of the symposium on multiple-use management of California's hardwood resources: pp. 59–63. USDA For. Serv. Gen. Tech. Rep. PSW-100.

Reifsnyder, W. E., L. P. Herrington, and K. W. Spalt. 1967. Thermophysical properties of bark of shortleaf, longleaf, and red pine. Yale University School of Forestry Bull. 70.

Reinhart, E. D., and K. C. Ryan. 1988. Eight-year tree growth following prescribed underburning in a western Montana Douglas-fir/western larch stand. USDA For. Serv. Res. Note INT-387.

Rice, R. R. 1973. The hydrology of chaparral watersheds. In Living with the chaparral: pp 27–33. Riverside, CA: Sierra Club.

Rich, L. R. 1962. Erosion and sediment movement following a wildfire in a ponderosa pine forest in central Arizona. USDA For. Serv. Rocky Mtn. For. and Range Exp. Sta. Res. Note 76. Fort Collins, CO.

Richards, L. W. 1940. Effect of certain chemical attributes of vegetation on forest flammability. J. Agr. Res. 60:833–38.

Riegel, G. M., R. F. Miller, and W. C. Krueger. 1992. Competition for resources between understory vegetation and overstory *Pinus ponderosa* in northeastern Oregon. Ecol. Appl. 2:71–85.

Riegel, G. M., B. G. Smith, and J. F. Franklin. 1992. Foothill oak woodlands of the interior valleys of southwestern Oregon. Northwest Sci. 66:66–76.

Roe, A. L., W. R. Beaufait, L. J. Lyon, and J. L. Oltman. 1971. Fire and forestry: A task force report. J. For. 68:464–70.

Romme, W. 1980. Fire history terminology: Report of the ad hoc committee. In Stokes, M. A., and J. H. Dieterich (eds.), Proceedings of the fire history workshop: pp. 135–37. USDA For. Serv. Gen. Tech. Rep. RM-81.

Romme, W. H. 1982. Fire and landscape diversity in subalpine forests of Yellowstone National Park. Ecol. Monogr. 52:199–21.

Romme, W. H., and D. Despain. 1989. Historical perspective on the Yellowstone fires of 1988. Bioscience 39:695–99.

Romme, W. H., and D. Knight. 1981. Fire frequency and subalpine forest succession along a topographic gradient in Wyoming. Ecology 62:319–26.

Romme, W. H., and M. G. Turner. 1991. Implications of global climate change for biogeographic patterns in the greater Yellowstone ecosystem. Conserv. Biol. 5:373–86.

Rose, C. W. 1966. Agricultural physics. New York: Pergamon Press.

Rothermel, R. C. 1983. How to predict the spread and intensity of wildfires. USDA For. Serv. Gen. Tech. Rep. INT-143.

Rothermel, R. C., and J. E. Deeming. 1980. Measuring and interpreting fire behavior for correlation with fire effects. USDA For. Serv. Gen. Tech. Rep. INT-93.

Rothermel, R. C., and G. C. Rinehart. 1983. Field procedures for verification and adjustment of fire behavior predictions. USDA For. Serv. Gen. Tech. Rep. INT-142.

Rowe, J. S. 1983. Concepts of fire effects on plant individuals and species. In Wein, R. W., and D. A. Maclean (eds.), The role of fire in northern circumpolar ecosystems: pp. 135–54. New York: John Wiley and Sons.

Rummell, R. S. 1951. Some effects of livestock grazing on ponderosa pine forest and range in central Washington. Ecology 32:594–607.

Rundel, P. W., D. W. Parsons, and D. T. Gordon. 1977. Montane and subalpine vegetation of the Sierra Nevada and Cascade Ranges. In Barbour, M. G., and J. Major (eds.), Terrestrial vegetation of California: pp. 559–600. New York: Wiley-Interscience.

Russell, J. J., and W. F. McCulloch. 1944. Slash burning in western Oregon. Oregon State Board of Forestry Bull. 10.

Ruth, R. H., and C. M. Berntsen. 1955. A four-year record of Sitka spruce and western hemlock seed fall. USDA For. Serv. Pacific Northwest For. and Range Exp. Sta. Res. Pap. 7.

Ruth, R. H., and A. S. Harris 1979. Management of western hemlock–

Sitka spruce forests for timber production. USDA For. Serv. Gen. Tech. Rep. PNW-88.

Ryan, K. C., and W. H. Frandsen. 1991. Basal injury from smoldering fires in mature *Pinus ponderosa* Laws. Int. J. Wildland Fire 1:107–18.

Ryan, K. C., and E. D. Reinhardt. 1988. Predicting postfire mortality of seven western conifers. Can. J. For. Res. 18:1291–97.

Sackett, S. S. 1980. Reducing ponderosa pine fuels using prescribed fire: Two case studies. USDA For. Serv. Res. Note RM-392.

Sackett, S. S. 1984. Observations on natural regeneration in ponderosa pine following a prescribed fire in Arizona. USDA For. Serv. Res. Note RM-435.

Saenz, L., and J. O. Sawyer, Jr. 1986. Grasslands as compared to adjacent *Quercus garryana* woodland understories exposed to different grazing regimes. Madrono 33:40–46.

Salih, M.S.A., F.K.H. Taha, and G. F. Payne. 1973. Water repellency of soils under burned sagebrush. J. Range Manage. 26:330–31.

Salisbury, H. E. 1989. The Great Black Dragon fire: A Chinese inferno. Boston: Little, Brown.

Sampson, A. W. 1944. Effect of chaparral burning on soil erosion and on soil-moisture relations. Ecology 25:171–91.

Sandberg, D. V. 1980. Duff reduction by prescribed underburning in Douglas-fir. USDA For. Serv. Res. Pap. PNW-272.

Sandberg, D. V. 1989. Future concerns and needs: Air resource management. In Hanley, D., et al. (eds.), The burning decision: Regional perspectives on slash: pp. 356–62. Inst. of For. Resour. Contr. 66. University of Washington, Seattle.

Sandberg, D. V., and F. N. Dost. 1990. Effects of prescribed fire on air quality and human health. In Walstad, J., et al. (eds.), Natural and prescribed fire in Pacific Northwest forests: pp. 191–218. Corvallis: Oregon State University Press.

Sapsis, D. B., and J. B. Kauffman. 1991. Fuel consumption and fire behavior associated with prescribed fires in sagebrush ecosystems. Northwest Sci. 65:173–79.

Sartz, R. S. 1953. Soil erosion on a fire-denuded forest area in the Douglas-fir region. J. Soil and Water Conserv. 8:279–81.

Sauter, J., and B. Johnson. 1974. Tillamook Indians of the Oregon coast. Portland, OR: Binsford and Mort.

Savage, M., and T. W. Swetnam. 1990. Early nineteenth-century fire decline following sheep pasturing in a Navajo ponderosa pine forest. Ecology 71:2374–78.

Sawyer, J. O., and D. A. Thornburgh. 1977. Montane and subalpine vegetation of the Klamath Mountains. In Barbour, M., and J. Major. (eds.),

Terrestrial vegetation of California: pp. 699–732. Calif. Native Plant Soc. Special Pub. 9. Sacramento, CA.

Sawyer, J. O., D. A. Thornburgh, and J. R. Griffin. 1977. Mixed evergreen forest. In Barbour, M., and J. Major (eds.), Terrestrial vegetation of California: pp. 359–81. Calif. Native Plant Soc. Special Pub. 9. Sacramento, CA.

Scheffer, V. 1959. Field studies of the Garry oak in Washington. University of Washington Arboretum Bull. 22:88–89.

Schiff, A. 1962. Fire and water: Scientific heresy in the Forest Service. Cambridge: Harvard University Press.

Schmidt, R. L. 1960. Factors controlling the distribution of Douglas-fir in coastal British Columbia. Quart. J. For. 54:156–60.

Schmidt, W. C., R. C. Schearer, and A. L. Roe. 1976. Ecology and silviculture of western larch forests. USDA Tech. Bull. 1520.

Schoenberger, M. M., and D. A. Perry. 1982. The effect of soil disturbance on growth and ectomycorrhizae on Douglas-fir and western hemlock seedlings: A greenhouse bioassay. Can. J. For. Res. 12:343–53.

Schroeder, M. 1961. Down-canyon afternoon winds. Bull. Amer. Meteor. Soc. 42:527–42.

Schroeder, M., and C. C. Buck. 1970. Fire weather. USDA Agr. Handbook 360.

Schultz, A. M., and H. H. Biswell. 1959. Effect of prescribed burning and other seedbed treatments on ponderosa pine seedling emergence. J. For. 57:816–17.

Schultz, A. M., J. L. Launchbaugh, and H. H. Biswell. 1955. Competition between grasses reseeded on burned brushlands in California. J. Range Manage. 5:338–45.

Schumaker, F. X. 1928. Yield, stand, and volume tables for red fir in California. University of California Agr. Exp. Sta. Bull. 456.

Scott, D.R.M. 1980. The Pacific Northwest region. In Barrett, J. W. (ed.), Regional silviculture of the United States: pp. 503–71. New York: John Wiley and Sons.

Scott, V. H. 1956. Relative infiltration rates of burned and unburned upland soils. Trans. Amer. Geophys. Union 37:67–69.

Scott, V. H., and R. H. Burgy. 1956. Effect of heat and brush burning on the physical properties of certain upland soils that influence infiltration. Soil Sci. 82:63–70.

Seabloom, R. W., R. D. Sayler, and S. A. Ahler. 1991. Effects of prairie fire on archeological artifacts. Park Sci. 11(1):1,3.

Seidel, K. W. 1986. Tolerance of seedlings of ponderosa pine, Douglas-fir, grand fir, and Engelmann spruce for high temperature. Northwest Sci. 60:1–7.

Selter, C. M., W. D. Pitts, and M. G. Barbour. 1986. Site microenvironment and seedling survival of Shasta red fir. Amer. Midl. Naturalist 115:288–300.

Serveen, C. 1983. Grizzly bear food habits, movements, and habitat selection in the Mission Mountains, Montana. J. Wildlife Manage. 47:1026–35.

Shepard, H. B. 1931. Supposedly harmless surface fires damage old growth forests. U. S. For. Serv., Pacific Northwest For. Exp. Sta. For. Res. Notes 7. Portland, OR.

Sheppard, P. R., J. E. Means, and J. P. Lassoie. 1988. Cross-dating cores as a nondestructive method for dating living, scarred trees. For. Sci. 34:781–89.

Shinn, D. A. 1980. Historical perspectives on range burning in the inland Pacific Northwest. J. Range Manage. 33:415–23.

Show, S. B. 1920. Forest fire protection in California. Timberman 21(3):88–90.

Show, S. B., and E. I. Kotok. 1924. The role of fire in the California pine forest. USDA Bull. 1294.

Shugart, H. H., Jr., and D. C. West. 1980. Forest succession models. Bioscience 30:308–13.

Silen, R. R. 1958. Silvical characteristics of Oregon white oak. USDA For. Serv. Silvical Ser. 10. Pacific Northwest For. and Range Exp. Sta., Portland, OR.

Singer, F. J., W. Schreier, J. Oppenheim, and E. O. Gorton. 1988. Drought, fires, and large animals. Bioscience 39:716–22.

Smathers, G. A., and D. Mueller-Dombois. 1974. Invasion and recovery of vegetation after a volcanic eruption in Hawaii. USDI Nat. Park Serv. Monogr. Ser. 5.

Smith, C. W. 1968. A review of the waiver program in the southern Oregon area. Northwest Forest Fire Council 1968: sec. 2.

Smith, F. W. 1981. Effects of competition for soil water on growth and water status of *Pinus contorta* on pumice soil. Ph.D. diss., University of Washington, Seattle.

Smith, J. H., and G. A. Kozak. 1967. Thickness and percentage of bark of the commercial trees of British Columbia. University of British Columbia Faculty For. Ser.

Smith, W. P. 1985. Plant associations within the interior valleys of the Umpqua River basin, Oregon. J. Range Manage. 38:526–30.

Snyder, J. R. 1984. The role of fire: Much ado about nothing? Oikos 43:404–5.

Soeriaatmadja, R. E. 1966. Fire history of the ponderosa pine forests of the

Warm Springs Indian Reservation, Oregon. Ph.D. diss., Oregon State University, Corvallis.

Spalt, K., and W. Reifsnyder. 1962. Bark characteristics and fire resistance: A literature survey. USDA For. Serv. South. For. Exp. Sta. Occ. Pap. 193.

Spies, T. A., J. F. Franklin, and T. B. Thomas. 1988. Coarse woody debris in Douglas-fir forests of western Oregon and Washington. Ecology 69:1689–1702.

Sprague, F. L., and H. P. Hansen. 1946. Forest succession in the McDonald Forest, Willamette Valley, Oregon. Northwest Sci. 20:89–98.

Sprugel, D. G. 1991. Disturbance, equilibrium, and environmental variability: What is "natural" vegetation in a changing environment? Biol. Conserv. 58:1–18.

Stage, A. R. 1976. An expression for the effect of aspect, slope, and habitat type on tree growth. For. Sci. 22:457–60.

Stahelin, R. 1943. Factors influencing the natural restocking of high altitude burns by coniferous trees in the central Rocky Mountains. Ecology 24:19–30.

Stark, N. 1973. Nutrient cycling in a Jeffrey pine ecosystem. Montana For. and Conserv. Exp. Sta., University of Montana, Missoula.

Starker, T. J. 1934. Fire resistance in the forest. J. For. 32:462–67.

State of Washington. 1991. Washington natural heritage plan. Olympia: Washington Natural Heritage Program, Dept. of Natural Resources.

Steele, R., S. V. Cooper, D. M. Ondov, D. W. Roberts, and R. D. Pfister. 1983. Forest habitat types of eastern Idaho–western Wyoming. USDA For. Serv. Gen. Tech. Rep. INT-144.

Steen, H. K. 1976. The U.S. Forest Service: A history. Seattle: University of Washington Press.

Stein, W. I. 1990. *Quercus garryana* Dougl. ex Hook, Oregon white oak. In Burns, R. M., and B. H. Honkala (tech. coords.), Silvics of North America, vol. 2: pp. 650–60. USDA Agr. Handbook 654.

Stender, R. H. (N.d.) Washington Smoke Management Program, 1988 annual report. Olympia: Washington Dept. of Natural Resources, Div. of Fire Control.

Stewart, G. H. 1986. Population dynamics of a montane conifer forest, western Cascade Range, Oregon, USA. Ecology 67(2):534–44.

Stewart, G. H. 1988. The influence of canopy cover on understory development in forests of the western Cascade Range, Oregon, USA. Vegetatio 76:79–88.

Stewart, O. C. 1951. Burning and natural vegetation in the United States. Geogr. Rev. 41:317–20.

Stickney, P. 1986. First decade plant succession following the Sundance forest fire, northern Idaho. USDA For. Serv. Gen. Tech. Rep. INT-197.
Stillinger, C. R. 1944. Damage to conifers in northern Idaho by the Richardson red squirrel. J. For. 42:143–45.
Stocks, B. J. 1987. Fire behavior in immature jack pine. Can. J. For. Res. 17:80–86.
Stohlgren, T. J. 1988. Litter dynamics in two Sierran mixed conifer forests, I: Litterfall and decomposition rates. Can. J. For. Res. 18:1127–35.
Stokes, M. A., and T. L. Smiley. 1968. Tree-ring dating. Chicago: University of Chicago Press.
Stone, E. C., and R. Vasey. 1968. Preservation of coast redwood on alluvial flats. Science 159:157–61.
Stuart, J. D. 1983. Stand structure and development of a climax lodgepole pine forest in south-central Oregon. Ph.D. diss., University of Washington, Seattle.
Stuart, J. D. 1984. Hazard rating of lodgepole pine stands to mountain pine beetle outbreaks in south-central Oregon. For. Ecol. and Manage. 5:207–14.
Stuart, J. D. 1987. Fire history of an old-growth forest of *Sequoia sempervirens* (Taxodiaceae) forest in Humboldt Redwoods State Park, California. Madrono 34:128–41.
Stuart, J. D., J. K. Agee, and R. I. Gara. 1989. Lodgepole pine regeneration in an old, self-perpetuating forest in south-central Oregon. Can. J. For. Res. 19:1096–1104.
Stuart, J. D., D. R. Geiszler, R. I. Gara, and J. K. Agee. 1983. Mountain pine beetle scarring of lodgepole pine in south-central Oregon. For. Ecol. and Manage. 5:207–14.
Stubblefield, G., and C. D. Oliver. 1978. Silvicultural implications of the reconstruction of mixed alder/conifer stands. In Briggs, D., et al. (eds.), Utilization and management of alder: pp. 307–20. USDA For. Serv. Gen. Tech. Rep. PNW-70.
Stuiver, M., and P. Quay. 1980. Changes in atmospheric carbon-14 attributed to a variable sun. Science 207:11–19.
Sugihara, N. G., and L. J. Reed. 1987a. Vegetation ecology of the Bald Hills oak woodlands of Redwood National Park. USDI Nat. Park Serv. Redwood National Park R&D Tech. Rep. 21. Orick, CA.
Sugihara, N. G., and L. J. Reed. 1987b. Prescribed fire for restoration and maintenance of Bald Hills oak woodlands. In Plumb, T. R., and N. H. Pillsbury (tech. coords.), Proceedings of the symposium on multiple-use management of California's hardwood resources: pp. 446–51. USDA For. Serv. Gen. Tech. Rep. PSW-100.
Sugihara, N. G., L. J. Reed, and J. M. Lenihan. 1987. Vegetation of the Bald

Hills oak woodlands, Redwood National Park, California. Madrono 34:193–208.

Sugita, S. 1990. Palynological records of forest disturbance and development in the Mountain Meadows watershed, Mt. Rainier, Washington. Ph.D. diss., University of Washington, Seattle.

Sullivan, T. P., and D. S. Sullivan. 1982. Barking damage by snowshoe hares and red squirrels in lodgepole pine stands in central British Columbia. Can. J. For. Res. 12:443–48.

Susott, R. A. 1982. Differential scanning calorimetry of forest fuels. For. Sci. 28:839–51.

Swain, A. M. 1973. A history of fire and vegetation in northeastern Minnesota as recorded in lake sediments. Quat. Res. 3:383–96.

Swanson, F. J. 1981. Fire and geomorphic processes. In Mooney, H., et al. (eds.), Fire regimes and ecosystem properties: Proceedings of the conference: pp. 410–20. USDA For. Serv. Gen. Tech. Rep. WO-26.

Swanston, D. N. 1974. Slope stability problems associated with timber harvesting in the mountainous region of the western United States. USDA For. Serv. Gen. Tech. Rep. PNW-21.

Swetnam, T. W. 1984. Peeled ponderosa pine trees: A record of inner bark utilization by Native Americans. J. Ethnobiol. 4:177–90.

Swetnam, T. W., and J. L. Betancourt. 1990. Fire–Southern Oscillation relations in the southwestern United States. Science 249:1017–20.

Swezy, D. M., and J. K. Agee. 1991. Prescribed fire effects on fine root and tree mortality in old growth ponderosa pine. Can. J. For. Res. 21:626–34.

Switzer, R. 1974. The effects of forest fire on archeological sites in Mesa Verde National Park, Colorado. Artifact 12:1–8.

Tackle, D. 1962. Infiltration in a western larch–Douglas-fir stand following cutting and slash treatment. USDA For. Serv. Intermountain For. and Range Exp. Sta. Res. Note 89.

Tandy, G. F. 1979. Fire history and vegetation patterns of coniferous forests in Jasper National Park, Alberta. Can. J. Bot. 57:1912–31.

Tansley, A. G. 1924. The classification of vegetation and the concept of development. J. Ecol. 8:118–49.

Tappeiner, J., J. Zasada, P. Ryan, and M. Newton. 1991. Salmonberry clonal and population structure: The basis for a persistent cover. Ecology 72:609–18.

Tarrant, R. F. 1956a. Effect of slash burning on some physical soil properties. For. Sci. 2:18–22.

Tarrant, R. F. 1956b. Changes in some physical soil properties after a prescribed burn in young ponderosa pine. J. For. 54:439–41.

Tarrant, R. F. 1956c. Effects of slash burning on some soils of the Douglas-fir region. Soil Sci. Soc. Amer. Proc. 20:408–11.

Taskey, R. D., C. L. Curtis, and J. Stone. 1989. Wildfire, ryegrass seeding, and watershed rehabilitation. In Berg, N. H. (tech. coord.), Proceedings of the symposium on fire and watershed management: pp. 115–24. USDA For. Serv. Gen. Tech. Rep. PSW-109.

Taylor, A. H., and C. B. Halpern. 1991. The structure and dynamics of *Abies magnifica* forests in the southern Cascade Range, USA. J. Veg. Sci. 2:189–200.

Taylor, A. R. 1970. Lightning effects on the forest complex. Tall Timbers Fire Ecol. Conf. 9:127–50.

Taylor, A. R. 1974. Ecological aspects of lightning in forests. Tall Timbers Fire Ecol. Conf. 13:455–82.

Taylor, K. L., and R. W. Fonda. 1990. Woody fuel structure and fire in subalpine fir forests, Olympic National Park, Washington. Can. J. For. Res. 20:193–99.

Taylor, R. J., and T. R. Boss. 1975. Biosystematics of *Quercus garryana* in relation to its distribution in the state of Washington. Northwest Sci. 49:49–57.

Teensma, P.D.A. 1987. Fire history and fire regimes of the central western Cascades of Oregon. Ph.D. diss., University of Oregon, Eugene.

Teensma, P.D.A., J. T. Rienstra, and M. A. Yeiter. 1991. Preliminary reconstruction and analysis of change in forest stand age classes of the Oregon Coast Range from 1850 to 1940. USDI Bureau Land Manage. Tech Note T/N OR-9. Portland, OR.

Thies, W. G. 1984. Laminated root rot: The quest for control. J. For. 82:345–56.

Thies, W. G. 1990. Effects of prescribed fire on diseases of conifers. In Walstad, J., et al. (eds.), Natural and prescribed fire in Pacific Northwest forests: pp. 117–21. Corvallis: Oregon State University Press.

Thies, W. G., K. W. Russell, and L. C. Weir. 1977. Distribution and damage appraisal of *Rhizina undulata* in western Oregon and Washington. Plant Disease Reporter 61:859–62.

Thilenius, J. F. 1964. Synecology of the white-oak (*Quercus garryana* Dougl.) woodlands of the Willamette Valley, Oregon. Ph.D. diss., Oregon State University, Corvallis.

Thilenius, J. F. 1968. The *Quercus garryana* forests of the Willamette Valley, Oregon. Ecology 49:1124–33.

Thomas, T. L., and J. K. Agee. 1986. Prescribed fire effects on mixed conifer forest structure at Crater Lake, Oregon. Can. J. For. Res. 16:1082–87.

Thornburgh, D. A. 1969. Dynamics of the true fir-hemlock forests of the west slope of the Washington Cascades Range. Ph.D. diss., University of Washington, Seattle.

Thornburgh, D. A. 1982. Succession in the mixed evergreen forests of northwestern California. In Means, J. E. (ed.), Forest succession and stand development research in the Northwest: pp. 87–91. Corvallis: Oregon State University For. Res. Lab.

Tiedemann, A. R. 1987. Combustion losses of sulfur from forest foliage and litter. For. Sci. 33:216–23.

Tiedemann, A. R., and G. O. Klock. 1973. First-year vegetation after fire, reseeding, and fertilization on the Entiat Experimental Forest. USDA For. Serv. Res. Note PNW-195.

Tiedemann, A. R., and G. O. Klock. 1976. Development of vegetation after fire, seeding, and fertilization on the Entiat Experimental Forest. Tall Timbers Fire Ecol. Conf. 15:171–92.

Tomback, D. F. 1986. Post-fire regeneration of krummholz whitebark pine: A consequence of nutcracker seed caching. Madrono 33:100–110.

Tomback, D. F., L. A. Hofmann, and S. K. Sund. 1990. Coevolution of whitebark pine and nutcrackers: Implications for forest regeneration. In Schmidt, W. C., and K. J. McDonald (comps.), Proceedings: Symposium on whitebark pine ecosystems: Ecology and management of a high-mountain resource: pp. 118–29. USDA For. Serv. Gen. Tech. Rep. INT-270.

Traylor, D., L. Hubbell, N. Wood, and B. Fiedler. 1979. The La Mesa Fire Study: Investigation of fire and fire suppression impact on cultural resources in Bandelier National Monument. Nat. Park Serv. Southwest Cultural Resources Center, Santa Fe, NM.

Trembour, F. N. 1979. A hydration study of obsidian artifacts, burnt vs. unburnt by the La Mesa forest fire. In Traylor, N., et al. The La Mesa Fire Study: Investigation of fire and fire suppression impact on cultural resources in Bandelier National Monument: app. F. Nat. Park Serv. Southwest Cultural Resources Center, Santa Fe, NM.

Trewartha, G. T. 1968. An introduction to climate. 4th ed. New York: McGraw-Hill.

Tsukada, M., S. Sugita, and D. M. Hibbert. 1981. Paleoecology of the Pacific Northwest, I: Late quaternary vegetation and climate. Verh. Int. Verein. Limnol. 21:730–37.

Tunstall, B. R., J. Walker, and A. M. Gill. 1976. Temperature distribution around synthetic trees during grass fires. For. Sci. 22:269–76.

Ugolini, F. C., and A. K. Schlicte. 1973. The effect of holocene environmental changes on selected western Washington soils. Soil Sci. 116:218–27.

Uhl, C., and J. B. Kauffman. 1990. Deforestation, fire susceptibility, and potential tree responses to fire in the eastern Amazon. Ecology 71:437–49.

Uresk, D. W., J. F. Cline, and W. H. Rickard. 1976. Impact of wildfire on three perennial grasses in south central Washington. J. Range Manage. 29:309–10.

USDA Forest Service. 1964–81. Wildfire statistics. Washington, DC: Div. Coop. Fire Control.

USDA Forest Service. 1987. Wood handbook: Wood as an engineering material. USDA Agr. Handbook 72.

USDI, National Park Service. Management policies. Washington, DC: Government Printing Office.

Ustin, S. L., R. A. Woodward, M. G. Barbour, and J. L. Hatfield. 1984. Relationships between sunfleck dynamics and red fir seedling distribution. Ecology 65:1420–28.

Vale, T. R. 1975. Presettlement vegetation in the sagebrush-grass area of the Intermountain West. J. Range Manage. 28:32–36.

Van Wagner, C. E. 1973. Height of crown scorch in forest fires. Can. J. For. Res. 3:373–78.

Van Wagner, C. E. 1977. Conditions for the start and spread of crown fire. Can. J. For. Res. 7:23–24.

Van Wagner, C. E. 1978. Age-class distribution and the forest fire cycle. Can J. For. Res. 8:220–27.

van Wagtendonk, J. W. 1972. Fire and fuel relationships in mixed-conifer ecosystems of Yosemite National Park. Ph.D. diss., University of California, Berkeley.

van Wagtendonk, J. W. 1974. Refined burning prescriptions for Yosemite National Park. USDI Nat. Park Serv. Occ. Pap. 2.

van Wagtendonk, J. W. 1983. Prescribed fire effects on understory mortality. Fire and Forest Meteorology Conf. 7:136–38.

van Wagtendonk, J. W. 1985. Fire suppression effects on fuels and succession in short-fire-interval wilderness ecosystems. In Lotan, J. E., et al. (eds.), Proceedings: Symposium and workshop on wilderness fire: pp. 119–26. USDA For. Serv. Gen. Tech. Rep. INT-182.

van Wagtendonk, J. W. 1986. The role of fire in the Yosemite wilderness. In Lucas, R. C. (comp.), Proceedings: National wilderness research conference: Current research: pp. 2–9. USDA For. Serv. Gen. Tech. Rep. INT-212.

Vasek, F. C. 1966. The distribution and taxonomy of three western junipers. Brittonia 18:350–72.

Veihmeyer, F. J., and C. N. Johnson. 1944. Soil moisture records from burned and unburned plots in certain grazing areas of California. Trans. Amer. Geophys. Union. Part 1:72–78.

Veirs, S. D. 1980. The influence of fire in coast redwood forests. In Stokes, M. A., and J. H. Dieterich (eds.), Proceedings of the fire history workshop: pp. 93–95. USDA For. Serv. Gen. Tech. Rep. RM-81.

Veirs, S. D. 1982. Coast redwood forest: Stand dynamics, successional status, and the role of fire. In Means, J. (ed.), Forest succession and stand development research in the Northwest: pp. 119–41. Corvallis: Oregon State University.

Viereck, L. A. 1973. Ecological effects of river flooding and forest fires on permafrost in the taiga of Alaska. In North American contribution: Permafrost, second international conference: p. 66. Washington, DC: National Academy of Science.

Vines, R. G. 1968. Heat transfer through bark, and the resistance of trees to fire. Austral. J. Bot. 16:499–514.

Viro, P. 1969. Prescribed burning in forestry. Metsantritkemuslartoken Julkaisuga 67.

Visibility Program Subcommittee. 1982. Air quality and prescribed burning: Strategies for protecting visibility in Class I areas. Olympia: Washington Dept. of Natural Resources.

Vlamis, J., H. H. Biswell, and A. M. Schultz. 1955. Effects of prescribed burning on soil fertility in second growth ponderosa pine. J. For. 53:905–12.

Volland, L. A. 1976. Plant communities of the central Oregon pumice zone. R-6 Area Guide 4-2. USDA For. Serv., Pacific Northwest Region, Portland, OR.

Wagle, R. F., and J. H. Kitchen, Jr. 1972. Influence of fire on soil nutrients in a ponderosa pine type. Ecology 53:118–25.

Wakimoto, R. H. 1990. National fire management policy. J. For. 88:22–26.

Walters, C. 1986. Adaptive management of renewable natural resources. New York: Macmillan.

Warcup, J. H. 1981. Effect of fire on the soil microflora and other nonvascular plants. In Gill, A. M., et al. (eds.), Fire and the Australian biota: pp. 203–14. Canberra: Australian Academy of Science.

Waring, R. 1969. Forest plants of the eastern Siskiyous: Their environmental and vegetational distribution. Northwest Sci. 43:1–17.

Waring, R., and J. F. Franklin. 1979. Evergreen coniferous forests of the Pacific Northwest. Science 204:1380–86.

Waring, R. H., and G. B. Pitman. 1980. A simple model of host resistance to bark beetles. For. Res. Lab. Res. Note 65. Oregon State University, Corvallis.

Washington Department of Forest Resources. 1973–81. Annual report, Smoke Management Program. Olympia: Div. of Fire Control.

Weast, R. C. (ed.). 1982. Handbook of chemistry and physics. Cleveland: CRC Press.

Weatherspoon, C. P. 1988. Preharvest prescribed burning for vegetation management: Effects on *Ceanothus velutinus* seeds in duff and soil.

In Proceedings: Ninth annual vegetation management conference: pp. 125–41. Redding, CA.

Weaver, H. 1943. Fire as an ecological and silvicultural factor in the ponderosa pine region of the Pacific slope. J. For. 41(1):7–15.

Weaver, H. 1951a. Fire as an ecological factor in the southwestern ponderosa pine forests. J. For. 49:93–98.

Weaver, H. 1951b. Observed effects of prescribed burning on perennial grasses in the ponderosa pine forests. J. For. 49:267–71.

Weaver, H. 1955. Fire as an enemy, friend, and tool in forest management. J. For. 53:499–504.

Weaver, H. 1957. Effects of prescribed burning in ponderosa pine. J. For. 55:133–37.

Weaver, H. 1959. Ecological changes in the ponderosa pine forest of the Warm Springs Indian Reservation in Oregon. J. For. 57:15–20.

Weaver, H. 1961. Ecological changes in the ponderosa pine forests of Cedar Valley in southern Washington. Ecology 42:416–20.

Weaver, H. 1974. Effects of fire on temperate forests: Western United States. In Kozlowski, T. T., and C. E. Ahlgren, Fire and ecosystems: chap. 9. New York: Academic Press.

Wells, C. G. 1971. Effects of prescribed burning on soil chemical properties and nutrient availability. In Prescribed burning symposium: pp. 86–99. USDA For. Serv. Southeastern For. Exp. Sta., Charleston, SC.

Wells, C. G., R. E. Campbell, L. F. DeBano, C. E. Lewis, R. L. Fredricksen, E. C. Franklin, R. C. Froehlich, and P. H. Dunn. 1979. Effects of fire on soil: A state-of-knowledge review. USDA For. Serv. Gen. Tech. Rep. WO-7.

West, N. E. 1969. Tree patterns in central Oregon ponderosa pine forests. Amer. Midl. Naturalist 81:584–90.

Western Forestry and Conservation Association. 1929. Cooperative forest study of the Grays Harbor area (Washington). Portland, OR: Western Forestry and Conservation Association.

Whisenant, S. G. 1990. Changing fire frequencies on Idaho's Snake River Plains: Ecological and management implications. In McArthur, E. D., et al. (eds.), Proceedings: Symposium on cheatgrass invasion, shrub dieoff, and other aspects of shrub biology and management: pp. 4–10. USDA For. Serv. Gen. Tech. Rep. INT-276.

White, A. S. 1985. Presettlement regeneration patterns in a southwestern ponderosa pine stand. Ecology 66:589–94.

White, E. M., W. W. Thompson, and F. R. Gartner. 1973. Heat effects on nutrient release from soils under ponderosa pine. J. Range Manage. 26:22–24.

White, P. S. 1979. Pattern, process, and natural disturbance in vegetation. Bot. Rev. 45:229–99.

White, P. S., and S.T.A. Pickett. 1985. Natural disturbance and patch dynamics: An introduction. In Pickett, S.T.A., and P. S. White, The ecology of natural disturbance and patch dynamics: chap. 1. New York: Academic Press.

White, R. 1980. Land use, environment, and social change: The shaping of Island County, Washington. Seattle: University of Washington Press.

White, S. E. 1920. Woodsman, spare those trees! Sunset 44(3):115.

Whittaker, R. H. 1953. A consideration of climax theory: The climax as a population and pattern. Ecol. Monogr. 23:41–78.

Whittaker, R. H. 1960. Vegetation of the Siskiyou Mountains, Oregon and California. Ecol. Monogr. 30:279–338.

Wilkes, C. 1845. Narrative of the United States Expedition during the years 1838, 1839, 1841, 1842. Vol. 5. Philadelphia: Lea and Blanchard.

Williams, C., and T. Lillybridge. 1983. Forested plant associations of the Okanogan National Forest. USDA For. Serv. Pacific Northwest Reg. R-6-Ecol-132b-1983.

Williams, C. K., T. R. Lillybridge, and B. G. Smith. 1990. Forested plant associations of the Colville National Forest. USDA For. Serv., Pacific Northwest Region, Portland, OR.

Williams, J. T., R. E. Martin, and S. G. Pickford. 1980. Silvicultural and fire management implications from a timber type evaluation of tussock moth outbreak areas. Fire and Forest Meteorology Conf. 6:191–96.

Williams, P. A. 1981. Aspects of the ecology of broom (*Cytisus scoparius*) in Canterbury, New Zealand. New Zealand J. Bot. 19:31–43.

Wills, R. D., and J. D. Stuart. In prep. Fire history and stand development of a Douglas-fir/hardwood forest in northern California.

Wischnofske, M. G., and D. W. Anderson. 1983. The natural role of fire in the Wenatchee Valley. USDA For. Serv., Pacific Northwest Reg., Wenatchee National Forest, Wenatchee, WA.

Wolbach, W., I. Gilmour, E. Anders, C. J. Orth, and R. R. Brooks. 1988. Global fire at the Cretaceous-Tertiary boundary. Nature 334:665–69.

Wollum, A. G., II, C. T. Youngberg, and F. W. Chichester. 1968. Relation of previous timber stand age to nodulation of *Ceanothus velutinus*. For. Sci. 14:114–18.

Woodard, P. M. 1977. Effects of prescribed burning on two different-aged high-elevation plant communities in eastern Washington. Ph.D. diss., University of Washington, Seattle.

Woodard, P. M., J. A. Bentz, and T. Van Nest. 1983. Producing a prescribed crown fire in a subalpine forest with an aerial drip torch. Fire Manage. Notes 44(4):24–28.

Woodard, P. M., and G. Cummins. 1987. Engelmann spruce, lodgepole pine, and subalpine fir seed germination success on ashbed conditions. Northwest Sci. 61:233–38.

Woodmansee, R. G., and L. S. Wallach. 1981. Effects of fire regimes on biogeochemical cycles. In Mooney, H., et al. (eds.), Fire regimes and ecosystem properties: Proceedings of the conference: pp. 379–400. USDA For. Serv. Gen. Tech. Rep. WO-26.

Wooldridge, D. D., and H. Weaver. 1965. Some effects of thinning a ponderosa pine thicket with a prescribed fire, II. J. For. 63:92–95.

Worf, W. 1984. Wilderness management: A historical perspective on the implications of human-ignited fire. In Lotan, J. E., et al. (eds.), Proceedings: Symposium and workshop on wilderness fire: pp. 276–82. USDA For. Serv. Gen. Tech. Rep. INT-182.

Wright, E., and R. Tarrant. 1957. Microbiological soil properties after logging and slash burning. USDA For. Serv. Pacific Northwest For. and Range Exp. Sta. Res. Note 157.

Wright, E., and R. Tarrant. 1959. Occurrence of mycorrhizae after logging and slash burning in the Douglas-fir type. USDA For. Serv. Pacific Northwest For. and Range Exp. Sta. Res. Note 160.

Wright, H. A. 1985. Effects of fire on grasses and forbs in sagebrush-grass communities. In Sanders, K., et al. (eds.), Rangeland fire effects: pp. 12–21. Boise: USDI Bureau of Land Management.

Wright, H. A., and A. W. Bailey. 1982. Fire ecology. New York: John Wiley and Sons.

Wright, H. A., and J. O. Klemmedson. 1965. Effects of fire on bunchgrasses of the sagebrush-grass region of southern Idaho. Ecology 46:680–88.

Wright, H. A., L. F. Neuenschwander, and C. M. Britton. 1979. The role and use of fire in sagebrush-steppe and pinyon-juniper plant communities: A state-of-the-art review. USDA For. Serv. Gen. Tech. Rep. INT-58.

Wright, K. H. 1971. Sitka spruce weevil. USDA For. Serv. Pest Leaflet 47.

Wyant, J. G., P. N. Omi, and R. D. Laven. 1983. Fire effects on shoot growth characteristics of ponderosa pine in Colorado. Can. J. For. Res. 13:620–35.

Yamaguchi, D. K. 1986. The development of old-growth Douglas-fir forests northeast of Mount St. Helens, Washington, following an A.D. 1480 eruption. Ph.D. diss., University of Washington, Seattle.

Young, J. A., and J. D. Budy. 1979. Historical use of Nevada's pinyon-juniper woodlands. J. For. Hist. 23:113–21.

Young, J. A., and R. A. Evans. 1981. Demography and fire history of a western juniper stand. J. Range Manage. 34:501–5.

Young, J. A., R. A. Evans, R. E. Eckert, Jr., and B. L. Kay. 1987. Cheatgrass. Rangelands 9:266–70.

Youngberg, C. T. 1966. Silvicultural benefits from brush. Soc. Amer. For. Proc. 1965:55–59.

Youngberg, C. T., and A. G. Wollum II. 1976. Nitrogen accretion in developing *Ceanothus velutinus* stands. Soil Sci. Soc. Amer. Proc. 40:109–12.

Zavitkowski, J., and M. Newton. 1968. Ecological importance of snowbrush, *Ceanothus velutinus*, in the Oregon Cascades. Ecology 49:1134–45.

Zeigler, R. S. 1978. The vegetation dynamics of *Pinus contorta* forest, Crater Lake National Park. M.S. thesis, Oregon State University, Corvallis.

Zinke, P. H. 1981. Floods, sedimentation, and alluvial soil formation as dynamic processes maintaining superlative redwood groves. In Coats, R. (ed.), Watershed rehabilitation in Redwood National Park and other Pacific coastal areas: pp. 26–49. Berkeley, CA: Center for Natural Resource Studies, John Muir Institute.

Zinke, P. H. 1988. The redwood forest and associated north coast forests. In Barbour, M. G., and J. Major (eds.), Terrestrial vegetation of California: pp. 679–98. Calif. Native Plant Soc. Special Pub. 9. Sacramento, CA.

Zobel, D. B., A. McKee, G. M. Hawk, and C. T. Dyrness. 1976. Relationships of environment to composition, structure, and diversity of forest communities of the central western Cascades of Oregon. Ecol. Monogr. 46:135–56.

INDEX

C

M

N

ABOUT THE AUTHOR

James K. Agee is professor of forest ecology in the College of Forest Resources at the University of Washington, Seattle. He recently completed a five-year term as chair of the Division of Ecosystem Science and Conservation, and he continues to teach and conduct research on forest and fire ecology. Before coming to the University of Washington, he was a forest ecologist and research biologist for the National Park Service in Seattle and San Francisco. Agee received his Ph.D. from the University of California, Berkeley, in 1973. He is the author of more than 100 technical reports and professional papers in forest and fire ecology, and he has extensive experience with fire research and management in the Pacific Coast states. He has been a trustee for the Washington chapter of The Nature Conservancy, currently chairs the Washington Natural Heritage Council, and is an associate editor of *Northwest Science*.

ISLAND PRESS BOARD OF DIRECTORS